Terror in the Name *of* God

ALSO BY JESSICA STERN

The Ultimate Terrorists

Terror in the Name of God

Why Religious Militants Kill

JESSICA STERN

An Imprint of HarperCollins*Publishers*

 For information, address HarperCollins Publishers Inc., 10 East 53rd Street, New York, NY 10022.

HarperCollins books may be purchased for educational, business, or sales promotional use. For information, please write: Special Markets Department, HarperCollins Publishers Inc., 10 East 53rd Street, New York, NY 10022.

FIRST EDITION

Designed by Kris Tobiassen

Library of Congress Cataloging-in-Publication Data

Stern, Jessica, 1958–

Terror in the name of God : why religious militants kill / Jessica Stern.—1st ed.

p. cm.

Includes index.

ISBN 0-06-050532-X (alk. paper)

1. Terrorism—Religious aspects—Christianity. 2. Violence—Religious aspects—Christianity. I. Title.

BL65.T47S74 2003

303.6'25—dc21

2003048508

03 04 05 06 07 BVG/RRD 10 9 8 7 6 5 4 3 2 1

For: Evan and Jeff

ACKNOWLEDGMENTS

I would like to thank the following individuals for their assistance with the manuscript. My editors Jason Epstein and Dan Halpern persuaded me to take the story of the interviews out of the footnotes and put them in the text. Together with my remarkable agent, Martha Kaplan, they pushed me to grow into a very different writer from the one I started out.

A virtual army of students served as research assistants. I would like to thank Nicole Simon, Hassan Abbas, Darcy Bender, Laila el-Haddad, Alex Fox, Edward Flinter, Matthew Gershwin, Susanna Krueger, Michael Rindner, and Jason Sanchez. I am grateful in particular to Assaf Moghadam, who was an extraordinary research assistant through much of the project. I would also like to thank Michael Johnson, my faculty assistant, who is particularly skilled at finding legal documents, writing polite letters of inquiry, and maintaining a sense of humor.

Colleagues who read parts of the manuscript include Greg Barton, Sugata Bose, Neta Crawford, Laura Donohue, Jeffrey Frankel, James Gilligan, Richard Hackman, Robert Hefner, Charles Heckscher, Judith Herman, Michael Ignatieff, David Lazer, Mark Moore, Barbara Pizer, Sam Popkin, Michael Reich, Hilary Stern, Monica Toft, Ivan Toft, Kenneth Winston, and Sundeep Waslekar. Mark Moore gathered together a group if ethicists to help me think about the ethical issues. Robert J. Lifton's study group on mass violence, and Jacqueline Bhabha and Michael Ignatieff's study group on terrorism and human rights, provided helpful comments. Ajai Sahni was a particularly helpful reader of three chapters. I would like also to thank the Advisory Group to the Harvard Project on Religion in Politics.

I would like to thank Leslie Cockburn, who first persuaded me to go to Pakistan; Ahmed Rashid, who took care of me when I arrived; and Muzamal Suherwardy, with whom I learned about Pakistan's jihadi groups. As a result of the project we started together, Muzamal made the jihadi groups a major focus of his own work as a reporter, now for the *Friday Times*. He also administered questionnaires in Pakistan and Afghanistan. Magnus Ranstorp accompanied me on my first trip to Gaza and provided a driver in Beirut. The late Ehud Sprinzak introduced me to a number of Jewish extremists. Konrad Huber, a former student, persuaded me to study Indonesian jihadi groups; and Diwakar, another former student, helped in India. Steve Kull accompanied me to Texas, and Steve Goldstein to Florida. David Stewart, a partner at Ropes and Gray, provided legal assistance when my research notes were subpoenaed by the U.S. government.

I am grateful to the following organizations that helped to fund this project: the Center for Public Leadership, the Council on Foreign Relations, the Ploughshares Fund, the Smith Richardson Foundation, the Weatherhead Center for International Affairs, the Women's Leadership Board, and the U.S. Institute of Peace.

This book is written with the many victims of terrorism in mind and heart.

CONTENTS

Religious terrorism arises from pain and loss and from impatience with a God who is slow to respond to our plight, who doesn't answer. Its converts often long for a simpler time, when right and wrong were clear, when there were heroes and martyrs, when the story was simple, when the neighborhood was small, when we knew one another. When the outside world, with its vulgar cosmopolitanism, didn't humiliate us or threaten our children. When we did not envy these others or even know about them. It is about finding a clear purpose in a confusing world with too many choices. It is about purifying the world. The way forward is clear: kill or be killed. Kill and be rewarded in heaven. Kill and the Messiah will come. It is about seeing the world in black and white. About projecting all one's fears and inadequacies on the Other. Why is my life not going as well as it should? The answer is America. The answer is affirmative action. The answer is the Jews. The answer is the Dome of the Rock. A devilish cabal controls the banking system and the press through globalization and world government, through the Council on Foreign Relations, or the Arab oil sheiks. My people are in the majority. This is the temple's wall. The wall where his horse stood tied. It is clear from the Bible that this land is legitimately ours. Archaeologists show. History proves. My ancestors' bones. My people are suffering. Without this piece of land or this temple, I am not whole. My people are not whole. We are spiritually dead. We are dry bones cast about the earth. This is where our Messiah will rule. This is where our prophets walked. This is the furthermost place of his nocturnal ride, where miracles happened, where He made us the chosen people, where loaves became fishes, where He comforted the afflicted, where He rose to heaven, where the angel Gabriel's handprints remain. This, in short, is where bloodbaths begin.

INTRODUCTION

Any creative encounter with evil requires that we not distance ourselves from it by simply demonizing those who commit evil acts. In order to write about evil, a writer has to try to comprehend it, from the inside out; to understand the perpetrators and not necessarily sympathize with them. But Americans seem to have a very difficult time recognizing that there is a distinction between understanding and sympathizing. Somehow we believe that an attempt to inform ourselves about what leads to evil is an attempt to explain it away. I believe that just the opposite is true, and that when it comes to coping with evil, ignorance is our worst enemy.

—KATHLEEN NORRIS

I teach a course called "Terrorism" at Harvard University's Kennedy School of Government. I have been studying terrorism for many years in various capacities—as a government official, a scholar, and as a university lecturer. A few years ago I decided to do something scholars rarely do: I decided to talk with terrorists.

People have always told me their secrets. Taxi drivers tell me about their dreams, their relationship with their bosses or their siblings, their affairs. A professional acquaintance once declared that he had killed someone in self-defense and had never before told anyone. I will also confess that I am intensely curious, especially about spiritual and emotional matters, and I suppose secret-sharers sense this. This personality quirk has been useful in talking to terrorists, as important, I believe, as my academic training and my experience working in national security agencies.

All of us at various points in our lives experience spiritual longing. I started this project in such a phase. I visited synagogues, churches, and mosques. Not attending services, or anyway not often. I wanted to be in rooms saturated with prayer, to feel prayer rugs under my bare feet, to hear the sound of hymns and chants sung by believers. I longed to be able to say that I *know* God, that I feel His presence every moment, everywhere, even typing in front of this computer screen. I envy people whose parents trained them to believe, who don't have to battle intellect to make room for faith.

In March of 1998, I had my first extended conversation with a religious terrorist. He is an American who had been released from prison and was living in a Texas trailer park at the time we spoke. I called him in connection with an earlier project on terrorists' potential to use weapons of mass destruction. Although I had been studying and working on terrorism for many years by that time, none of what I had read or heard prepared me for that conversation, which was about faith at least as much as it was about violence.

Kerry Noble had been second-in-command of a violent apocalyptic cult active in the 1980s, whose members were convicted of murder, firebombing a synagogue and a church that accepted homosexuals, conspiracy to assassinate federal officials, and other crimes. They had stockpiled cyanide with the aim of poisoning major city water supplies and, like Timothy McVeigh ten years later, plotted to bomb the Oklahoma City Federal Building. Their political goals included racially cleansing the United States; bringing down the U.S. "Zionist occupied" government and replacing it with a Christian one; destroying multilateral institutions such as the United Nations and the World Bank, the same institutions that Al Qaeda would come to describe as instruments of Western domination; and stopping the creation of a new world order based on "humanism" and "materialism."[1] They believed that ridding the world of Jews, blacks, and sinners would facilitate the Apocalypse and the Messiah's return. "The major cities to us were like Sodom and Gomorrah, like the Tower of Babel," Noble explained. "Who would be judged? The homosexual; the liberal, idolatrous preachers; those officials in high places; the merchants of trade and usury; and all those who refused the word of the Lord. They were the enemy. And so they would have to die. . . . We wanted peace, but if purging had to precede peace, then let the purge begin."[2]

During our first conversation, Noble told me he spends a lot of time in meditation and prayer. He is an accomplished student of the Scriptures—he knows whole chapters of the Bible by heart. And he feels he has a personal relationship with God. He is still 100 percent certain that there is a God and that God is good, even though at an earlier period in his life he listened to God and ended up living on an armed compound in rural Arkansas, doing things he now feels were wrong. Much to my consternation, I found myself feeling envious of Noble's faith, even as I was horrified by his cult's plots and crimes. I wanted to keep talking with him. I wanted to understand how a person so obsessed with good and evil, with such strong faith, could be led so far astray.

I was the "Superterrorism Fellow" at the Council on Foreign Relations in Washington at the time I began this project. I started pestering a psychologist whose office was near mine in Washington. Steve Kull works as a pollster, but he is interested in spirituality and has sometimes assisted people who have had frightening mystical experiences. This man tells me he has seen God's hand in visions, I report to Steve. He has heard God's voice. He has experienced revelations. He has spoken in tongues. But he thought he could persuade the Messiah to return more quickly by killing people. Could it be that he really was having spiritual experiences, but misinterpreted them? Or is he simply mad? After meeting with Noble, Steve told me that in his view Noble was not mentally ill. This is a group phenomenon. Once inside an organization whose goals include killing, ordinary people can commit seemingly demonic acts.[3] According to psychiatrist Robert J. Lifton, who has studied Nazis and other violent, fanatical groups, cult members become two people: the self they were, and the new, morally disengaged killer self.[4] Some people are more susceptible to such doubling than others, often in response to trauma. Certain professions, including medicine, psychiatry, military work, and research, encourage doubling, at least to some degree. A surgeon incapable of suppressing his capacity to empathize would have trouble slicing open his patient's chest. This book is in part about how terrorist organizations foster extreme doubling, extinguishing the recruit's ability to empathize with his victim, encouraging him to create an identity based on opposition to the Other.

Steve said several things to me back then that I had trouble absorbing, some of which now make perfect sense. He listened to me discuss my conversations with Noble. I explained how exhausting it was to talk to Noble because, to make him feel at ease, I felt that I had to try hard to see the world through his eyes. I had to suspend judgment, to try to understand his view that killing mixed-race couples, homosexuals, blacks, and Jews was a way of worshiping God. Steve observed my struggle and told me, you won't be able to explain terrorism to others until you can completely empathize with the pain and frustration that cause it. You need to picture yourself joining the groups you study. This can't be a superficial feeling, he said. At least during the period you are speaking to them, you need to feel yourself ready to join their cause. You need to sustain that feeling—go into it completely—but at the same time trust that you will recover yourself at the end of the conversation.

This struck me as an impossible task at the time. Coming back to myself wasn't the hard part—it was the possibility of empathy that seemed impossibly difficult. The individuals I had met or studied up to that point seemed so irrational, and the crimes they had committed so evil. How could I, as a Jew, imagine becoming a Nazi? How could I believe—even for a moment—that killing was a form of worshiping God?

This is tricky moral ground. I realize that the reader may be curious about my own moral position in regard to these questions, so I provide a summary argument at the end of this chapter. Despite this trickiness, however, I held fast to the view that informing myself about what leads to evil, as Kathleen Norris suggests, "from the inside out" is the best way to fight it.[5]

It is important to point out that empathy does not necessarily imply sympathy. To empathize is "to understand and to share the feelings of another," without necessarily having feelings of pity or sorrow for their misfortunes, agreeing with their sentiment or opinions, or having a favorable attitude toward them—the feelings that define sympathy.[6] It is a kind of vicarious introspection.[7] Although empathizing with a religious-extremist killer is difficult, I discovered that it can be learned. It is possible to understand and vicariously share the feelings that give rise to terrorism—if only briefly—and still maintain that the terrorist's actions are immoral, or even evil.

The hardest part to deal with, I told Steve, was the religious aspect. Although I was brought up in a secular household, I had a prejudice in favor of religion when I started this project. My image of religion was based on two formative experiences: my reading of Simone Weil in high school, and my exposure to a nun named Sister Miriam Therese. Simone Weil struck me (and undoubtedly many teenage girls) as an extremely romantic figure. She was a brilliant French Jewess, born to a family of secular intellectuals. She read the paper aloud to her family by the time she was five years old and mastered Greek and several modern languages by her early teens.[8] Her determination to understand the nature of suffering led her to relinquish temporarily her profession as a philosophy teacher to work variously as a factory hand, as a farmworker, and as a cook on the front lines of the Spanish Civil War. Although her physical awkwardness prevented her from excelling at any of these tasks (her most often cited accomplishment as a cook was to spill boiling water on her leg, wounding herself), the work strengthened her compassion and her interest in helping the less fortunate. She struggled to believe in God, eventually finding faith through music and poetry. She became a Christian after listening to monks singing Gregorian chants and reading the poetry of John Donne and George Herbert. Had she accepted baptism, it is likely she would have been canonized for her work with the poor and for her philosophical writings.[9]

Sister Miriam Therese was my grandmother's best friend. I saw that her faith fueled her desire to help the poor and abandoned in my grandmother's town of New Rochelle, New York. She considered my grandmother to be her "Jewish mother" and often invited my grandparents, my sister, and me to her convent—which struck me as a kind of "goodness spa," where, if a person would only spend enough time in contemplation, she would become perfectly good.

It seemed to me, in short, that faith made people better—more generous, more capable of love. Meeting Noble made me reconsider this position. Noble and other Christian terrorists I had spoken with are profoundly religious. They spend a lot of time in meditation and prayer. They are interested in good and evil, even if, from my perspective, they are confused about which is which. Some are intellectuals. How is it that people who profess strong moral values, who, in some cases, seem truly to be motivated by those values, can be brought to do evil things? Is there something inherently dangerous about religion? How can it be that the same faith in God that inspired Michelangelo, Mozart, Simone Weil, and Sister Miriam Therese also inspires such vicious crimes? Why, when they read religious texts, do these terrorists find justification for killing innocents, where others find inspiration for charity?

These are the questions that inspired this book—and once they occurred to me, I couldn't let them go. I couldn't stop this quest. My curiosity compelled me to travel, far beyond Kerry Noble's Texas trailer park, to Lebanon, Jordan, Palestine, Israel, India, Indonesia, and Pakistan. I can't pretend to have answered these questions definitively, but I have learned something about them, which I hope to share with others interested in the topic of terrorism.

I soon realized that the grievances Noble described were similar to those of religious extremists around the world. Al Qaeda's complaints about the new world order sound remarkably similar to Kerry Noble's, for example. Ayman Zawahiri, Osama bin Laden's second-in-command, accuses Western forces of employing international institutions such as the United Nations, multinational corporations, and international news agencies as weapons in their "new crusade" to dominate the Islamic world. The new world order is "humiliating" to Muslims, he writes.[10] Religious extremists see themselves as under attack by the global spread of post-Enlightenment Western values such as secular humanism and the focus on individual liberties. Zawahiri accuses the "new crusaders" of disseminating "immorality" under the slogans of progressiveness, liberty, and liberation.[11] Many see America's way of life as motivated by evil, "Satan," "bad for the human being," and overly materialistic. "Globalization," a Hezbollah militant told me, "is just another word for McDonaldization."

They often reject feminism in favor of "family values," whether their families are in Oklahoma or Peshawar. They see themselves as defending sacred territory or protecting the rights of their coreligionists. They view people who practice other versions of their faith, or other faiths, as infidels or sinners. Because the true faith is purportedly in jeopardy, emergency conditions prevail, and the killing of innocents becomes, in their view, religiously and morally permissible.[12] The point of religious terrorism is to purify the world of these corrupting influences.

But what lies beneath these views? Over time, I began to see that these grievances often mask a deeper kind of angst and a deeper kind of fear. Fear of a godless universe, of chaos, of loose rules, and of loneliness—fears that we all have to one degree or another. The religious extremists' angst is familiar, as is their fear. What surprised me most was my discovery that the slogans sometimes mask not only fear and humiliation, but also greed—greed for political power, land, or money. Often, the slogans seem to mask wounded masculinity. This book is about those deeper feelings—the alienation, the humiliation, and the greed that fuel terrorism. And it is about how leaders deliberately intensify those feelings to ignite holy wars.

The book takes two cuts at the problem. First we look at the issue of religious terrorism from the perspective of individuals. What are the grievances that lead individuals to join holy-war organizations? And once they join such organizations, what makes them stay? Why do they risk their lives in support of a purported public good and not ride free on the "soldiering" of others? Second, we look at organizations. What does a leader need to run an effective terrorist organization? How have terrorist leaders structured their organizations in response to the challenges and opportunities posed by globalization and technological change?

I studied these issues in several ways. I visited the schools that recruit cannon fodder for "jihads." I talked to leaders, public-affairs officers, trainers, and operatives. I talked with terrorists in jails, in their homes, and at their training complexes. I talked with government officials and religious leaders—both sympathizers and opponents of the terrorist groups I studied. I arranged to have locals administer detailed questionnaires, querying the terrorists about their motivations.

Before we examine what the terrorists say about themselves, it may help to discuss briefly how terrorism will be defined in these pages, and the ethics of interviewing terrorists and of terrorism itself.

WHAT IS TERRORISM?

The student of terrorism is confronted with hundreds of definitions in the literature.[13] Some definitions focus on the perpetrator, others on his or her purpose, and still others on the terrorist's technique. But only two characteristics of terrorism are critical for distinguishing it from other forms of violence. First, terrorism is aimed at noncombatants. This characteristic of terrorism distinguishes it from some war-fighting. Second, terrorists use violence for dramatic purpose: instilling fear in the target audience is often more important than the physical result. This deliberate creation of dread is what distinguishes terrorism from simple murder or assault.[14]

In this book terrorism will be defined as an act or threat of violence against noncombatants with the objective of exacting revenge, intimidating, or otherwise influencing an audience. This definition avoids limiting perpetrator or purpose. It allows for a range of possible actors (states or their surrogates, international groups, or a single individual) and all putative goals (political, religious, or economic).[15] This book is concerned only with terrorists who claim to be seeking religious goals, i.e., religious terrorism. It is limited to three monotheistic religions: Christianity, Islam, and Judaism. As we shall see, most religious terrorists promote a mixture of religious and material objectives, for example, acquiring political power to impose a particular interpretation of religious laws or appealing to religious texts to justify acquisition of contested territory.

The characteristics of terrorism, as we have listed them, in turn raise additional questions. How do we define *noncombatants?*[16] The term is controversial. A soldier on a battlefield is unquestionably a combatant, but what if his country is not at war, and he is sleeping, for example, in a military housing complex in Dhahran, as nineteen U.S. soldiers were in June 1996 when they were killed by a bomb? What if he is riding a bus also carrying civilians, as happens regularly in Israel, when a suicide bomber attacked? What if the soldier's country is not at war, but he is working at a Defense Department office, as was the case for the 125 killed when Al Qaeda attacked the Pentagon on September 11, 2001?[17]

Under these circumstances, many would claim that the soldier is not a combatant. But what if troops are sent into a country on a humanitarian mission? And what if those troops are perceived to be partisan? This question is likely to arise whenever U.S. leaders send the military on humanitarian missions.

A second thorny issue is the perpetrator of the violent act. Can a state commit acts whose purpose is to intimidate noncombatants, acts that might be labeled terrorism? The answer is yes. States can and do unleash terrorist violence against their own civilians, as Saddam Hussein did with chemical weapons against Iraqi Kurds; as Stalin did, in acts of random violence against Soviet citizens; and as the Guatemalan government did for nearly forty years against its own people. And states have also used terrorism as an instrument of war, by deliberately attacking civilians in the hope of crushing enemy morale. Although states frequently engage in terrorism, I am concerned in this book only with substate actors.

Two religious terrorist organizations from history are of particular interest for our purposes: the Zealots-Sicarii and the Assassins. The first was active around the time of Jesus Christ, the second during the eleventh to the thirteenth century. The technologies they employed were primitive: their weapons were the sword and the dagger. Nonetheless, these groups, inspired by religious conviction, were highly destructive and were active internationally.[18]

A Jewish group, the Zealots-Sicarii, survived only twenty-five years, but profoundly influenced the history of the Jews. The Zealots murdered individuals with daggers and swords. Later they turned to open warfare. Their objective was to create a mass uprising against the Greeks in Judea and against the Romans that governed both Greeks and Jews. The revolt had unforeseen and devastating consequences, leading to the destruction of the Temple and to the mass suicide at Masada. Later revolts inspired by the Zealots-Sicarii led to the extermination of the Jews in Egypt and Cyprus, the virtual depopulation of Judea, and the Exile itself, which, David Rapoport explains, became central features of the Jewish experience over the next two thousand years. "It would be difficult to find terrorist activity in any historical period which influenced the life of a community more decisively," he observes,[19] though the influence was not what the terrorists had intended.

The Assassins, or Ismailis-Nizari, operated over two centuries, from 1090 to 1275. Their aim, like that of Islamist extremists today, was to spread a "pure" version of Islam. They stabbed their victims at close range in broad daylight. Under these circumstances, escape was nearly impossible. Like contemporary suicide bombers, they considered their own lives to be sacrificial offerings. Unlike today's suicide bombers, the Assassins murdered particular individuals—prominent politicians or religious leaders who refused to accept the new preaching. Despite their primitive technique, the Assassins seriously threatened the governments of several states, including those of the Turkish Seljuk Empire in Persia and Syria.[20]

The twenty-first century is seeing a resurgence of holy terror—the kind practiced by the Zealots-Sicarii and the Assassins.[21] Unlike their predecessors, however, today's terrorists attack randomly, targeting people whose only crime is to be in the wrong place at the wrong time. Religious terrorist groups are more violent than their secular counterparts and are probably more likely to use weapons of mass destruction.[22] It is for these reasons that I decided to focus exclusively on religious terrorism in this book, in addition to my intense curiosity about why people who are obsessed with good and evil end up murdering innocents, somehow slipping into becoming more evil than the evil they aim to fight.

How should we respond to this evil? One approach to evil insists that we look the other way to avoid being contaminated. Another, according to philosopher Susan Nieman, insists that morality demands that we make evil intelligible. It is, of course, the latter approach that I am adopting here.[23]

THE ENCOUNTER WITH EVIL

When I was asked to take a stand on the "evil" of terrorism, my first response was that I'm not a priest. I had no hesitation saying that terrorists are morally wrong. It doesn't matter how compelling their grievances, or how familiar their pain, it's terribly wrong to kill innocents. But I have come to think Hannah Arendt's conception of evil certainly applies—the unthinking evil of the person who follows rules that are morally wrong—and *wrong* is too weak a word. The person who commits atrocities. That is what they do—they commit atrocities. I decided it was important to learn something about evil in order to take a stand.

Theologians, psychologists, and moral and political philosophers, among others, have various perspectives on what constitutes evil, its causes, and how to fight it. Philosophers traditionally identify three kinds of evil: moral evil—suffering caused by the deliberate imposition of pain on sentient beings; natural evil—suffering caused by natural processes such as disease or natural disaster; and metaphysical evil—suffering caused by imperfections in the cosmos or by chance, such as a murderer going unpunished as a result of random imperfections in the court system. The use of the word *evil* to describe such disparate phenomena is a remnant of pre-Enlightenment thinking, which viewed suffering (natural and metaphysical evil) as punishment for sin (moral evil).[24]

If we look to literature and the Bible for our understanding of moral evil, we find evil men acting deliberately, often out of envy, sometimes in a fit of rage or apparent possession. Cain murders his brother out of envy that the Lord took more pleasure in Abel's sacrificial offerings. Iago persuades Othello that his wife was unfaithful. Othello is eventually driven mad by Iago's lies and murders his beloved wife. Iago's evil arises from his disappointment in his own professional failures and because he envies the Moor's goodness. Men may become evil by giving in to selfishness, despair, or ennui, as is the case for Stavrogin in *The Possessed.* The heroes of the Marquis de Sade's novels are perfect villains who planned their crimes in detail and took sensual pleasure from their victims' pain. But at the trial of Adolf Eichmann, Hannah Arendt observed another kind of moral evil: men who comply, unthinkingly, with evil rulers, regulations, or unfair systems, perpetrating unspeakably cruel acts. In this "banal" form of evil, perpetrators shut off the knowledge that their victims are human beings. It is this kind of evil that I observe in the terrorists described in these pages. The Evil One does not possess them. They love their families, they give alms to the poor, they pray. A guest, even a stranger espousing offensive religious or political views, is likely to be treated with respect and generosity. But they have lost the ability to empathize with their victims. This book is partly about how leaders bring themselves and their followers to the point where their empathy for victims is gone. Over time, some operatives, who may begin their terrorist careers as evil in Arendt's sense, will become accustomed to inflicting pain. They may even begin to take pleasure from atrocity in the name of "purification." One of the problems with employing unjust means for (subjectively determined) just ends is that violence and crime can become second nature. By resorting to terrorism, a man whose ends are undeniably just becomes a criminal.[25]

Another view, subscribed to by some psychoanalysts, is that evil arises from trauma. When the pain of trauma is so great that the victim cannot sustain feeling, he too becomes susceptible to propagating further evil, and evil thus proliferates.[26] In this case, suffering can lead to sin, rather than—as pre-Enlightenment philosophers believed—sin leading to suffering.[27] Absent intervention, victims of genocidal wars may raise tortured children who, in turn, are more susceptible to harm their own children psychologically.[28] Male children raised in cultures of violence are more likely to become delinquents or violent criminals.[29] Not surprisingly, many of the terrorists described in this book grew up in failed or failing states where violence was commonplace.

For Jung, evil was inherent, not only in every human being, but also in God. He viewed evil as an archetypal Shadow, an aspect of the unconscious that cannot be controlled, but can be integrated. When it is integrated, it becomes a source of creativity. When it is repressed, it can lead to overt acts of evil such as terrorism. All of these approaches to evil seem to me to be important, not only for understanding terrorism, but also for developing an effective response.

Some terrorism is evil in a straightforward way. The September 11 hijackers, for example, plotted their attack for years. They may have felt themselves grievously wronged by U.S. policies, but their victims were not responsible for creating or implementing them. The hijackers issued no ultimatum. Many of the victims were not American. Malice and forethought, the classic components of evil intentions, have "rarely been so well combined," philosopher Susan Nieman observes.[30]

Before September 11, we had grown used to complex villains, whose evil was less immediately apparent. We were in the habit of thinking about evil in Arendt's terms—ordinary people contributing, like cogs in a wheel, to evil outcomes. "Wall Street seemed determined to show us that everything could be bought and sold, the Pentagon bent on renewing the pre-Socratic belief that justice means helping your friends and hurting your enemies," Nieman writes. "Those whose conceptions of evil were always simple and demonic were happy to see them confirmed," she tells us. But for those of us whose conceptions of evil had been shaped not by Hollywood but by Vietnam, Cambodia, and Auschwitz, this "single-mindedly thoughtful evil" caught us entirely unprepared.[31]

Few of the terrorists described in these pages are single-mindedly thoughtful villains like those who masterminded the September 11 attacks. In some cases, determining the ethical basis of their actions is complicated. Many are followers, not leaders. Some fight militaries at least some of the time.[32] In rare cases, action that would otherwise be defined as terrorism could be construed as just, for example, in self-defense or to defend others from imminent death, where no other options are available. In such hypothetical cases, "just terrorism" could be consistent with just-war teachings.

Although none of the terrorism described in this book can be described as morally acceptable, at least in my view, the pro-life doctor killers probably come the closest and are worth examining in detail for that reason.[33] Unlike the September 11 hijackers, the doctor killers are discriminating: they target individuals who, in their view, are in the business of murder. If we accept their assumption that a fetus is a human being, it is easy to follow the moral logic that leads doctor killers to conclude that killing abortion providers is "justifiable homicide," even if we condemn their actions.

Doctor killers assert that "ensoulment" begins at conception. While many of us feel uncertain about precisely what ensoulment entails, if it exists, viability can be tested empirically. It is reasonable to assume, given the direction and pace of medical advances, that it will soon be possible to sustain and grow a fertilized egg outside the womb. So the assertion that a fertilized egg is a human being is difficult to reject out of hand, even for those of us who utterly condemn the doctor killers' actions.

Doctor killers see themselves as the moral equivalent of the abolitionists in the period before the American Civil War.[34] The abolitionists recognized the humanity of the slaves and felt that God too recognized their humanity—the same argument used by the doctor killers in regard to the unborn. Most people around the world accept that slavery is morally wrong. But the violent abolitionists went one step further than condemning slavery and working to stop it. They felt that the slaves' situation was so dire that terrorism was warranted to secure the slaves' release. Like John Brown and other violent abolitionists, the doctor killers believe that the risk to unborn children is sufficiently grave to warrant murder.

Even if we accept the view that abortion is morally wrong, if only for the sake of moral exploration, that doesn't mean that killing doctors who provide abortions is morally acceptable. Abraham Lincoln argued again and again that the institution of slavery was wrong. And yet he also argued that it was wrong to be "so impatient of it as a wrong as to disregard its actual presence among us and the difficulty of getting rid of it suddenly in a satisfactory way."[35] There were "constitutional obligations thrown about" the institution of slavery, and these could not be ignored without putting the Union at risk, he said.[36] For these reasons, Lincoln supported punishing the terrorist abolitionists, even though he concurred entirely with their cause. Lincoln's argument applies equally to the case of the doctor killers.

The religious terrorist's moral error is partly his impatience, to use Lincoln's word—or put another way, his zealotry. By taking the law into their own hands, the doctor killers—and other terrorists who claim to be motivated by moral concerns—make a grave moral error by putting at risk institutions that are a critically important part of our moral world. When they murder, pro-life killers step over the line from activists to criminals.[37]

The tendencies to focus on a single value to the exclusion of others, to use morally unacceptable means to address genuine grievances or achieve defensible goals, and to turn to violence when other means are available for achieving the same goal (even if more slowly) are common among religious terrorists all over the world. Thus, an additional question explored in these pages is this: Why and how do people who may be particularly sensitive to the suffering of others or to spiritual wrongs, who are motivated—at least initially—by a desire to purify the world of political and spiritual corruption, evolve from activists into murderers?

Writing this book has helped me to understand that religion is a kind of technology. It is terribly seductive in its ability to soothe and explain, but it is also dangerous. Convents such as the one I visited as a child may make good people better, but they don't necessarily make bad people good. They might even make bad people worse.

Religion has two sides—one that is spiritual and universalist, and the other particularist and sectarian. We should not turn away from this dangerous aspect of religion in an attempt to remain uncontaminated. We must recognize the seductiveness of sectarianism to understand the extent of the danger.

The philosopher Martha Nussbaum tells a wonderful story about her mentor, John Rawls, who felt that the dangerousness of the Wagnerian view could be comprehended only if one understood its appeal.

"I recall a conversation with him about Wagner's *Tristan,* when I was a young faculty member," she writes. "I made some Nietzschean jibes about the otherworldliness of Wagnerian passion and how silly it all was. Mr. Rawls, with sudden intensity, said to me that I must not make a joke about this. Wagner was absolutely wonderful and therefore extremely dangerous. You had to see the danger, he said, to comprehend how bad it would be to be seduced by that picture of life, with no vision of the general good."[38]

Religious terrorism attempts to destroy moral ambiguities. But we should be wary of succumbing to the extreme dualist view that the perpetrator is a manifestation of pure evil, rather than a suffering human beleaguered, as we are, by unmet aspirations, negation, and despair. In *The Origins of Satan,* Elaine Pagels explores the evolution of Satan from his roots in the Hebrew Bible. Pagels tells us that the evolving image of Satan served "to confirm for Christians their own identification with God and to demonize their opponents—first other Jews, then pagans, and later dissident Christians called heretics."[39] "The use of Satan to represent one's enemies lends to conflict a specific kind or moral and religious interpretation," she argues, "in which 'we' are God's people and 'they' are God's enemies, and ours as well. . . . Such moral interpretation of conflict has proven extraordinarily effective throughout Western history in consolidating the identity of Christian groups; the same history also shows that it can justify hatred, even mass slaughter," she observes.[40] This is the way religious terrorists view the world. Their commitment to a religious idea or a religious group leads them to dehumanize their adversaries to a degree that they become capable of murder. They start out with the intention to purify the world of some evil, but end up committing evil acts. Pagels's words teach us not only about the terrorists, but also about ourselves, and our own capacity to become counterterrorism zealots—dehumanizing our enemies, putting innocent civilians at risk. It is an approach we should strive to avoid if we aim to succeed in countering them.

What is so deeply painful about terrorism is that our enemies, whom we see as evil, view themselves as saints and martyrs. As such, religious terrorism is more than a threat to national security. It is psychological and spiritual warfare, requiring a psychologically and spiritually informed response. We cannot hope to develop such a response without analyzing the terrorists' methods, including skillful marketing of grievances as spiritual complaints and targeted charitable giving to generate support.

A psychologically and spiritually informed response demands that we understand that religious terrorists aim not only to frighten their victims in a physical sense, but also to spread a kind of spiritual dread, to shift their own existential dread of cultural and spiritual defeat onto their victims. Thus, fighting religious terrorism also requires examining not only our propensity to overreact in the face of such fears, including by demonizing the perpetrators and their supporters or coreligionists, but also how our actions and reactions play into their hands.

Although we see them as evil, religious terrorists know themselves to be perfectly good. To be crystal clear about one's identity, to know that one's group is superior to all others, to make purity one's motto, and purification of the world one's life's work—this is a kind of bliss. This is the bliss offered to those who join religious terrorist groups.[41] Participants in the Crusades, the Inquisition, and the kamikaze suicide-bombing raids all understood the appeal of purifying the world through murder. It is a bliss I have seen among the terrorists described in this book. This powerful yearning for bliss cannot be denied if we are to fight terror in the name of God, the gravest danger we face today.

A METHODOLOGICAL NOTE

Many people ask why terrorists have been willing to talk to me. I was acutely aware that the men I interviewed for this book must have been attempting to use me for something. The Pakistani jihadis, for example, seemed to feel they had not received enough attention from the U.S. government, and initially thought I was working for the CIA. They were probably right that the CIA was ignoring them at the time we started speaking, although they can hardly feel neglected now. Often, my interviewees hoped that I would broadcast their message to the world. Sometimes they spoke to me out of loneliness. That applies especially to terrorists in prison, for whom the experience of speaking with a woman who hangs on their every word, with no interruptions, was obviously a rare pleasure. Many of my interlocutors hoped to change my mind; they often asked me to join their cause or convert to their religion. Sometimes, they denied killing civilians—either because they were lying; because they considered enemy civilians to be potential soldiers, as Hamas sees Israeli children; or because they truly did focus on military targets (which in my view, makes them paramilitaries or mercenaries, not terrorists).

But the flip side is that I was using them too. I wanted to understand how they view the world and how they feel, in order better to understand how to stop them. On rare occasions I tried to persuade young men to turn back. But mostly, I took advantage of their desire to be heard. I did not share my own views. It was not a normal human encounter.

It is also important to point out that my interviewees lied to me. Sometimes, but undoubtedly not always, I knew when they were lying. Their lies revealed not only what they considered particularly sensitive information, but also their fantasies about what would impress me, frighten me, or make me sympathetic to their cause. In some cases, I suspected that intelligence agencies had briefed operatives in advance about what was appropriate to tell me or even arranged for me to encounter individual operatives. Readers should be alert to possible lies.

Another problem is that I had limited access to terrorists; my sample was far from random. Some terrorists refused to speak with me; and some I was afraid to approach. Although I spoke with jihadis from many of the groups that are member organizations in Osama bin Laden's International Islamic Front, I did not try to talk to members of Al Qaeda (other than one that was already in U.S. custody) because I was afraid.

A rigorous, statistically unbiased study of the root causes of terrorism at the level of individuals would require identifying controls, youth exposed to the same environment, who felt the same humiliation, human rights abuse, and relative deprivation, but who chose nonviolent means to express their grievances or chose not to express them at all. A team of researchers, including psychiatrists, medical doctors, and a variety of social scientists would develop a questionnaire and a list of medical tests to be administered to a random sample of operatives and their families.

Furthermore, to identify more systematically the attributes of terrorist organizations correlated with "success" (however defined), it would be important to examine a large variety of groups with a variety of purported goals, and attempt to tease out the causative factors. I am hopeful that the "risk factors" I identify in chapter 10 can serve as the basis of more systematic research by future scholars.

All subjects involved in this study were informed of how the material would be used. Where possible, I asked subjects to review not only my notes of the interviews but in some cases, the chapters in which they were mentioned.

In preparing to systematize my interviewing of terrorists, including developing a questionnaire administered to subjects in Pakistan and in India, I had to present the project to the Standing Committee on the Use of Human Subjects at Harvard University. All universities undertaking grant activities have such rules, which were originally established to prevent abuses in medical and psychological experiments. Normally, researchers are required to obtain signed consent forms from the subject. In this case, the Board allowed me to write a script explaining who funded the project and how the results would be used. I made clear to the Human Subjects Board that in the unlikely event any information I obtained could save someone's life, I considered my responsibility to possible victims to be more important than my responsibility to the Committee on Human Subjects or the subjects themselves.

Part I

Grievances That Give Rise to Holy War

Part 1 of this book explores the kinds of grievances that give rise to terrorism in the name of God. We learn in the first half of this book how leaders exploit feelings of alienation and humiliation to create holy warriors; and how demographic shifts, selective reading of history, and territorial disputes are used to justify holy wars.

Part 1 addresses the question: Why do some people respond to these religious grievances by joining terrorist groups, and once they join, what makes them stay? We learn, through the terrorists' stories, that the benefits they receive are partly spiritual, partly emotional, and partly material. Terrorism involves a collective-action problem, in the sense that only those who contribute incur the costs, but a broader collective shares the benefits. The theory of collective action suggests that people tend to "free ride" on others' contributions to collective goods. It suggests, for example, that it is irrational to pay taxes if there is no enforcement mechanism because we can reap the benefits of others' contributions whether or not we write a check to the government.[1] It is possible to encourage collective action through positive incentives (rewards or payments) or penalties for noncompliance (corporal punishment, incarceration, or fines).

When Jewish extremists attempt to lay a cornerstone for the Third

Temple they hope to build, all like-minded messianic Jews (and messianic Christians) benefit. Only the participants pay: When they ascend the Temple Mount, they incur risks to their person, livelihood, freedom, and families. Given this, the extremist should be asking himself: Why bother participating? Why not let others do the work and take the risks?

Participation in terrorist violence can be seen as kind of tax paid to redress the collectives' grievances. Those who contribute their lives, their money, or their support are paying their taxes; those who do not are free riders. The metaphor may sound far-fetched, but an Al Qaeda member has used precisely this language to chastise non-violent Muslims who don't contribute to Al Qaeda's goals. Ramzi bin al-Shibh, a mastermind of the September 11 attacks, describes violence as "the tax" that Muslims must pay "for gaining authority on earth." He says that "it is imperative to pay a price for Heaven, for the commodity of Allah is dear, very dear. It is not acquired through rest, but [rather] blood and torn-off limbs must be the price." The moral "obligation of jihad" is equally important as the duties of prayer and charity, he says. He urges Muslims to "grasp this understanding," claiming that the punishment awaiting those who neglect the obligation to pay their "taxes" by waging jihad will be "painful and harsh."[2]

Terrorist leaders encourage operatives to participate in terrorist violence by holding out the promise of heavenly rewards or the threat of heavenly retribution. Some operatives participate because they fear being punished in the afterlife, as Ramzi bin al-Shibh suggests, or because they desire to be virtuous (in their view) for its own sake. But leaders also offer material and emotional incentives—both rewards and punishments. They provide cash payments for successful operations. They provide money to "martyrs'" families. Recruiters in Kashmir coerce families into donating their sons by demanding large payments or the use of a child. One Al Qaeda recruit told his interrogators that the atmosphere at the training camps was one of intense psychological pressure enforced by the torture of those who did not embrace the violent code.[3]

Some operatives will admit they got involved in terrorism out of a desire for adventure.[4] Many join out of friendship or through social networks. In some cases, the desire to be with friends turns out to be more important, over time, than the desire to achieve any particular goal. Others are attracted to the "glamour" of belonging to a militant group. One operative told me about the appeal of living outside normal society under extreme conditions, on a kind of permanent Outward Bound. Some get involved in violent groups out of a sense of alienation and anomie. Once part of a well-armed group, the weak feel strong and powerful, perhaps for the first time in their lives. Some admit that they find guns and violence appealing. For such individuals, there are clear emotional benefits to belonging to violent groups. In short, fun and profit—status, glamour, power, prestige, friendship, and money—provide powerful incentives for participating in terrorist groups.

But fun and profit do not explain the whole picture. Foot soldiers are likely to receive no monetary compensation. They are often recruited from extremist religious seminaries where they are indoctrinated from an early age about the spiritual importance of donating their lives to a holy war. The September 11 hijackers apparently were not paid. Fun and profit also do not explain how an organization begins. "Why and how . . . the group committed from the start to fundamental transformation of the structure of power . . . remains one of the mysteries of our time," sociologist Charles Tilly famously observed in regard to social movements.[5] And yet revolutions and violent social movements do come about, much to the puzzlement of rational choice theorists. Something other than fun and profit appears to be at play.[6]

In real life (as opposed to elegant, parsimonious theory), people have mixed motives for everything they do. We may desire to do the right thing, but we may want our efforts to be noticed and rewarded—perhaps by God, perhaps by other people. Terrorists, similarly, have mixed motives. They see themselves as purifying the world. They believe that murdering the group's "enemies" is a way to "do good" or to "be good."[7] As some terrorists define it, virtue may be its own reward. But operatives may be influenced simultaneously by more pragmatic incentives, possibly including money for themselves or their families.

What seems to be most appealing about militant religious groups—whatever combination of reasons an individual may cite for joining—is the way life is simplified. Good and evil are brought out in stark relief.

Life is transformed through action. Martyrdom—the supreme act of heroism and worship—provides the ultimate escape from life's dilemmas, especially for individuals who feel deeply alienated and confused, humiliated or desperate.

When religious terrorist groups form, ideology and altruism play significant roles. Commitment to the goals of the organization, and the spiritual benefits of contributing to a "good cause" are sufficient incentives for many operatives, especially in the initial phase of the organization. Over time, in some cases, cynicism takes hold. Terrorism becomes a career as much as a passion. What starts out as moral fervor becomes a sophisticated organization. We will find in the pages that follow that grievance can end up as greed—for money, political power, or attention.

Astute leaders take advantage of the variety of motives that lead operatives to become terrorists. They do not rely on terrorists' (mistaken) moral convictions alone to sustain the group over time. They offer friendship, status, adventure, "glamour," and jobs. In commander and cadre-style organizations, leaders also realize they need a variety of recruits, some of whom will require material incentives in addition to moral, spiritual, or emotional ones.[8]

Although each chapter is named after a single grievance (alienation, humiliation, demographic shifts, historical wrongs, and claims over territory), multiple grievances play a role in the religious conflicts highlighted in these chapters, not just the one mentioned in the chapter's title. The goals of the terrorists discussed in part 1 also vary along three dimensions: from spiritual to temporal, from instrumental to expressive, and from ideological to profit-driven.

All the terrorists discussed in part 1 claim to be motivated by religious principles, but most pursue a mixture of spiritual and political goals. At the extreme religious end of the spectrum are the groups seeking eternal, spiritual goals such as redemption or helping to bring on the apocalypse and the Endtimes predicted in biblical texts. The Covenant, the Sword, and the Arm of the Lord, the Christian cult discussed in chapter 1, and the Jewish Underground, one of the terrorist groups discussed in chapter 4, probably come closest to this ideal type. Both were interested in influencing the process and timing of the apocalypse. Neither was seeking political power, at least when they started out. At the opposite end of the spectrum are the groups that are mainly pursuing temporal, pragmatic goals on this earth. They may propose to impose religious laws, but their principal interests are obtaining political power or expanding their territory. For example, some of the worst religious violence in Indonesia, discussed in chapter 3, has arisen in areas where indigenous groups living in natural-resource rich regions are seeking greater autonomy or independence. Hamas, discussed in chapter 2, claims to be protecting coreligionists from assault by other religious groups, but is largely focused on achieving political power and asserting control over the whole of Israel. Jewish extremist Avigdor Eskin invoked an ancient mystical prayer to bring about the death of Israeli Prime Minister Rabin. Despite his fascination with mysticism, however, Eskin is mainly interested in altering the situation in this world. He wanted Rabin to die because he was giving away "Jewish" territory to Muslims. Eskin and others discussed in the chapter are raising money to create a "genuine" right-wing party in Israel.

Terrorists also vary in their desire to accomplish something. Sometimes they are businesslike in their pursuit of objectives. The objective could be to frighten the enemy or damage an economy. It could be to force the enemy to overreact, thereby demonstrating his ruthlessness or weakness. It could be to impose religious laws. But sometimes the purpose is expressive rather than instrumental. The aim is to convey rage or to exact revenge with little thought to long-term consequences. For whom is the message intended? Usually we think of the audience for terrorism as the victims and their sympathizers. But attacks sometimes have more to do with rousing the troops than terrorizing the victims. Bin Laden, for example, appears to believe that spectacular attacks make him more appealing to his followers. In his words, people follow the strong horse, and abandon the weak one.

Terrorists groups also vary in terms of the extent to which ideology matters. Some terrorist organizations transform themselves, over time, into profit-driven organizations for which crime is an end rather than a means. These groups switch from grievance to greed.

The groups discussed in part 1 also vary in size, organizational sophistication, and potential to cause mass casualties. Some engage in terrorism essentially full time; while others engage in terrorism as a kind of hobby. Some are mainly involved in fighting enemy troops, resorting to terrorism (targeting noncombatants) only occasionally. For still others, the charitable and political wings of the organization are equally as important as the military ones.

ONE

Alienation

This chapter tells the story of a group of alienated individuals who joined a religious fellowship in rural Arkansas. After the leader received a "revelation" that the Endtimes had begun, the cult began "fusing together in one body" as directed by a prophetess living on the compound. They burned family photographs, sold their wedding rings, pooled their earnings, and destroyed televisions and other "reminders of the outside world's propaganda." They also began stockpiling weapons to prepare for the "enemy's" anticipated invasion. But the Apocalypse—and the battle between good and evil forces—failed to materialize on the appointed hour. Each failed prophecy was followed by a revised forecast. Instead of giving in to despair that their dream of the Endtimes might not materialize, cult members' confidence grew stronger. They intensified their military training, acquired more powerful weapons, and purified themselves to prepare to vanquish the forces of evil.

By examining this cult, we learn how leaders develop a story about imminent danger to an "in group," foster group identity, dehumanize the group's purported enemies, and encourage the creation of a "killer self" capable of murdering large numbers of innocent people. This chapter focuses on the evolution of a cult member named Kerry Noble. We observe how the leader cunningly capitalized on Noble's need to feel important inside the group, and how, over time, Noble was transformed from a gentle but frustrated pastor seeking transcendence to a terrorist prepared to countenance "war" against the cult's enemies—blacks, Jews, "mud people," and the U.S. government.

On April 19, 1985, two hundred federal and state law-enforcement agents staged a siege at a 240-acre armed compound in rural Arkansas inhabited by a Christian cult called the Covenant, the Sword, and the Arm of the Lord (CSA).[1] The cult had long been expecting an enemy invasion, and members had laid land mines around the periphery of the property. They had stockpiled five years' worth of food. James Ellison, the commander of the cult, wanted to shoot it out with the feds. Danny Coulson, head of the FBI's Hostage Rescue Team, eventually persuaded Ellison that the cult would lose such a battle. Coulson said he had a Huey helicopter, just over the hill, which would level the place if a cult member fired a single shot. He also said that an aircraft circling the property was equipped with heat-seeking devices. "We can watch your every move, day or night," he said. He told cult members that he had an armored personnel carrier around the bend, and weapons so advanced and new that the military didn't have them yet. "Your organization is considered by the government to be the best-trained civilian paramilitary group in America. That's why we're here. We're only sent against the best," he told the cult's second-in-command, Kerry Noble, who had been sent to negotiate with the enemy.[2]

The FBI asked the Reverend Robert Millar, a leading cleric of the American racist right, to help negotiate with the cult. Millar reports that he saw 150 men in camouflage, plus FBI and ATF agents, a SWAT team, and "a few Mossad agents," scattered in the woods around the compound, whom he blamed for provoking a "tense and dangerous confrontation."[3] "If it comes to a fight, hand me a gun, show me how to use it, and I am with you," he says he told Ellison.[4]

Three days after the siege began, the Covenant's "Home Guard" surrendered. The Reverend Millar was disappointed. "It ended with the whole group walking out, the womenfolk carrying their Bibles and singing, the men handing over their carbines."[5] When government officials searched the compound, they found a large cache of weapons, including fifty hand grenades; seventy-four assault weapons; thirty machine guns; six silencers; an M-72 antitank rocket; a World War II–era antiaircraft gun; three half-pound blocks of C-4 plastic explosives; an unfinished, homemade armored personnel carrier; and a large drum of cyanide, which cult members intended to use to poison major-city water supplies.

The cult hoped to hasten the return of the Messiah by "carrying out God's judgments" against unrepentant sinners.[6] They believed that humanists, communists, socialists, and Zionists had taken over the U.S. government. They knew for a fact that Jews, Satan's direct descendents, were working closely with the Antichrist, whose forces included the United Nations, the IMF, the Council on Foreign Relations, the Illuminati, and the "One-Worlders."[7] They had discovered, through their intelligence channels, that the aim of this cabal was to create a world government, a clear sign that the forces of Satan were at work.[8] The cult planned to poison residents of major cities—far more people than any modern terrorist group has killed before. They had joined forces with other right-wing groups in the hope of destroying what they called the Zionist Occupied Government (ZOG). Cult members and their coconspirators were eventually tried for sedition, but in the end, the government lost the case.[9]

Kerry Noble was a "God-anointed elder" of the cult and, by the end, its second-in-command. I first contacted him by telephone in March 1998. He was living in Texas, now released from prison. He told me that the group had started preparing for "war" because there were signs of Armageddon. "We believed that once those signs were there, it was time for us to act, to make judgments against those who were doing wrong or who refused to repent," he says. "The original timetable was up to God, but God could use us in creating Armageddon. That if we stepped out, things might be hurried along. You get tired of waiting for what you think God is planning."[10]

Kerry said he regrets his involvement in leading the cult, especially its paramilitary activities. He is active in anticult programs and helped the FBI investigate the Oklahoma City bombing. He tells me he got involved in CSA because he listened to God, and that God encouraged him to stay with the cult even after it became violent. He told me, "God directs us in everything we do. God is in control of everything." He believes that God speaks to individuals and provides direction, although individuals can misinterpret what they hear. I am instantly reminded of the famous warning that anyone who believes God speaks to him directly should be careful—it might actually be the devil speaking.

I want to know more about this faith. Kerry defines faith as "believing and obeying God when He tells you to do something or when He leads you into a situation. When you believe that God is directing you, it also follows that He will provide [for] and protect you."[11] He says that faith in God is key to every action he has ever taken, "whether relocating, living communally, casting out demons, or having a baby by natural childbirth."[12] Faith must also have played a role, then, when he drove to Kansas City with plastic explosives and a .22 pistol to shoot some "queers and niggers," bomb a pornographic bookstore, and blow up a church frequented by homosexuals.[13] In the end he discovered that he didn't have the stomach for killing at close range, but other cult members did. That same God also leads him now, he tells me, in his efforts to help deprogram youth who have joined violent cults.

We spoke a number of times, but I was dissatisfied, irritated, and, frankly, confused. This was early in my study of religious terror, and I was still of the view that faith in God makes people better human beings. How is it possible that Kerry could retain faith in a God that he believes directed him to do things that were evil? Things that he now sees as morally wrong?

I e-mailed Kerry to ask him more questions. He wrote me back, "Hi, Jessica. It always blesses me when someone asks questions concerning spiritual matters, as you did, in trying to understand people. I can't tell you how much our last conversation meant to me. I will be glad to help you any way possible."[14] He also told me, "You are welcome to come down to visit; we would be honored to have you here. Just let me know when." I decided to take him up on his offer in June 1999.

Kerry lives in Burleson, Texas. The Burleson, Texas, Web site calls the town "more than a whistle stop. It's a visitor's delight!" But there are no hotels in Burleson, so I stay in a Holiday Inn in a neighboring town.

Kerry invites me to dinner. After mulling it over awhile, I conclude that one should bring a gift to one's dinner host, even if he is a former terrorist. A box of chocolates? A cake? Finally I settle on a bottle of wine. On a whim, I select a substantial merlot, which a flyer hanging from the shelf informs me was highly rated by Mr. Robert Parker. It is the most expensive bottle of wine I have ever bought.

I follow Kerry's detailed instructions, eventually finding my way to a quiet neighborhood consisting mostly of mobile homes. Kerry, his wife, Kay, and their children live in a mobile home behind his parents' house. Kerry's grandparents bought the mobile home for Kay and the children when Kerry was in prison. The mobile home has four bedrooms. It is twenty-eight feet wide and seventy-six feet long.

Kerry meets me at the door. He is a big man, six feet tall, and stout. He has puffy, unhealthy-looking skin, and a neatly trimmed gray beard. The crown of his head is covered with a thin, gray fuzz. Tracks of worry are visible on his pale brow. Former minister, penitent neo-Nazi, regretful attempted bigamist, repentant terrorist. Convicted of conspiracy to possess unregistered weapons and of receiving stolen property. The prodigal son, now living on his parents' land. Six children, three granddaughters. He moves slowly, apologetically, as though weighed down by a childlike shame, and perhaps an embarrassing resentment at having gotten caught. He wants to please, to be forgiven. Is that my role? I wonder whether I am the first Jew to have entered this mobile home.

He thanks me for the wine, which he immediately takes into the kitchen and puts in the refrigerator. I notice through the kitchen door that the table is already laid. Everything is ready. Everything is neat and clean.

I ask to use the bathroom. It is spotless.

Kerry invites me into the living room. He sits on a recliner, where he tells me he sat for nearly three days straight reveling in the quiet and space of the mobile home after his release from solitary confinement. I sit on a couch. In front of me is a glass of iced tea sitting on a dust-free coffee table. I notice the mingled smells of "fresh scent" antibacterial soap on my now very clean hands, lemon Pledge, and something like fear, someone's fear, but I'm not sure whose. Kay joins us. We make small talk. How was your drive? Hot weather in Texas. Hot weather in Washington too. I hear locusts humming out the window.

Kay invites us to sit down to dinner. Kerry takes the wine out of the refrigerator. He offers some to me and pours a tiny amount for himself. This is the first time they have had wine with dinner, Kerry tells me. Kay pulls a perfectly white Corning Ware casserole out of the oven. She has made us Kerry's favorite dish, she says, Mexican chicken: taco shells and precooked boneless chicken pieces baked in Campbell's cream-of-mushroom soup.

What was it like living on the compound? I ask Kay. "There was a secure, Christian atmosphere," she says. "We helped one another—taking care of each other's babies and children. Canning, gardening. It was a big communal thing. It was nice."

"Back to basics, living in the woods," Kerry adds. It was "heaven" for the children, he says—no worries, lots of other children to play with, and peaceful surroundings. Many of the children had a hard time adjusting to life after the cult dispersed. Kerry's eldest daughter had nightmares for several years after the siege, having to adjust to a society she didn't understand, Kerry says.

Cult members built their own houses, each one equipped with a bunker or a nearby foxhole in preparation for the enemy's imminent invasion. James Ellison, the leader of the cult, lived in a stone house. But the Nobles lived in a wood-frame house with no insulation, no electricity, and no running water. Kay washed the family's laundry by hand. Their outhouse was on the porch, and in the winter they could see their breath even while standing right next to the woodstove.

When the Nobles joined the cult, it was a "religious fellowship," not a violent cult. The group believed in "clean" living: beer drinking, smoking, recreational drugs, and cussing were not allowed at the camp. "At the time of the siege," Kerry says, "I had agents coming up to me during the standoff saying you had such a nice way of living. I wish I could live in something like this. Too nice to mess it up. We knew that at our best, we worked hard, cared about people. Tight community, kids didn't have to worry about drugs, kidnappings . . ."

Later, I speak with a former government agent who asked me not to reveal his name. I'll call him Keith. Keith quit his job because he was so upset by what happened with CSA. He admired the cult, he admits. "I had a problem to start with when this case was first assigned to me because of my understanding of what CSA represented," he tells me. "I came from a Baptist background so I was accustomed to fire and brimstone [and] I didn't see anything wrong with what Ellison was saying or doing. I got some heat for holding these views. But then I kept hearing from informants that it was not just a religious organization. When you look behind [the religious facade], you realize that this guy is not really what he holds himself out to be." It was almost as if James Ellison were two people, he says. A religious leader, but also a thug. He was extremely charismatic, Keith says: "I can tell you right now, if he walked from where he is into the next state, he would gather a lot of followers." Terrorists often strike people who know them as two different people: the family man and the killer.[15]

In the beginning, the group held Bible meetings almost every night. Kerry was the main Bible teacher. At these meetings, members sometimes "prophesied" about fellow members' "pride." When this happened, the person guilty of pride would sit in a special chair, and the others would lay hands on him and speak in tongues to "cure" him. The chair was a "symbol in which to humble oneself in front of the group before asking for prayer, openly confessing what one needed," Kerry explains.[16] Public shaming of members is one of the hallmarks of a cult.

Sometimes the prophecies were about the group's future path. At one session, a "prophetess" named Donna told the group that "we would have to be willing to sacrifice much for the good of the group, that only by our fusing together in one body could He accomplish His will in us," Kerry recalls.[17] Ellison instructed all the men to shave their beards and cut their hair short. Members would no longer receive payment for their logging work; the group would pool all its earnings. Members were encouraged to burn any remaining vestiges of their precult identities: photographs, keepsakes, and high school yearbooks. They destroyed televisions and radios and other "reminders of the outside world's propaganda." They sold their wedding rings. They received little or no information from the outside world.[18]

Cutting off information from the outside world and destroying personal possessions or anything that reminds members of their precult lives is another common practice among cults. A French fascist told Robert J. Lifton that he felt that by joining the SS (a cultlike organization), he was entering a religious order that required that he "divest himself of his past" to be reborn as a person capable of what Himmler called heroic cruelty.[19]

The creation of a new self, which Lifton calls doubling, helps to explain how "banal" operatives, in Hannah Arendt's sense, come to kill innocent civilians.[20] William James considered the potential for doubling to be inescapable, although he considered it most likely to occur in extremis, for example, when a person faces his own or a loved one's death.[21] Doctors report the need to develop a "medical self," which is capable of slicing open a patient's body while remaining relatively inured to the patient's pain and even death.[22]

In a study of the psychology of killing, military psychologist David Grossman found that without desensitivity training, most soldiers will not fire at enemies at close range. Nearly 80 percent of riflemen neglected, declined, or omitted to fire at an exposed enemy in World War II, even to save their lives or the lives of their compatriots. He found a similar tendency in earlier wars. After extensive desensitivity training, however, the nonfiring rate in Vietnam was only 5 percent.[23] Desensitivity training requires learning to see the enemy as less than human. Terrorists also employ this technique, partly by referring to the "enemy" as subhuman. Neo-Nazi hate groups refer to nonwhites as the "children of darkness," and Jews as the "destroying virus." Aryans are described as "pure," "the Chosen," and "the children of light."[24] As a terrorist's assignments gradually become more violent, his capacity for moral revulsion is worn down. Psychologists find that even ordinary, decent people can be trained to do extraordinarily cruel things.[25]

People tend to find and stay in positions that satisfy their psychological needs, so there is probably some degree of self-selection in who joins and stays inside terrorist movements and violent cults. Lifton argues that Nazi doctors who worked in the death camps were more ideologically committed to the cause but may have had, in addition, "greater schizoid tendencies, or been particularly prone to numbing and omnipotence-sadism, all of which also enhance doubling."[26] But these characteristics suggest only greater susceptibility; "normal" people are also capable of extreme disassociation and inflicting great harm, under certain conditions.

It is also common for cult leaders to demand that adherents donate all earnings, and in some cases possessions, to the group or its leaders. Shoko Asahara, for example, the leader of the Aum Shinrikyo cult best known for its use of chemical weapons on the Tokyo subway in 1995, demanded that members donate all their earnings to the group's cause.

CSA came to accept the teachings of Identity Christianity, which sees Anglo-Saxons as the "true Israel," America as a sacred land, and the Declaration of Independence and the Constitution as a God-inspired, Christian inheritance.[27] Ellison told the cult that Christians had turned away from Old Testament laws and were allowing enemies of Christ to rule the land. It was time to take that sacred land back from God's enemies, he said. As we will see, many religious terrorists imbue material objectives and objects (such as land) with a spiritual dimension, making it impossible to compromise because the land now has spiritual content.

Like more mainstream Protestant fundamentalists, adherents of Christian Identity take the Scriptures literally and focus a great deal on the Endtimes. A key difference is the understanding of when Jesus appears during the tribulation. "Pretribulation" fundamentalist Protestants believe that Jesus will save them from experiencing the Apocalypse through a "divine rapture," the simultaneous ascension to heaven of all good Christians.[28] Followers of Christian Identity expect to be present during the Apocalypse. Christian militants who subscribe to "posttribulation" beliefs consider it their duty to attack the forces of the Antichrist, who will become leader of the world during the Endtimes. He will offer the people a false religion and a single world government. The strength of international institutions promoting world government, including the United Nations and the international banks, are indications that the Antichrist is already here, they believe. Identity Christianity has become the dominant religion of the racist right in America. Adherents include Gordon Kahl, a leader in the Posse Comitatus movement who died in a shoot-out with the FBI in 1983 and became the first Identity "martyr"; Randy Weaver, whose wife and son were killed in a government siege at his house in 1992; and Timothy McVeigh, who was executed for killing 168 people in the 1995 bombing of the Oklahoma City Federal Building.

Like other apocalyptic sects in the past, some Christian Identity adherents believe that they are now experiencing the tribulations as described in Revelation, that America is currently the equivalent of the corrupted and depraved Babylon. Others are not sure, nor are they certain how long the period of tribulation will last. I called up the Reverend Robert Millar, the leading cleric of the American racist right, to ask his views. "Armageddon could come anytime in the next thirty, forty, fifty years, it might come in 2160, or it might even be five years from now," Pastor Millar tells me. "We are not really hung up on the date."[29]

In late 1979, after Kerry Noble had been living at the compound for over a year, Ellison called the group together. "The Jews have declared war on our race, promoting race-mixing and thereby polluting the pure seed of God," Ellison explained to cult members. "This ZOG, this Zionist Occupied Government, is killing our white babies through abortion! It is destroying white minds with its humanistic teachings of evolution! I tell you this—niggers may be descended from apes, but my ancestors never swung from trees by their tails. In order to preserve our Christian heritage and race, it is our right, our patriotic duty, to overthrow the Antichrist government!"[30] He continued, "Prepare war, O Israel! Wake up the mighty men! Let all the men of war come near. Beat your plowshares into spears and your pruning hooks into swords. Let the weak say, 'I am strong!' "[31] Kerry says that the men, dressed in camouflage and armed with rifles and pistols, shouted, "I am strong!"[32]

Terrorists frequently invoke the notion that they are protecting the in-group from pollution by impure outsiders. Ellison's speech also makes clear that he hopes to make his converts feel like new men—partly to foster the development of a killer self who wears combat fatigues, but also to make members feel strong inside the group, enhancing their commitment to the cult. Outside the cult, Kerry tells us, he felt weak and repeatedly humiliated. Inside, Ellison made him feel needed and strong.

The cult sold the pigs they had earlier raised for meat and, believing that they were the true children of Israel, started celebrating Passover. During their first Passover celebration, the men dressed in military uniforms. They prayed that the angel of death would come and kill the firstborn of all the "Egyptians" still living in America. "We were disappointed when we awoke the next morning not to find a plague having struck the nation," Kerry recalls.[33]

Ellison also subscribed to the "sonship" doctrine, which teaches that Jesus is only the head of God's Christ, or Anointed One. The body of Christ is composed of a number of people—possibly the 144,000 people mentioned in the Book of Revelation. Eventually the people that make up the body of Christ will achieve spiritual perfection. Although they retain their mortal bodies, these prophets become incapable of sin.[34]

This interpretation requires "spiritualizing" the biblical text.[35] To spiritualize a scripture is to see a deeper meaning than the literal meaning or the one commonly taught, Kerry explains.[36]

Cult leaders tend to create their own religions as they go along, freely "spiritualizing" texts, sometimes picking and choosing from a variety of religions, sects, or ideologies. The leader often has a "vision" that becomes an organizing myth, often related to hopes and fears about the end of the world as we know it, to be replaced by a new, fairer, better world. Shoko Asahara, the leader of Aum Shinrikyo, mixed Christianity, Hinduism, and Buddhism to legitimate his autocratic rule and to inculcate total commitment among his followers. He chose Shiva the Destroyer as the principal deity for his cult, but he also emphasized the Judeo-Christian notion of Armageddon. He used his yoga skills to impress adherents and to establish schools, raising money for the cult. He used longing for the new, post-Apocalypse world to urge his followers to hurry it along through violence, including with weapons of mass destruction.

Ellison believed that he and other cult members were among the people that made up the body of Christ. When he felt himself to have achieved the requisite level of perfection, he took a second wife. He felt that whatever he chose to do, he was no longer capable of sinning, so nothing was off-limits. Kerry planned to do the same, but his fiancée broke off the engagement just before the marriage was to take place. Cults frequently engage in unusual sexual practices, whether abstinence, free love, or polygamy. Shoko Asahara also took on multiple sexual partners among adherents.

Ellison employed essentially all the techniques for enhancing commitment that cults traditionally favor. These include sharing property and/or signing it over to the group upon admission, limiting interactions with the outside world, employing special terms for the outside world, ignoring outside newspapers, speaking a foreign language or special jargon, requiring free love, polygamy, or celibacy, no compensation for labor, communal work efforts, daily meetings, mortification procedures such as confession, mutual surveillance and denunciation, institutionalization of awe for the group and its leaders through the attribution of magical powers, the legitimization of group demands through appeals to ultimate values (such as religion) and the use of special forms of address.[37] Most terrorist groups employ at least some of these mechanisms.

Kay explains that cult members felt they needed weapons because they had obligations to take care of one another. It is common for members of terrorist groups to begin taking on a group identity and to feel that the need to protect group members is as strong—or stronger—than the need to protect their own lives. "We bought guns as protection—against people who wanted to come and steal our food," she says. "Part of it was antigovernment." The compound was supposed to be a place where God-fearing Christians could escape when the tribulation occurred. Ninety percent of the world's population would die within the first hour of the Apocalypse, "leaving an elect people to rebuild society and usher in the millennial rule of Christ."[38] "We believed we were to house, feed, and clothe those who came to us," Kerry explains.[39]

From the very beginning Ellison held out the possibility that, by joining the cult, members would receive early news about the coming Apocalypse. There are still prophets and apostles in the church, Ellison said, and God would warn those prophets before Armageddon began.

The cult received early intelligence about an imminent Apocalypse several times, but their disappointed hopes did not lessen their faith in the least. In 1978, they received word that the judgments of God would begin on August 12, the ninth of Ab on the Jewish calendar, when Jews commemorate the destruction of both the First and Second Temples. Their source, whom they considered reliable, warned them of natural catastrophes "beyond measure," including flooding on all sides of the United States in a band two hundred miles wide. When the prediction turned out to be premature, Kerry says, "We were disappointed it didn't occur, but still anticipated its arrival as being near."[40]

The cult repeatedly prepared for the coming economic and social collapse, which was certain, Ellison told them again and again, to arrive the following season.[41] When the judgments started, Ellison told them, "It will get so bad that parents will eat their children. Death in the major cities will cause rampant diseases and plagues. Maggot-infested bodies will lie everywhere. Earthquakes, tidal waves, volcanoes, and other natural disasters will grow to gigantic proportions. Witches and satanic Jews will offer people up as sacrifices to their gods, openly and proudly; blacks will rape and kill white women and will torture and kill white men; homosexuals will sodomize whoever they can. Our new government will be a part of the one-world Zionist Communism government. All but the elect will have the mark of the Beast."[42] The deliberate inculcation of apocalyptic fears often precedes violence in cults that are cut off from society. It is also common for cults to believe even more strongly in the world's imminent end after prophecy fails.[43]

To prepare, the men started practicing military maneuvers. Every male member of the cult was issued a rifle, a pistol, and full military gear.[44] They started stockpiling food. Kerry says that at one point he had over three thousand pounds of food in storage.[45] They raised money by stealing from department stores and committing arson for profit. They built factories for manufacturing grenades and silencers and sold weapons at National Rifle Association gun shows.[46] They offered classes in "Christian martial arts" at a school they called the Endtime Overcomer Survival Training School. They charged $500 for the full course, which included shooting cardboard cutouts of blacks and Jews.[47] Terrorist groups often raise money through criminal activities.

They published a series of books, including *Witchcraft and the Illuminati, Christian Army Basic Training Manual, The Jews: 100 Facts,* and *Prepare War!* They sold racist and survivalist literature as part of their official book list, including such titles as *The Protocols of the Learned Elders of Zion, The Negro and the World Crisis, Who's Who in the Zionist Conspiracy, The Talmud Unmasked,* and *A Straight Look at the Third Reich.*[48]

How did you feel when the group turned violent? I ask Kay.

"Well, it came about slowly," she says. "Over a couple of years. We got pulled into it, it became a way of life.

"One of the doctrines that was heavily relied on was that women submit to their husbands," she continues. "That was what we believed. We didn't question. And we weren't directly involved. Acts of violence were usually kept among the men; the women really didn't know much of what was going on." Women called their husbands "lord" as a sign of respect, in imitation of the way the biblical Sarah referred to her husband, Abraham.

"End of times, I've heard of that for years as a child in the Baptist church. Everyone is left to fend for themselves," Kay says. "And James [Ellison] was like a saint. I mean, he could just like hypnotize you." She is crying now.

Ellison enters a trancelike state when he concentrates, Keith, the former government agent, confirms to me later. "There are not too many things he would not do to establish his goals, once he puts his mind to something," he says.[49]

Do you still believe Armageddon is imminent? I ask Kay.

"I don't worry about it now. If it happens, it happens," she answers.

What was happening in your life before you joined CSA? I ask Kerry.

"I desperately wanted to be valedictorian of my high school class," he tells me. "In my sophomore year, there was no doubt that I would be valedictorian, my grades were so outstanding," but the family kept moving, so he was ineligible. Kerry then volunteered to go into the military, but was rejected because of an earlier illness. "I got disenchanted again," he says. "At that point I had no direction and I was just going with the crowd. Then I got called to the ministry."

What does it mean to be called to ministry? I ask.

"I had a spiritual experience, I guess," he says. "In March of 1972, I was living alone, and one night, after smoking a joint, I went to bed. During the night I found myself all of a sudden standing before God, even though all I could see was His arm and hand. Then He spoke; His voice seemed to echo throughout the entire place. Then I saw a book on a table about my life and why He had done what He did in those events. It totally changed my understanding about my life. God began to tell me what He had planned for me and what He wanted me to do. He said He had given me the gift of teaching and pastoring. The place I was in was so peaceful and so full of light, yet somehow dark at the same time. Anyway, I woke up and felt a power in my life I had never felt before. I went to a Christian bookstore and bought a pocket-size Bible to carry to work. Then I went to see a Baptist preacher at a church down the street and told him about my experience. He simply said I had been called to the ministry. It wasn't until three years later that I realized I had experienced the baptism of the Spirit as well. When I got to work, I told everyone what happened. They said it was just the marijuana. But I ignored them and began to read the Bible every day, every chance I got. I read it through, underlining verses and memorizing scriptures, and then I would read it again, cover to cover, over and over. I couldn't get enough of it.[50]

"Everyone I knew just stood back and said, 'Well, Kerry's a religious fanatic now.' So, I decided to move to Lubbock. I started going to college. Needed some kind of degree to graduate with. My first class, religion class, was very liberal, main denomination. The very first day I go to class, this Baptist preacher got up and said, 'There's no devil or hell, no calling, no God, Moses didn't part the Red Sea.' I mean, he goes through the whole 'spiel' in the first class. Everything that we had been taught as Southern Baptists growing up, here's this Baptist getting up and saying this wasn't true. And I'm thinking, 'Well then, what is true?' I mean, how can you be teaching at this Southern Baptist school and saying it's not true? That just threw me. I just had no vision for the college thing." Leaders of new religions take advantage of people's frustration with mainstream religions, which many people believe do not help them deal with the problems of contemporary life.

"Then I started working as a telephone counselor for the *700 Club* with Pat Robertson," Kerry tells me. People would call in to ask for prayers. At the end of the evening, all the telephone counselors would pray together and 'believe for a healing,' Kerry says. Kerry was curious about whether the prayers worked, so he suggested calling people back to see how they were doing. "That's a big no-no. We were just supposed to proclaim our healing and believe it and then leave it. I don't see that, I'm sorry. I saw this as a conflict at the time and I quit doing that," he says.

"I was also working as an associate pastor. You don't get paid. It's more of a training thing." But Kerry interpreted the texts in his own way, infuriating his superiors. "And then I find out that the deacons are talking about me."

So you were in trouble with the church and you had this awful job you didn't like, and then you heard that your friends were moving up to this place in Arkansas? I ask.

"Yeah," Kerry says. "While we were in Dallas, we lived with our friends Tom and Barbara in the same house for a year. So, that was my first experience with community living. To me it was a high ideal. Two families living in a house and we never had a problem amongst us. Traditional patriarchal hierarchy where the women did all the chores and all that kind of stuff and the men went off to work. But to me, it was the best I had seen of what Christianity should be."

The two couples had regular prayer meetings. They sometimes conducted healing sessions that included laying on of hands. During one of these sessions Kerry says he left his body. He was praying to be released of "religious pride," and he suddenly found himself outside his body, watching the scene from above. "It was so powerful. . . . I can still picture it today," Kerry recalls.[51] "When we went to Arkansas to visit [Tom and Barbara, who by that time had joined the cult], it was just a bigger picture of what that community living could be."

You told me once that Ellison liked to recruit people who were young and vulnerable, and that he was good at figuring out what they needed, I say. You told me he was somewhat rattled by people who are intelligent. What did he figure out you needed? I ask.

"He has a gift of knowing what people needed. He knew enough about me to know what I was looking for. That's what makes for good cult leaders. You've got to almost be good at psychology, have to have that feel for what people want, what they are thinking, what is missing," Kerry says. "Jim [Ellison] told me he needed me to be the Bible teacher, there's this ego thing. He really needed me. He realized that he wasn't that good at teaching, only preaching. He didn't know the Scriptures that well. So he needed someone to teach."

And the fact that God had appeared to Ellison "perked my attention," Kerry adds.

You believed Ellison when he told you this? I ask.

"Yeah." Kerry was skeptical of Ellison at first, he admits. "But as soon as I met him, it wasn't like he had this huge mansion and everyone else was working and he was just sitting up there. He worked harder than everybody else. I saw him do things compassionate-wise with people when people had done him wrong and he would still show mercy to them, forgive them. I mean, he had a big heart."

Kerry had suffered from chronic bronchitis as a child, and his mother had discouraged him from exerting himself. He was weak, he tells me, with little endurance. In first grade he was forced to attend the girls' physical education classes because he couldn't keep up with the boys. "I don't know if I ever got over the shame and humiliation of not being able to keep up with the other boys—or even with some of the girls," he says. "Other boys often picked on me or hit me. But I never fought back. My mother taught me that violence and fighting never solved anything."

Kerry says that he desperately wanted to be a "rugged, hardworking man" with the confidence of his stepbrother and stepfather. He also remembers his brother-in-law telling him he would "never amount to a damn" because of his reluctance to exert himself and his physical weakness.[52] Even his sister was more willing to fight the neighborhood bullies than Kerry was.

Now he is ready to answer my question more directly. "So I'm disillusioned. I've got fears and insecurities 'cause my mom taught me to be antiviolent. I had a lot of fights when I was in school, a lot of bullies picking on me. I had a fear of man in me. In the paramilitary, for the first time I felt as if I could protect myself. For me, I needed this. I never had this growing up.

"But I also needed to know that there was somebody who walked with

God, or at least who I perceived walked with God. This was in '77; it's been five years since I'd been called to ministry. And here was a group where everybody was seeking the same thing—hearing God, knowing God, and talking to Him and Him talking back. I had never seen this before. I had heard about it, but not seen it."

I tell Kerry that I've noticed that one thing that distinguishes religious terrorists from other people is that they know with absolute certainty that they're doing good. They seem more confident and less susceptible to self-doubt than most other people.

"Sure," Kerry says. "They believe they are directed by a higher authority. They see themselves as chosen, anointed by God. So you're in a group of people who understand the world the way you do, who see themselves as chosen to carry out a particular mission. And not only are you part of a group of people that understand you, but you think maybe you can do something.

"Out of all the multitudes in the Book of Acts, there were five hundred in the first part that follow Jesus Christ. Out of these five hundred, He chose twelve to be His disciples. Three were given more responsibilities than the rest: Peter, Paul, and John. One scripture says that Jesus loved John—with the implication that there was something deeper there, a special relationship. So, God loves a multitude of people, but only some are chosen. At a certain point, it becomes easy to believe you are chosen."

Did any of you think you were a John? I ask.

"I think Ellison did. I saw myself as a John in terms of my relationship with Ellison, not God," Kerry says.

Terrorists often suffer delusions of grandeur. They come to believe that their actions are of intense interest to everyone, especially their enemies. These characteristics, together with profound suspicion of the government and premonitions of doom, are symptoms of what psychiatrist Jerry Post calls "politicized paranoia," which often leads activist organizations to turn violent.[53]

Ellison left the compound to attend a meeting of the Aryan Nations Congress in July 1983. At that meeting, a number of leaders realized they had a common goal—to overthrow the Zionist Occupied Government (ZOG)—and they decided to pool their resources.[54] To facilitate communication, they would establish a nationwide computer system linking right-wing organizations.[55] They would assassinate federal officials, politicians, and Jews; sabotage gas pipelines and electric power grids; and bomb federal office buildings, including the Alfred P. Murrah Federal Building in Oklahoma City.[56] And they would poison municipal water supplies with cyanide. The ultimate purpose was to spark the Second American Revolution, creating an all-white state.

The most violent members of the participating organizations were recruited to form a new group called The Order, named after a fictional terrorist cell in William Pierce's *The Turner Diaries.* This book, which inspired the Oklahoma City bombing, also describes the use of radiological, nuclear, chemical, and biological weapons. Until he died in 2002, Pierce was the head of the neo-Nazi organization National Alliance, with headquarters in West Virginia. Timothy McVeigh was a fan of the book. He sold it at gun shows and reportedly slept with a copy under his pillow.

Kerry says that Ellison returned from the Congress with more energy and excitement than he had seen in him for a "long, long time." Ellison called the Elders together and told them about plans to finance the right-wing movement through counterfeiting, robbing armored cars or banks, stealing from stores, or "whatever it takes." "If the left wing could do it in the sixties," he told the Elders, "the right wing can do it in the eighties." They also talked about forming small cells, which would never meet with other cells, and "silent warriors," who would operate on their own, to minimize the risk of leaks. Ellison told Kerry that he now understood that it was his destiny to become famous, "to go down in the history books in a major way." He informed Kerry that he saw himself as a founding father of the Second American Revolution.[57]

Soon after the Congress, CSA paramilitary forces detonated an explosive device along a pipeline that supplied natural gas to much of the midwestern United States, ending in Chicago. The attack failed. The pipeline was damaged but natural gas deliveries were not interrupted.[58] "It was winter," Kerry explains. "We thought people would freeze, that they might start riots."[59] They also detonated an explosive device on an electrical transmission line at Fort Smith, Arkansas.[60]

In 1988, the U.S. government took the historically unusual step of accusing the fourteen people who had plotted to overthrow the government of sedition.[61] Federal prosecutors argued that they had returned home from the Congress determined to "wage war" against the government.[62] The government lost the case, but it succeeded in destroying The Order in a violent raid.[63] A decade later, in 1998, another group named after Pierce's fictional fraternity emerged in rural Illinois. The group, which called itself the New Order, was arrested for plotting a series of attacks, including bombing public buildings, assassinating several individuals, and poisoning water supplies with cyanide.[64]

I ask Kerry how the group "spiritualized" away the Sermon on the Mount, which extols the virtues of pacifism. Christians typically ascribe the qualities of "light" and "love" to God and try to manifest those qualities in their own lives, Kerry concedes. "But the Scriptures describe another aspect of God: 'The Lord God is a man of war' " (Exodus 15:3). And in Deuteronomy, the Lord says, "If I whet my glittering sword, and my hand take hold on judgment: I will render vengeance to my enemies, and will reward them that hate me. I will make my arrows drunk with blood and my sword shall devour flesh" (Deuteronomy 32:41, 42). They wanted to mimic this more violent aspect of God, Kerry says, a practice common to Identity Christian groups.

Informants, in some cases hoping to reduce their own sentences, told the FBI that the cult was kidnapping children, stockpiling illegal, military-style weapons, conducting paramilitary exercises, and burying land mines around the perimeter of the compound. They stole cars whenever Ellison felt they were "needed in furtherance of the Lord's work," then altered or disguised them.[65] They said that Ellison had taken two wives.[66] They warned the FBI that cult members had been assigned sniper positions in the event of a government raid. All of these allegations were eventually confirmed, with the exception of the kidnapping of children.

Kerry Noble was convicted of conspiracy to possess unregistered weapons in 1985 and sentenced to five years. Ellison was convicted and sentenced in September 1985 to twenty years in prison on federal racketeering and firearms violations charges. In return for a reduced sentence, Ellison told the government about other people in the movement and also provided details about the conspiracy to destroy the government.[67] He completed his probation three days before the 1995 bombing of the Alfred P. Murrah Federal Building in Oklahoma City. He moved to Elohim City, another Identity Christian compound in Oklahoma, run by Pastor Robert Millar. By this time Ellison's wives had divorced him or gone into government witness programs, and he married Pastor Millar's granddaughter.[68] Elohim City came to national attention when the FBI revealed that Timothy McVeigh had phoned the compound while plotting to attack the Murrah Federal Building—the same building that Ellison had wanted to attack a decade earlier.

Richard Snell, Ellison's chief accomplice in the CSA plot to blow up the Murrah Building, was sentenced to death for killing a pawnbroker, whom he mistook for a Jew, and an Arkansas state trooper, who was black.[69] His death was scheduled for April 19, 1995. Snell "repeatedly predicted that there would be a bombing or an explosion on the day of his death," according to Alan Ables, a prison official.[70] The day of Snell's execution, ten years to the day after the siege that ended CSA, Timothy McVeigh bombed the Alfred P. Murrah Federal building in Oklahoma City.[71] One hundred sixty-eight people died. Snell was still alive when the building was bombed and reportedly spent his last day watching television coverage of the bombing and laughing to himself.[72] Kerry believes that McVeigh was inspired by his contacts with CSA members, but the FBI has reportedly found no proof.

I ask Keith whether Ellison remains dangerous. He tells me, "If he envisions the Endtime rolling around again, there is no telling what he could do. He could easily become agitated or excited to the extent that he believes the Apocalypse is coming. Any sign that he sees could make him turn violent."

Do you think that he might try to get his hands on more sophisticated weapons than cyanide if he were to become violent again? I ask.

"I don't think there is anything anywhere at this point that he would not have access to, or that some member of a radical group would not have access to. They have such an intelligence network that would have knowledge about any kind of weapons of mass destruction," Keith says.

What kind of intelligence network? I ask Keith. Over the Internet?

"The Internet and word of mouth. CSA had numerous 'prophets' that would drop by that would carry messages about technical matters related to weapons. Even before computer communication they were knowledgeable about all that stuff, even then. They know the intricacies of warfare. These prophets that travel around from one group to another are quite knowledgeable about any number of weapons," Keith says.[73]

But it is not the people we hear about that we should fear, Kerry tells me. "The guys that are taking in the spotlight and giving speeches—they're in it for the fame and glory. Leaders aren't radicals—they have too much to lose." It's the people we don't hear about who should concern us, he says.

Before talking to Kerry Noble, I had read that when prophecy fails, people can come to believe even more strongly in a false Messiah or that the prognosticated events will eventually come to pass. In the seventeenth century, for example, a handsome, charismatic young man named Shabbtai Tzvi was banished from Turkey for outlandish behavior. He came to Jerusalem. A well-known seer who lived in Gaza became convinced that Shabbtai Tzvi was the Messiah. Tzvi was prepared to play the part. He announced that the time of redemption had come and revealed a plan to rebuild the Temple. The Jews of Gaza fell entirely under his sway, and the movement spread quickly to the rest of the Jewish world. Word of the miracles performed by the prophet and his Messiah spread worldwide. The pope sent a delegation to Jerusalem to investigate. The two were expelled from Jerusalem and were gone by the time the pope's emissaries arrived. When the sultan demanded that Shabbtai Tzvi convert to Islam on pain of death, he chose conversion, telling his followers that this was a stage in the redemption process. Many of his followers converted with him. Decades after his death, followers continued to believe that Shabbtai Tzvi was the Messiah.

It is one thing to read about how this happened among seventeenth-century followers of a charismatic false Messiah. It is another to hear the story from a person who himself fell under the sway of apocalyptic prophecy, who admits that his faith grew stronger when prophecy failed.

Learning about Noble's evolution from a mild-mannered pastor to a "soldier" taught me how cult leaders can harness alienation and anomie to construct a group identity, eventually creating killers out of lost souls. The leader has to be a psychologist, Noble tells us. He has to have a gift for knowing what people need, what they want, what is missing from their lives. We will see many examples of leaders catering to followers' needs in the stories that follow.

We also learn through Noble's story about the importance of "sacred territory." The cult came to believe that America is sacred land and that it was time to take the land back from God's enemies. As we shall see, once material objects, such as land, are imbued with a spiritual dimension, it becomes impossible to compromise. We also learn from Noble about the importance of selective interpretation of texts to justify violence.

In the next chapter we continue to explore how leaders can harness perceived humiliation to create support for a terrorist movement. The humiliation that Noble talked about is personal, however. He alone was forced to take the girls' gym class. In the chapter that follows, we explore the tragic effects of repeated humiliation of a whole people.

TWO

Humiliation

In this chapter we explore how real or perceived national humiliation of the Palestinian people by Israeli policies, and often by Israeli individuals, has given rise to desperation and uncontrollable rage. Terrorist leaders have learned to harness this sense of outrage to encourage youth to murder Israeli civilians, creating a vicious cycle of atrocities on both sides.

As we will see in the pages that follow, Palestinians are engulfed in an epidemic of despair, to the degree that mothers proclaim on television that they are joyful that their sons and daughters have committed murder-suicide. On a per capita basis, Israelis and Palestinians have suffered multiple September 11–scale attacks. The effects of trauma on the general population are visible on both sides, drawing increasing attention from the medical community. But murder-suicide is not just an expression of individual hopelessness. In most cases, terrorist groups with clear political aims organize and facilitate the suicide bombings. We also learn in this chapter how terrorist groups use charities not only to garner support for their movement, but also to buy the quiescence of mothers, whose children have donated their lives to murder Israeli women and children.

Hamas and the other terrorist groups explored in this chapter use religion to justify their aspirations for political power and to recover Palestinian territory from Israeli occupation.[1] Part of this land is sacred to Muslims but also to Jews and Christians, as we will see in chapter 4. To achieve their ends, some of which are accepted as legitimate by much of the world, Hamas and the other terrorist groups discussed in this chapter are committing atrocities against Israeli citizens, oftentimes injuring or killing innocent Palestenians as well. The terrorist leaders deliberately inculcate the idea that "martyrdom operations" are sacred acts, worthy of both earthly and heavenly rewards. Mainstream Islamic scholars are increasingly voicing their view that suicide-bombing attacks against civilians are not acts of martyrdom but suicide and murder, both of which are forbidden by Islamic law.[2]

In the summer of 1999, the commander of Jordan's Special Forces invites me to Amman to visit Jordan's prisons, where he tells me he will arrange for me to interview incarcerated terrorists. I decide to take him up on his offer and fly to Amman in late July of that year. The day after I arrive, an official from the prison authority calls up to my room early in the morning, awakening me. She informs me that she has come to take me to visit some prisons. I rush to dress—a long skirt, long sleeves, a scarf—and hurry downstairs. A white Mercedes with police lights escorts us on the highway, forcing slow-moving vehicles out of our way. We visit Al Jweda jail, including the women's division. I meet murderers, prostitutes, and drug pushers. My guides tell me that a number of the women are living in the jail solely for their own protection—because they have been raped or have purportedly committed adultery, and authorities fear that they are vulnerable to "honor killings," murders perpetrated by male relatives to protect the family's "honor." The officials confirm that in many cases perpetrators of honor killings get off scot-free. Some of the women that I meet in the jail were involved in car accidents. They have been incarcerated to protect them from the wrath of the victims' families. I find there are no terrorists in Al Jweda jail, however.

Our next stop is Swaga, Jordan's largest prison, an hour's drive from Jweda. Security at Swaga is much tighter than at Jweda. The prison is surrounded by high walls and barbed-wire fences.

The manager of Swaga meets me with great fanfare, as though I were a visiting dignitary. He invites my guide and me to his office, where he serves us glasses of hot, sweet tea with cardamom. He then escorts us throughout the grounds, which are substantial. He shows us workshops where the prisoners learn woodworking, chemistry, sewing, and cooking. Midsummer in Jordan is punishingly hot, and submitting to the manager's enthusiasm about his prison and his hospitality takes fortitude. Again, I meet many criminals, but no terrorists. When I ask the manager where the terrorists are incarcerated, he tells me that no one is incarcerated at Swaga for political crimes. I am puzzled, but don't want to offend my hosts, whom I have gradually come to realize are under the false impression that I am an authority on prisons.

The manager invites us back to his office for lunch. Several of his deputies join us. My hosts are excited because the best chef among the prisoners has prepared our meal. A whole roasted lamb lies on a bed of rice, surrounded by fresh herbs and pieces of the lamb's liver. The rice is seasoned with cardamom, almonds, and dried fruits. My hosts encourage me to partake of the liver, a delicacy. They ask me about prisons in America. How are they different from those I've seen in Jordan? I tell them about drugs, weapons in the prison, homosexual rape, AIDS, fights among guards and prisoners, and occasional escapes. I tell them about what I observed on death row in Florida: metal-detecting equipment so sensitive that visitors have to remove their shoes. I tell them about fear and guards with guns. None of these, with the exception of rape, they tell me, occurs in Jordan. Not a single Jordanian prisoner has ever escaped.

The terrorists, I discover, are incarcerated in another prison, which is located far from Amman. It was built during the British mandate and is in bad shape. The prison officials tell me they can't take me there, presumably because the conditions are not fit for foreign observers.

I am disappointed, but still hopeful that I can meet with leaders of Hamas, the "Islamic Resistance Movement." Some leaders of Hamas were living in Amman at the time. When I phone Ibrahim Ghosheh, chief spokesperson of Hamas, he demands to know my name. He refuses to meet with anyone named Stern, a Jewish name, and tells me that none of his colleagues will meet me either. I resolve to try talking with the Hamas leadership in Gaza instead, whom I will attempt to meet without revealing my name in advance.

I travel by bus to Jerusalem and from there by car to Gaza. Israelis have warned me that a car with Israeli license plates would be stoned, so I hire a taxi to take me to the border crossing at Erez. I walk on hot tarmac to the crossing reserved for Israelis and foreigners. At the checkpoint on the Palestinian side, two border guards stand behind a rickety desk. One of them politely requests my passport. When he sees that I am American, he smiles and says, "Welcome to my country." He wants to be welcoming even though he has no country here, just an overcrowded city dotted with Israeli settlements and military outposts.

I have hired a young Palestinian woman named Amira to translate for me, and she meets me at the border.[3] She is an undergraduate at a top-ranked American university, spending the summer with her mother in Gaza. An official with the Palestinian Authority (PA), Palestine's interim self-government, has offered to give us a tour of the city.[4] The PA was established in accordance with the Gaza-Jericho Agreement signed in Cairo on May 4, 1994. The official is General Osama al-Ali, introduced to me as Abu-Zeid (the name he uses with friends and family).

The general picks us up in an air-conditioned, black SUV. A tough-guy's car. He drives with apparent pleasure, but safely, bureaucratically. He brings us to the office of the DCO—the joint Israeli-Palestinian command. The office is in a barrackslike building. A servant brings tea. The general begins a talk that I can see he has given before. He shows us a wall-sized map of the Gaza Strip, pointing out the Jewish settlements, and also the larger "settlement areas" encircling the settlements. Under the Oslo agreement these settlement enclaves would be governed by Israel even after the PA takes control over most of the Gaza Strip. He is especially angry that the Israelis are not complying with their part of the agreement, which in any case favors the Israelis, he tells us.

Abu-Zeid invites us back into the SUV and takes us for a drive. Two things irritate him intensely: military outposts of the Israel Defense Forces (IDF) and the greenhouses on the settlements. He drives us past numerous military posts and greenhouses so that we get the picture. These IDF structures are not allowed under the Oslo Accords and they should be removed immediately, he tells us. The greenhouses use an unfairly large fraction of the region's water supply. There are hundreds of them. Many of them have been built in the last few months, we are told, perhaps in anticipation of a peace treaty.

Israelis build settlements where there is water. The settlement next to the Khan Younis refugee camp, for example, is built above the coastal aquifer. The six thousand settlers in the Gaza Strip use 70 percent of Gaza's water resources, which is available to them at subsidized prices.[5] Although Palestinians living in Gaza don't have enough potable water, some of the water is shipped to Israel through a pipeline built in 1994.[6] "The settlers could not survive without subsidized water and Palestinian labor," Abu-Zeid scoffs. "And they treat Palestinian workers unfairly. They expose the workers to unsafe levels of dangerous pesticides, and the workers often end up with damaged lungs. They hire five hundred workers on one day and twenty the next. The workers have no job security." He wants to put a stop to this, he says. If he could find a way to prevent the Palestinian laborers from working in the greenhouses, it would force the settlers to shut them down, he muses. The settlers have also tried bringing in migrant workers from Thailand. He wants to stop this as well.

As we drive around the city, Abu-Zeid points out buildings that he has erected next to Israeli military outposts, with the main goal of annoying the IDF. "They provoke me with their outposts, I provoke them with my buildings," he says. Why don't you set up security outposts right next to the IDF's? I ask. "We are the rabbit," he says. "They are the elephant. The rabbit will not be able to strangle the elephant no matter how hard he tries."

Gaza Strip is known as one of the most overcrowded places on earth. It is a small area—around twice the size of Washington, D.C. Three-quarters of the 1.2 million Palestinians living in the 147-square-mile area are refugees, half of them living in camps. Under the Oslo accords Israel retains 42 percent of the land, most of it reserved for the six thousand settlers (0.5 percent of the Gazan population).[7] Still, we pass huge tracts of privately held open land, owned by wealthy Palestinians, seemingly neglected. Most of the land is littered with garbage and junked cars.

Later, we walk through the city. The sidewalks are uneven and covered with garbage. Mingled smells—of sewage, sweat, spices, and rotting meat—assault us as we walk. There is an inescapable feeling of depression here, of utter humiliation and despair. The city itself looks much like other third-world cities around the world. But something is missing. There is none of the unabashed consumerism or entrepreneurial spirit you often feel in the third world. It's as though the smog was made of despair.

The settlers live in a different world, which a passerby can glimpse through chain-linked fences and barbed-wire entanglements. A world with pristine white villas, gardens, and manicured lawns. The settlers burn the ancient olive groves to make room for their lawns and pools and consume, on average, five times as much water as their Palestinian neighbors. The passage from one world to the other is dizzying. For Gazans, a world of relentlessly humiliating occupation by a vastly superior military power. For settlers, a southern-California-style oasis, kept up by Palestinian laborers.

Tawfiq Abu-Ghazaleh, a renowned Palestinian lawyer, has invited us to lunch. He tells us that both Jews and Palestinians are profoundly hurt as a result of the difficulty of achieving peace. His wife interrupts him to reject this view. "The hurt is on our side," she says. "Until you have been forced out of your own home, until you have watched the police beat your own child, you can never understand the Palestinians' pain." Later I discover that her son fell while being chased by Israeli soldiers during the first Intifada (uprising). He fell so hard that he broke his leg and was unable to move. A friend dragged him home. He had already been accepted to Northeastern University in Boston and had to travel to Massachusetts in a wheelchair.

Since the occupation began, Palestinians have been entirely at the mercy of the Israeli Civil Administration "in every sphere of economic life," respected Israeli reporters Ze'ev Schiff and Ehud Ya'ari, explain. "Each requirement for a permit, grant, or dispensation entailed an exhausting wrestle with a crabbed bureaucracy of mostly indifferent but sometimes hostile clerks and officials—a veritable juggernaut of four hundred Jewish mandarins managing thousands of Arab minions bereft of all authority."[8] A Palestinian student told me that it was at border crossings where she first experienced humilitation at the age of nine, while traveling with her parents and two siblings to Gaza. "Our trip from Amman to Gaza through the Allenby Bridge border crossing should have taken us no longer than three hours. Instead, it would last more than twenty hours," she said. "I will never forget those days, that seemed all the more difficult for a nine-year-old child. The toilets that were piled to the roof with excrement. The endless lines of other travelers and children, waiting for the unwelcoming and belligerent faces of their occupiers to place a simple stamp in their travel document giving them approval to return to their home; or to arbitrarily interrogate them; imprison them; or deny them entry. The strip searches."[9]

Hamas leaders recognize that poverty and hopelessness increase support for them. "Hardship always brings people back to God. It is like sickness," Sheik Younis al-Astal, another Hamas leader, explains. "[A] believer should never be afraid of being poor but of being rich. When you become rich, you think only of things. This kills your soul. Islam distinguishes us in that it prepares people to die for the sake of Allah. They are always ready to die for Allah."[10] Hopelessness, deprivation, envy, and humiliation make death, and paradise, seem more appealing. "Look around and see how we live here," an elderly resident of Jenin told a visiting reporter. "Then maybe you will understand why there are always volunteers for martyrdom. Every good Muslim understands that it's better to die fighting than to live without hope."[11]

Since the Second Intifada began in late September 2000, the economic situation in Gaza has worsened significantly. Since then, unemployment has risen 11 percent, to about 40 percent. The United Nations estimates that one in three Palestinians lives on less than $2.10 a day; an estimated two-thirds live below the poverty line. UNRWA,[12] the UN organization in charge of providing relief and works assistance to Palestinian refugees in the Middle East, reports that the population of Palestinian refugees is growing at 3.1 percent annually.[13] Half the population is under age fifteen.

After lunch, Amira and I go to see Ismail Abu Shanab, a Hamas leader who is also head of the Society of Engineers in Gaza. He knows only that I am a visitor from Harvard University. He studied engineering at the University of Colorado and is completely comfortable—even excited—to talk with an American. When I give him my card, he seems entirely unfazed by my name. Perhaps he is more accustomed to talking with Jews than his counterparts in Amman.

"The Palestinian issue must be understood from its origins," he says. "It started when Jews began to immigrate to Palestine in 1917 or even before, and continued when the Jews evicted Palestinians from their homes in 1948. The Jews took advantage of the hospitality of Palestinian people and settled here under encouragement of the British mandate. They developed their own army in 1948. They forced the Palestinians from their homes to neighboring Arab countries. This is the starting point of the problem. The six hundred thousand Palestinians who were evacuated have become four and a half million refugees today.

"In 1965, the Palestine Liberation Organization [PLO] established the Palestinian struggle. Those of us inside Gaza and the West Bank remained docile until 1987, when the [first] Intifada arose. It was the culmination of Palestinian frustration and suffering. You must understand: we were living under occupation. Occupation is prohibited under international law," Abu Shanab asserts, with the tone of a teacher accustomed to the frustrations of attempting to instruct mentally disadvantaged students.[14]

It is not just a matter of law, Abu Shanab says, but also of religion. "It is a duty for Muslims to struggle against occupation. It is our duty to defend the land for the sake of God. For Jews, the issue is the 'Promised Land.' For us, it is not a question of something promised—it *is* our land. We believe it is a natural law that power deters power. Without power there is no deterrence. We believe in talks, but to carry out talks we must be armed with power."

Do you see any psychological differences between those who join the military wing and those who don't? I ask Abu Shanab.

"They are more religious than typical. Often they are angry—they may have seen someone being hurt. It's also a question of the general atmosphere they live in."

This line of questioning reminds Abu Shanab of a theory he has developed about the correlation between militants' personalities and the weapons they choose. "While I was in prison, I tried to figure out whether there is any particular personality type that gets involved in various kinds of military operations. I found that those who use knives tend to have nervous personalities. Usually they become violent as a direct reaction to an incident. The person who uses a gun is well trained. The person who explodes a bomb does not need a lot of training—he just needs to have a moment of courage."

I am surprised that Abu Shanab is speaking so openly. He seems to have forgotten that he is talking to an American who will scrutinize his every word for clues. He has told me, in effect, that while in prison, he realized that suicide bombers are a cost-effective weapon.

A suicide bomber should be someone in whom the organization invests only minimal training—the minimum required to get the job done. Some operations, like the September 11 attacks, are complex. The leaders would have to be reliable experts. But for an ordinary suicide-bombing attack in an Israeli shopping mall, all that is required is a bomb, a detonator, and a moment of what Abu Shanab calls courage. "Courage" is the scarce resource. Hamas's job, then, is to find youth with the capacity to feel this "courage," and then to find ways to nurture it. This requires understanding the psychology of Palestinian youth, and the variety of spiritual, emotional, and financial incentives that will make them willing to be martyr-murderers.

"This is the genius of the Intifada," he says. "People acquire the courage to carry out attacks from having seen something terrible—some kind of atrocity. Islam says an eye for an eye. We believe in retaliation. When someone is killed in jihad, it is a joyful day."

Who are the combatants in your dispute with Israel?

"There are no civilians in Israel because every citizen is required to serve in the army," he replies. "We are at war with Israel. Americans are helping Israelis . . ." He seems suddenly to remember that he is speaking to an American. He tells me, smiling, "We distinguish between the American government and the American people."

Would the Israelis' withdrawal to the June 4, 1967, borders, i.e., those existing prior to the 1967 war, satisfy Hamas? I ask.

"If the Israelis withdraw to the 1967 borders, we would consider that a truce, not the end of the war."

How do you feel about globalization? I ask Abu Shanab.

"Globalization is just a new colonial system. It is America's attempt to dominate the rest of the world economically rather than militarily. It will worsen the gap between rich and poor. America is trying to spread its consumer culture. These values are not good for human beings. The problem with pursuing capitalism as an end in itself is that the name of the game is the dollar. In the West, money really does talk. This is bad for the human being. It leads to disaster for communities."

Why are you involved in the political wing of Hamas, rather than the military wing? I ask. "In 1989 I was put in prison for directing the Intifada. I was in Ashqelon prison for eight years, so I didn't have a chance to be in the Qassam brigades [the military wing of Hamas]. But I think every Palestinian should serve as a soldier."[15]

The most important element of Hamas's success is its social welfare activities, he says. "We started getting involved in charity before Hezbollah did. Our obligation as Muslims is comprehensive. This is the meaning of the phrase 'Islam is the solution.' The PA doesn't understand this. They don't provide social welfare. They are completely corrupt. Our discipline and lack of corruption are part of our appeal." Arafat's officials are widely reported to be running illegal import-export businesses, demanding kickbacks, and pocketing money sent as foreign aid.

"Even before Hamas came into being, in 1976, there were two organizations that were engaged in social welfare functions: al-Jam'iya al-Islamiyah and al-Mujjama'. In those days the priority was to work on social, educational, and welfare programs. After 1980, there were three such organizations, including the Islamic Benevolence Society. They had no connection with politics, even during the occupation. I founded Jam'iya al-Islamiyah, but I cut my connection with them when Hamas was first established. That happened on fourteen December 1987, during the Intifada."

Charitable giving is an important aspect of Islam. *Zakat,* the obligatory giving of alms to the poor, is one of the five pillars of Islam. The word *zakat* means both "purification" and "growth." Islam teaches that by providing alms to the needy, one purifies one's possessions. Radical Islamist groups use the concepts of benevolence and self-sacrifice to spread their movements in regions where the government has failed to provide social welfare, especially for the poor.

Later, I learn that Islamists are hardly the originators of the idea of using charitable works to recruit adherents. Early Christians also employed this technique, with enviable success in terms of conversions. For example, during the reign of Marcus Aurelius, beginning in 165, a plague swept over the Roman Empire.[16] The Christians ministered to the sick and dying, both Christians and pagans, including by preparing the dead for burial at great risk to their own lives. The Romans were highly suspicious of the Christians' motives. They believed that the Christians engaged in good works only to spread their religion, a policy that the Romans were dismayed to discover was highly effective. Two centuries later the emperor Julian attempted to institute pagan charities that would rival the Christian ones. Sociologist Rodney Stark explains that by Julian's day, in the fourth century, the Romans could no longer compete with the Christians in terms of providing social welfare. The seeds for this successful policy can be found in the doctrine itself, which emphasizes charity.[17]

Although the Christians did not practice the combination of martyrdom and murder that has become so common in Palestine, there is something to be learned about the role of martyrdom—and the way the Church encouraged it—from the period when the Christian movement was perceived as a dangerous threat to the ruling Roman elite. Elaine Pagels writes that the Christian movement challenged converts to put their allegiance to fellow Christians before any other commitment, including not only the corrupt Roman elite, but even their families.[18] Perpetua, perhaps the most famous Christian martyr, wrote in her diary that the governor beat her father with a rod to try to persuade her to deny her beliefs in order to save her life. She felt sorry for her father, she wrote, as though she herself had been beaten, but refused to deny her faith, despite the pain it would cause her father.[19] Similarly, Hamas encourages suicide bombers in training to focus on the *ummah,* the Muslim community, not the demands of corrupt Muslim rulers or the emotional loss of their parents. Hamas attempts to soften the blow for families, however, by providing financial assistance to those left behind.

Early Christian martyrs, like Palestinian suicide bombers, received many rewards for their sacrifices, including material, emotional, and spiritual ones. In the period leading up to a martyr's death, fellow Christians would often shower the martyr designate with gifts of food and clothing as well as attention. The martyr was promised not only eternal life in the next world, but also posthumous fame in this one. The letters of Ignatius, for example, make clear that he was "reaching for glory," in the words of sociologist Rodney Stark, "both here and beyond. He expected to be remembered through the ages and compares himself to martyrs gone before him, including Paul, 'in whose footsteps I wish to be found when I come to meet God.' "[20] Christians would hold celebrations dramatizing the martyr designate's forthcoming test of faith. These celebrations served several functions, perhaps the most important of which was to establish a kind of social contract between the martyr designate and fellow Christians, to minimize the risk that he would recant at the last moment. Videotapes taken of suicide bombers would seem to fulfill a similar role, publicizing the *shaheed*'s commitment to sacrifice his life for the purported good of the community.

Like contemporary suicide-bombing campaigns, which receive wide coverage in the press, Christian martyrdom was a kind of theater—always in public, always with the aim of demonstrating faith and recruiting new followers. A witness to Perpetua's murder wrote: "On the day before, when they had their last meal, which is called the free banquet, they celebrated not a banquet but rather a love feast. They spoke to the mob with the same steadfastness, warned them of God's judgment, stressing the joy they would have in their suffering, and ridiculing the curiosity of those that came to see them. Saturus said: 'Will not tomorrow be enough for you? Why are you so eager to see something that you dislike? Our friends today will be our enemies on the morrow. But take careful note of what we look like so that you will recognize us on the day.' Thus everyone would depart from the prison in amazement, and many of them began to believe. The day of their victory dawned, and they marched from the prison to the amphitheater joyfully as though they were going to heaven, with calm faces, trembling, if at all, with joy rather than fear. Perpetua went along with shining countenance and calm step, as the beloved of God, as a wife of Christ, putting down everyone's stare by her own intense gaze."

Perpetua "screamed as she was struck on the bone; then she took the trembling hand of the young gladiator, and guided it to her throat. It was as though so great a woman, feared as she was by the unclean spirit, could not be dispatched unless she herself were willing."[21] Christian martyrs' refusal to back down was seen as an important testament to the power of Christian faith and the appeal of the sect, and thus an important recruitment tool for the Christian movement.[22] The Roman officials realized that the Christians' "inflammatory views, accompanied by passionate religious fervor, could catch fire among the disaffected and the restless, especially among subject nations and slaves. Thus Rome showed no toleration for these dangerous Christians," Pagels explains.[23] The martyrs provoked the Romans to react in a way that increased the Christians' appeal to the public at large.

Hamas provokes Israel to overreact for the same reason: to mobilize support. Martyrdom (including suicide bombing) is a cheap form of psychological warfare.

To be clear, I do not mean to suggest a moral equivalence between the Christian martyrs and suicide bombers. In my view, murder-martyrdom raises far more serious ethical and legal concerns; while suicide may be forbidden by most religions, murdering innocents is forbidden by all. But the example of Christian martyrdom helps to elucidate how organizations can provide spiritual, financial, and emotional incentives to persuade individuals that it is rational to sacrifice their lives for the good of a religious organization.

I want to learn about how the Palestinian Authority (PA) views Hamas. Amira arranges for me to meet with Brigadier General Nizar Ammar of the Palestinian General Security organization. The offices are in the Saraya security compound, where the Palestinian Authority's prison is also located. An aide leads us through seemingly endless, grime-encrusted hallways. The general's office is similarly dingy, as if he intends to broadcast the message "We are overworked, underpaid, with few resources," even though the general perception of the PA on the streets of Gaza is that the officials are all on the take, funneling moneys meant for the Palestinian people to their private bank accounts. The walls are stained and the windows blackened with dust.

Amira seems even more distressed by the filth than I am. She periodically wrings her hands with an antibiotic lotion. Last summer, she tells me, she got sick when she visited Gaza. The septic system in Gaza is barely functional. In some places, raw sewage is dumped directly on the sand dunes with no treatment. Soaking pits and septic tanks frequently overflow onto the streets and into people's homes.[24] The water makes everyone sick, Amira tells me. The general himself seems utterly oblivious of his surroundings, however. He greets us energetically. I scrutinize his face, looking for clues about how he feels talking to the two of us. Amira is a beautiful young woman, raised in privilege in Saudi Arabia, now completing an expensive American education. I am an American academic and former government official, now teaching counterterrorism at Harvard. Undoubtedly he will feel obligated to give us a particular impression, but what will it be?

The general tells us that Fatah (a precursor of the PLO) emerged out of the Muslim Brotherhood, which was founded by an Egyptian schoolteacher named Hassan al-Banna in 1928. Hassan al-Banna was strongly influenced by revolutionary totalitarian movements from the far left as well as the far right, including glorification of the military and a fascination with violence, a cult of martyrdom, and the Russian revolutionaries' idea of the "propaganda of the deed."[25] By the late 1930s, revolutionary junior officers in the Egyptian army, including those affiliated with the Brotherhood, had established links with Nazi Germany. Although the Brotherhood had started out as a charitable and cultural organization, it soon had a paramilitary wing, which took on fascistlike slogans and practices. From the very beginning, one of its explicit goals was to counter liberal democratic principles.[26]

Banna was assassinated in 1949, and Sayyid Qutb—considered by many to be the father of modern Islamist extremism—became the Brotherhoods' chief spokesperson (and its liaison with the communists). Qutb was an early advocate of Islamic holy war as a legitimate response to regimes that claim to be Islamic but whose implementation of Islamic law is found wanting. Like Banna, Qutb was not an Islamic scholar by training. He worked as an inspector of schools and published literary criticism. In 1948 he left Egypt to study education in the United States. He found Americans' materialism and the freedoms that American men gave their wives deeply distressing, and he returned to Egypt in 1951 with profoundly anti-Western as well as anticapitalist views. Qutb described Americans as "violent by nature" and "having little respect for human life." In his eyes, American churches were "not places of worship as much as entertainment centers and playgrounds for sexes." When an American female college student told him that the sexual issue "was not ethical, but merely biological," he concluded that Americans were "primitive in their sexual life."[27]

Qutb was most critical of Arab leaders, whom he described as arrogant, corrupt, Westernized princes and autocrats. He considered them the equivalent of Jahili Arabs, who practiced paganism prior to the birth of Muhammad and the revelation of the Koran. Qutb became convinced that the most important enemies of Islam were the secular leaders of the Arab world and advocated that a jihad be waged against them. He found support for his views in the writings of Ibn Taymiyya, a thirteenth-century theologian and jurist who wrote that jihad against Muslim unbelievers was a legitimate means for protecting the purity of the faith. Qutb described internal jihad as a necessary component of the permanent revolution of the Islamic movement.[28]

After the Egyptian revolution of 1952–54, the military government, which promoted secularism, became the Brotherhood's chief enemy. President Gamal Abdel Nasser suppressed the Brotherhood in 1954, and many of its members went into exile in Algeria, Saudi Arabia, Iraq, Syria, and Morocco, from where they established a network of adherents in religious schools and universities.[29] Qutb was imprisoned and, in 1966, executed. But an offshoot of the Muslim Brotherhood, Egyptian Islamic Jihad, murdered President Anwar Sadat, who had initially courted the Brotherhood as a counter to the Communists.[30] The Muslim Brotherhood and the writings of Qutb inspired not only Egyptian Islamic Jihad, the group responsible for Sadat's murder, but many of the Islamist terrorist groups active today. Al Qaeda and Hamas are perhaps the most prominent examples. As we will see in chapter 9, Egypt's strong anti-Islamist policies induced the Egyptian Islamic Jihad to focus on international targets, including, most famously, the first World Trade Center bombing of 1993 and, after it merged with Al Qaeda, the September 11 strikes.

Some Islamic scholars argue that the Islamism these groups promote has more to do with totalitarianism than with Islam.[31] Islamist terror is "first and foremost an ideological and moral challenge to liberal democracy," the historians Boroumand and Boroumand argue. It is an eminently modern practice "thoroughly at odds with Islamic traditions and ethics," they claim.[32] The problem is that the Islamists are able to persuade their followers that they are preaching Islam, even if they are reading the texts selectively. All religious terrorists engage in hermeneutics (interpreting texts), as we shall see. But Islamists seem to be able to spread their message to a larger group of followers, in part because of the organizational tools they employ.

Palestinians living in Gaza at the time of the first Intifada talk about the social pressure to participate, even for youth not living in the camps. It was just what everyone did, one young man told me.[33] Interviewees in a study overseen by psychiatrist Jerrold Post also talked about social pressure, and the feeling that they would be ostracized if they didn't participate in the violence.[34] One said a friend recruited him to join Hamas, but that joining was just "the normal thing to do, as all young people were enlisting. With my Islamic leanings and the social pressure from the Islamic center, it is only natural that I joined in Hamas activities in the camp."[35] Another reported, "My entire spiritual, cultural, and social world revolved around the movement, and it was natural for me to join Hamas. . . . All the religious men in the area joined Hamas."[36]

In 1991, Hamas carried out its first act of terrorism inside Israel, an attack on a Tel Aviv bus. In December 1992, Israel deported 415 members of various Islamic organizations, including Hamas and Palestinian Islamic Jihad, to the Lebanese no-man's-land on the border of Israel and Lebanon, where they remained until 1993.[37]

"This was a big mistake on Israel's part, because Hamas started cooperating with Hezbollah," General Ammar recalls, referring to the Shia organization established in the aftermath of the 1982 Lebanon war. Hezbollah attempted to adopt Iranian revolutionary doctrine into Lebanon. Since its inception, Hezbollah fought the Israeli occupation of a self-declared "security zone" in southern Lebanon, and the group's combination of guerrilla and terrorist tactics contributed much to Israel's decision to withdraw its forces from Lebanon in May 2000.

Israel's deportation of the 415 Islamists "utterly transformed Hamas," the general continues. Hezbollah and the Iranian revolutionary guards taught Hamas members to carry out suicide-bombing campaigns during this period of exile. "Leaders like Dr. Rantissi and Dr. Zahar emerged during that period. They recruited intellectuals and academics for other leadership positions, following the pattern the PLO had used in 1965."

During the period of exile, the PLO was preparing to renounce terrorism, and Hamas became a more professional terrorist organization. As a result, Arafat's Palestinian Authority began to curb Hamas's unrelenting militancy. Once the PA started cracking down, Hamas divided itself into several wings—a political wing, a charitable organization, and the military wing. "But the difference between the wings is often a fiction," the general explains. "We learned through interrogations that some of the people involved in operations inside Israel had been in the political wing only forty-eight hours before the operation. This is a big problem for the PA interrogators because people jump between the political and military wings at a moment's notice.

"This is how it works," he tells us, seemingly more comfortable now. "Say Mohammed Deif [head of Hamas's military wing] is planning an operation in Tel Aviv. He plans all aspects of the operation—he transports the explosives and specifies the target. But he needs two volunteers to carry explosives to the target. He calls the person in charge of Hamas at the university, and that person calls the head of the student union, or someone familiar with the students, and they will select two student volunteers. Usually they are troubled youths. Forty-eight hours before the initiation of the operation, the two youths would be sent to meet with Mohammed Deif."

The central leadership is not located in Gaza or the West Bank. Khaled Mashaal, the man whom the Mossad tried to poison and whose life was saved by the late King Hussein, is the overall leader. Mousa Abu Marzook, who had moved to the United States but was deported to Jordan in May 1997, is now based in Damascus. They send the funding and provide overall direction from their safe houses elsewhere in the Middle East.

Where do they get the money? I ask. "There are tens of channels," the general tells us. "Some money comes from Palestinians living in Saudi Arabia, the Gulf, or the United States."

A large fraction of Hamas's money comes from Muslim charities in the West, including the United States. The Holy Land Foundation, for example, the largest Muslim charity in America, is a significant contributor, according to the U.S. government.[38] About 60 percent of Hamas's budget goes to social welfare—schools, libraries, youth clubs, athletic teams, mosques, orphanages, and clinics, which are spread throughout the territories.[39] Running charities that actually support the poor is a great cover for fund-raising for terrorism; but charitable giving is also a critically important aspect of Hamas's appeal.

The charitable wing, known as Dawa, plays a significant role in increasing support for the other two wings. It offers apartments to students at reduced rates. It provides families of suicide bombers lifetime annuities. The bombers are recruited in Hamas classrooms and in Hamas sports clubs. A Hamas activist explains, "These guys kill Israelis, but they also secure their families from poverty."[40]

Iran is the only government that funds Hamas directly, the general tells us. Emad al-Alami, a senior Hamas official, provides the link between Hamas and Iran.[41] The Iranians fund a variety of Islamic organizations, especially military wings. "In the beginning they considered Palestinian Islamic Jihad the most important Palestinian organization, but they have started funding Hamas as well. Now they support both," the general says. Iran reportedly provides $20–$30 million a year to Hamas.[42] "Saudi Arabia and the Gulf States also support Hamas," the general continues. "But it is not government support. Nongovernment organizations provide the money so that governments can't be blamed. But the governments know what is going on; the governments oversee the money. Money comes in from Qatar, UAE, and Kuwait. When Sheik Yassin visited the Gulf, he collected millions of dollars for Hamas," he tells us.

Here is how it works. The Saudi government administers a foundation that provides funds to the families of suicide bombers. The Saudi embassy in Washington issued a press release in January 2001 claiming that the Saudi Committee for Support for the Al-Quds Intifada, chaired and administered by Interior Minister Prince Nayef bin Abdulaziz, distributed $33 million in support of wounded and handicapped Palestinians and to "the families of 2,281 prisoners and 358 martyrs." The press release also reported that the committee "pledged a sum of SR 20,000 ($5,333) cash to each family that has suffered from martyrdom."[43]

I want to know the Palestinian Authority's view of who becomes a suicide bomber. What makes suicide bombing an appealing tactic to Hamas? How does the organization mold the beliefs and actions of recruits, and how do they keep them from changing their mind before detonating the bomb? How does Hamas identify a likely candidate? The general recounts for me, tick by tick, the profile of the typical Palestinian suicide bomber prior to the Second Intifada, and before Mohammad Atta's September 11, 2001, attack:

Young, often a teenager.

He is mentally immature.

There is pressure on him to work.

He can't find a job.

He has no options, and there is no social safety net to help him.

He would try to work for the PA but he doesn't get a job because he has no connections.

He tries to get into Arafat's army, but again, he doesn't have the right connections. He doesn't have "vitamin W." (*Vitamin W* is an expression for *wasta* in Arabic, which refers to political, social, and personal connections.)

He has no girlfriend or fiancée.

On the days he's off, he has no money to go to the disco and pick up girls (even if it were acceptable).

No means for him to enjoy life in any way.

Life has no meaning but pain.

Marriage is not an option—it's expensive and he can't even take care of his own family.

He feels he has lost everything.

The only way out is to find refuge in God.

He goes to the local mosque.

It's not like in the United States where they just go to church on Sundays. He begins going to the mosque five times a day—even for the 4 A.M. prayers. (An average devout Muslim will not attend the early-morning prayer.)

Hamas members are there and notice him looking anxious, worried, and depressed and that he's coming every day. It's a small society here—people tend to know each other. They will ask about him, discover his situation.

Gradually they will begin to recruit him.

They talk to him about the afterlife and tell him that paradise awaits him if he dies in the jihad. They explain to him that if he volunteers for a suicide bombing, his family name will be held in the highest respect. He'll be remembered as a *shaheed* (martyr, a hero). He'll become a martyr and Hamas will give his family about $5,000, wheat flour, sugar, other staples, and clothing. The most important thing is that his family's status will be raised significantly—they too will be treated as heroes. The condition for all this: he is not allowed to tell anyone.

They will take him away from home forty-eight hours before the operation so there is no chance for him to reconsider. During this period he will write his last letters and sign his will, making it difficult to turn back.

Ariel Merari, a leading Israeli authority on suicide terrorism, visited Harvard during the 1999–2000 academic year. In a lecture he gave to my class at my request, he added some detail to the general's assessment. In interviews he conducted, he found that despite their coming disproportionately from refugee camps, suicide bombers tended to be of average economic status. More than half had spent time in Israeli prisons.[44] The most important factor is the organization: almost nobody does this as an individual; candidates are almost always trained.[45] An organization provides logistics and planning. After the prospective *shaheed* is recruited, he will be referred to as a "living martyr." In the last days before the operation, he writes letters to family and friends, explaining his decision and his expectation of paradise. Often, audio- or videotapes are made of the candidate to be used as his final farewell. Photographs are taken of him in heroic positions, and the photos are then used to make recruitment posters and calendars to be disseminated after the *shaheed*'s death. Sometimes additional footage is spliced to farewell videos to make them more effective recruiting advertisements. The idea, Ariel Merari and other experts on suicide bombing explain, is to create points of no return, to make it nearly impossible for the candidate to back out of his commitment. Once these tapes and photos are made, it would be humiliating to change one's mind out of fear.[46]

After the al Aqsa Intifada of 2000 and the September 11 attack of 2001, it has become clear that developing a single profile of suicide bombers is nearly impossible. Candidates are not necessarily poor; they may in fact be wealthy. Nor are they necessarily uneducated. Women are now getting involved. Women have been responsible for over a third of the suicide bombings carried out by the Liberation Tigers of Tamil Eelam in Sri Lanka, and over two-thirds of those perpetrated by the Kurdish Workers' Party PKK.[47] But until recently, female suicide bombers were considered rare among Muslims.[48] Hamas no longer needs to recruit suicide bombers; they are swamped with volunteers requiring little indoctrination.[49]

Islam explicitly forbids suicide *(intihar)*. The Koran instructs Muslims, "And do not kill yourself, for God is indeed merciful to you." In another verse the Koran states, "And do not throw yourself into destruction with your own hands." There is also a widely accepted tradition *(hadith)* that warns Muslims, "Whoever kills himself with a knife will be in hell forever, stabbing himself in the stomach; whoever kills himself by drinking poison will eternally drink poison in the hellfire; and whoever kills himself by falling off a mountain will forever fall in the fire of hell."[50] But terrorist leaders have for some time been arguing that suicide-bombing attacks are not suicides but acts of martyrdom *(istishhad)*. Although God punishes the suicide, he rewards the martyr. The Koran states: "Think not of those who are slain in the cause of God as dead. Nay, they are live in the presence of the Lord and are granted gifts from him" (3, 169).[51]

Soldiers are trained to risk their lives for their country; but a suicide bomber goes into the operation assuming not that he *might* die, but that he *will* die. The more training a soldier receives, the more skilled he is at avoiding death, whereas the opposite is true for a suicide bomber. When such a person makes a cost-benefit analysis about the value of his life versus the value of his death, he attaches greater value to death—both for his country and for himself. This suggests that something is terribly wrong—either with him, his training, or with his situation.

Ordinary suicide has been shown to spread through social contagion, especially among youth.[52] Studies have shown that a teenager whose friend or relative attempts or commits suicide is more likely to attempt or commit suicide himself.[53] Not surprisingly, ordinary suicide is more common among youths who are depressed or exposed to intense social stress. Suicide bombing is different from ordinary suicide: it entails a willingness not only to die, but also to kill others. Often, an organization takes charge of planning the suicide operation, and the terrorist may be on call for weeks or, in the case of the leaders of the September 11 attacks, years.[54] But suicide bombing has some things in common with ordinary suicide.

The situation in Gaza suggests that suicide-murder can also be spread through social contagion, that at some tipping point a cult of suicide-murder takes hold among youth. Once this happens, the role of the organization appears to be less critical; the bombing takes on a momentum of its own. "Martyrdom operations" have become part of the popular culture in Gaza and the West Bank.[55] For example, on the streets of Gaza, children play a game called *shuhada,* which includes a mock funeral for a suicide bomber. Teenage rock groups praise martyrs in their songs. Asked to name their heroes, young Palestinians are likely to include suicide bombers on the list.[56]

Suicide seems to spread more readily in subcultures heavily exposed to violence. In the United States, the local rate of suicide tends to be correlated with the local homicide rate, especially among youths.[57] Easy access to weapons plays a role. Teenagers are more likely to commit suicide if a gun is kept in their home. One study shows that youths who commit suicide are thirty-two times more likely to have lived in a house with a loaded gun than matched controls in the same community.[58] High school shooters, who are often suicide-murderers, are likely to have guns in their homes, and many spend a lot of time playing violent video games and watching violent films, again suggesting that exposure to violence is a risk factor.[59]

Although some Palestinian parents claim to be pleased when their children donate their lives to jihad, Palestinian mental health workers report that parents are seeking advice about how to prevent their children from martyring themselves.[60] A backlash began after three Palestinian students from Sheikh Radwan refugee camp, one fourteen years old and two fifteen, set out on a suicide mission against a settlement armed only with knives and makeshift bombs, trying to infiltrate the settlement. IDF troops guarding the settlement shot and killed them.[61] Shortly after that incident, a seventeen-year-old girl ran away, leaving her parents a note saying that she was going to blow herself up in Israel. But her father requested help from both Palestinian and Israeli security officials, who found the girl and returned her to her parents. Dr. Mahmud Sehawail, general director of the Treatment and Rehabilitation Center for Victims of Torture in Ramallah, explained that Israeli solders had killed the girl's cousin and she wanted "revenge," and that she required psychiatric treatment for anger and depression.[62] Dr. Elia Awaad, director of mental health at the Palestine Red Crescent in Beit Sahur, said, "It's horrible what has been happening, but a suicide bomber acts because of accumulated trauma, going back generations, in some cases back to 1948."[63] Mark Juergensmeyer sees suicide bombing as a means to "dehumiliate" the deeply humiliated and traumatized. "They become involved in terrorism not only to belittle their enemies but also to provide themselves with a sense of power," he argues.[64]

The organizations that recruit suicide bombers encourage youth to donate their lives, and their parents' quiescence, in a variety of ways. The parents are showered with gifts and attention, including substantial financial rewards offered by a variety of charities. They hold celebrations for the *shaheed* to celebrate his purported marriage in paradise. Death notices in Palestinian papers often take the form of wedding announcements. For example, a notice in *Al-Istiqlal,* the Palestinian Authority paper, read: "With great pride, the Palestinian Islamic Jihad marries the member of its military wing . . . the martyr and hero Yasser Al-Adhami, to the 'black-eyed' [virgins]."[65] A suicide bomber's will, published in *Al Risala,* the Hamas newspaper, urged his mother to "call out in joy" and "distribute sweets" because "a wedding to 'the black-eyed' awaits your son in Paradise." Al-Hutari had carried out a suicide bombing outside a disco in Tel Aviv, killing twenty-three people, most of them teenage girls.

Islamic scholars argue that the tradition makes clear that seventy-two virgins are the reward for every believer who is admitted to paradise, not only martyrs. But terrorist organizations emphasize that the seventy-two virgins are a special reward for martyrdom. Hamas leader Isma'il Abu Shanab explained to Agence France Presse, "Anyone who dies a martyr's death has a reward. If the martyr dreams of 'the black-eyed,' he'll get [them]." Although many Islamic scholars claim that the tradition is not entirely clear on whether getting married to the "black-eyed" entails having sex with them, suicide bombers seem to believe that it does. A sixteen-year-old Hamas youth leader told a visiting American reporter, "I know my life is poor compared to Europe and America, but I have something awaiting me that makes all my suffering worthwhile. . . . Most boys can't stop thinking about the virgins."[66] The Israeli Defense Forces report that one of the suicide bombers whose attack they managed to prevent had wrapped toilet paper around his genitals, apparently to protect them for later use in paradise.[67] In a review of the Egyptian press, the respected Egyptian journalist Hasanain Kurum wrote that the late author and reporter Muhammad Galal Al-Kushk caused a scandal when he wrote that "the men in paradise have sexual relations not only with . . . the 'black-eyed' but also with the serving boys," and that "a believer's penis is eternally erect."[68]

The evening after our interviews with Abu Shanab and the general, I consider staying in a hotel. Amira, my translator, invites me to stay with her family in their apartment in Gaza City. This strikes me as safer, so I take her up on her offer. We stop by the apartment in midafternoon. Amira's mother, who is a doctor, offers to make some calls for me. She offers to phone Dr. Abdel Aziz Rantissi, the spokesperson for Hamas in Gaza. He is a pediatrician and known to her circle of friends. She will make me appointments for tomorrow, she says. In the meantime, she prepares an extraordinary meal. Cucumbers and zucchini stuffed with ground meat. Chicken stewed with cinnamon, cloves, and cumin. Rice with dried fruits. Afterward, Amira and I walk to a hotel on the beach. The broken sidewalks are cooler now. Amira points out the movie theater that Hamas tried to shut down. The large square where students hold demonstrations. The liquor stores whose windows Hamas operatives have shattered.

When we return, I discover that Amira and I are to share a bed. The combination of sleeping with a person whom I don't know, the singing of the muezzin in a neighboring mosque, and the crowing of an enthusiastic, insomniac rooster keep me up most of the night.

The next morning, Amira and I visit Dr. Abdel Aziz Rantissi, one of the founders of Hamas and a member of its executive committee, who is under house arrest at his home in Gaza. There appear to be security personnel in every part of the house—on the front porch, in the foyer, in every room we can see. Many are wearing civilian clothes, but they have handguns. We are directed to a sitting room. An unidentified person—Rantissi's bodyguard? His servant?—brings us soft drinks and tea. After ten minutes, a young man brings us to Dr. Rantissi's office. Rantissi sits behind a desk, apparently entirely unaffected by the security officials stationed throughout his home. Posters of the Al Aqsa Mosque and the brilliant gold Dome of the Rock hang on the wall behind him.

Dr. Rantissi is not as friendly as my other Palestinian interlocutors were. He waits for me to ask questions, which I find intimidating, especially on so little sleep.

What is your view of globalization? I ask.

"The West offers the rest of the world a very valuable civilization. The Prophet said that we should take [the] best that other tribes have to offer, and leave the worst."

What is the best? I ask. "Technological advances, democracy, the information revolution, the industrial revolution, and elections—these are things that should be absorbed by Islam," he says, in that order. "But dancing, drinking, seductive behavior—these are forbidden by Islam. There should be no inappropriate mixing of sexes. Women are very highly regarded in Islam, and these things adversely affect them." He then adds, as if sensing that an educated American woman might not appreciate this kind of chivalry, "In Islam the man must support the woman. Even if she has a higher salary, even if she is a millionaire, he has to support their children."

Why are you involved in the political wing rather than the military wing? I ask.

He tells us that he considers himself a *mujaheed* even if he doesn't bear arms. "The political and military wings both are struggling. That is the meaning of jihad."

He tells us that after he had been imprisoned for two months, the Israeli High Court passed a resolution to release him. But the Mossad and the CIA did not want him released. Why? I ask. "I refused to recognize the state of Israel. Palestinians' rights cannot be returned on the negotiating table. They will only be returned through war. When the Germans occupied France during World War II, the French people did not hesitate to fight Germany. We don't like war. If it were possible to solve our problems without it, that would be better. But clearly war is our only option. You can't forget that there are generations of Palestinians that have been dispossessed for fifty years. A large part of our people are still under occupation—the worst form of slavery. The Jews have killed thousands of our people. They did not spare women and children. They declared themselves a state. The Israelis are also preparing for war. Why do they need those F-16s if they are really in favor of peace? They are for war.

"We are living in a time when we are at a weak point. It is very hard for us. But history and religion tell us that the stronger does not always remain strong. When we become strong, we'll tell the world that this land is our land. The world will not find this unusual because they already heard Jews make this claim."

As a pediatrician, are you bothered that your organization kills Israeli children? I ask.

Dr. Rantissi responds angrily, "Our religion condemns killing women, children, and civilians. The intention is to kill combatants. When children are killed it's collateral damage," he says, using the language used by the American military in explaining the loss of civilian life in war and by Israelis when they kill Palestinian children.

He thinks for a moment, then continues with a different line of argument. "All Israelis are combatants because they all participate in the army. All Israelis are the children of those who threw us off our homeland. If the Jew considers Israel to be his homeland because he was exiled two thousand years ago, then we will use the same logic: we were exiled half a century ago. It is our land."

The Jews killed over two thousand people during the first Intifada, he exaggerates, "and the vast majority of them were children.[69] They have killed tens of hundreds, mostly in mosques. They forced us to resist."

That evening we return to Jerusalem. Amira is not really supposed to be in Jerusalem at night, but she stays in my room at my hotel.

On a Saturday morning, Amira and I get up early to go visit the Dome of the Rock and Al Aqsa Mosque, beneath which many believe the remains of the first two Jewish temples are buried. We walk down Nablus Road to Damascus Gate and make our way toward the hill known as the Temple Mount to Jews, the Noble Sanctuary or *al-haram al-sharif* to

Muslims. At first we try going through the Jewish quarter, as instructed by the hotel. Tourists are required to pass through a metal detector to get into the Jewish quarter on Saturdays, and it turns out that Palestinians are not allowed through that gate at all. Amira could easily pass for an Israeli or an American if she took off her *hijab* (head scarf), but she doesn't. I argue with the guards, explaining that Amira is a student at a U.S. university and that we are together. They refuse to grant her access. We decide to try walking through the gate they have directed her to—the gate for Muslims. A Jewish soldier looks me in the eye and asks whether I am a Muslim. When I say no, he refuses me entry. I am forced to walk around. Amira waits for me near the Dome. While she is waiting for me she does something that for her is utterly unprecedented—she talks to an Israeli soldier. By the time I reach her, the two of them have been chatting for ten minutes. I watch her question him and am surprised to see her unconsciously mimicking the line of questioning she had translated for me when I was interviewing one of the leaders of Hamas. "Have you ever killed anyone?" she asks the soldier.

"No," he answers.

"Do you think you could?" He answers that he thinks he could but only in self-defense.

The Dome is extraordinarily beautiful, lined with soft mosaics and stained glass. In the center is "the rock"—a cave in which the prophet Muhammad is believed to have prayed, alongside earlier prophets, including Abraham and Jesus Christ. A guard urges everyone—including me—to leave the cave quickly to give others a chance to look. I want to stay here to see if I am susceptible to the power of this stone. I pretend to be in such a deep trance that I don't notice him shooing me out. Eventually he gives up on me and concentrates on talking to beautiful Amira. Some of the visitors pray here.

As we are leaving, the Israeli soldier beckons to Amira. The two of us walk toward him, surprised. He asks her about her plans for the week. Could he come pick her up some evening and take her out? he asks. She tells him she is busy.

The following week the Israeli government grants me permission to interview a senior Hamas operative. I hire a car to take me to the prison at

Ramla, where I meet with Hassan Salameh—perhaps the most important Hamas leader in an Israeli jail. In January 1996, Yahya Ayyash, known as the Engineer, was murdered with a booby-trapped mobile telephone. Following Ayyash's assassination, Hamas embarked on a retaliatory suicide campaign, which became the deadliest series of suicide bombings that Israel had known up to that point, causing over sixty deaths. Salameh, Ayyash's deputy, was responsible for organizing that campaign. He planned many successful suicide bombing attacks.

Hassan Salameh is twenty-eight years old. I ask him why he joined Hamas. He joined during the Intifada because he was attracted to the idea of fighting the Israeli government and he liked the way members of Hamas thought, the way they acted. "Many young people joined at that time," he tells me.

I ask whether part of his motivation was religious. "Yes, but not exclusively," he tells me. The most important motivation was the Israeli occupation; he saw Palestinians were oppressed and wanted to take action.

I ask whether he would ever consider carrying out a suicide bombing himself, or whether he sees himself exclusively as an organizer of suicide attacks. "The latter," he tells me. "This is an organization. Every person has his own role." This is fairly typical; terrorist leaders generally think of themselves as playing a different role from those they recruit as human bombs.

I ask whether he feels any remorse about the lives of the young men that were lost when they carried out suicide attacks against the Israelis. "The terrible things that have happened to the Palestinian people are far bigger and far stronger than feeling sorry or guilty," he tells me. "As a Palestinian, I feel that my people and I have been murdered in the soul by the Israeli occupation. This feeling stays with me in every situation. There is a big difference between murder and killing to defend his country—attacks against Israelis, even against Israeli citizens, are the latter kind of killing, not murder. All religions allow people the right to kill in self-defense, or to defend their land. Land has been taken from us with violence, and we have the right to take it back. You must understand the difference between Hassan the person and Hassan the Palestinian. I was born in a refugee camp near Gaza. The jail I am now incarcerated in is situated on what was our land."

He had one year of college prior to the Intifada, and that was the end of his education. His parents had essentially no education. His father had three wives and twenty-five children. The older children worked to help support the family. Everyone in the family worked as tailors. He is in solitary confinement and says that he is lonely. It is inhumane, he says. But he believes he is in jail because God decided he had to be here. Religion gives me a feeling of peace, he tells me. He spends his time reading the Koran and watching the news. The officers treat him pretty well because they want him to be quiet. They mostly give him whatever he wants, he claims. A few weeks ago he decided to go on a hunger strike to protest his solitary confinement and didn't eat for twenty-one days. A doctor was sent in to examine him every day. He lost ten kilograms. Four days ago he started eating again. You do not look emaciated, I tell him. He tells me he was fat when he began his hunger strike. He started eating again because they reached an agreement, the nature of which he does not reveal.

If you are let out of prison, will you continue to do the same thing? I ask him.

"I can't say," he says. "It's in God's hands."

I want to hear the Israeli counterterrorism office's perspective on how Hamas works. I hire a taxi to take me to the offices of the Ministry of Defense in Tel Aviv. What a change from Gaza. In Gaza you feel humiliation, confusion, and desperation. Here you feel power, certainty, and determination. A part of me feels at home in this setting, but coming here immediately after spending time in Gaza is disorienting. I heard only partial truths in Gaza. But the truth is not what counts for terrorists or for those who come to support them. It is perception and pain, not truth, that leads to terrorism.

In the counterterrorism office I meet with a leading expert on Hamas.[70] He tells me, "Hamas's raison d'être is the symbiotic relationship with the Palestinian people. They believe they have to change the behavior of the people prior to an Islamic revolution, but at the same time they have to persuade the Palestinian people that only they should be vested with the authority to lead. Hamas has a very good public image, and that is critical to its success."

What are the principal challenges Hamas faces today? I ask.

"For one thing, sources of authority." He describes how the founders of Hamas, including Abu Shanab and Dr. Rantissi, the engineer and pediatrician, respectively, I interviewed in Gaza, as well as Hamas's spiritual leader, Sheik Ahmed Yassin, are from poor refugee camps. "They feel a duty to the Palestinian people, and the Palestinian problem is the issue that drives them. Living in poverty in refugee camps is what gave the original leaders legitimacy," the expert says. "But in 1989, we arrested the founders of Hamas: Zahar, Abu Shanab, Rantissi, and Yassin. After that, new leaders emerged: Musa Abu Marzook, Ibrahim Ghosheh, and Khaled Mashaal." These new leaders, he adds, never lived in Gaza or the West Bank. "Their authority does not stem from their having lived in Gaza, but from their money. What gives Marzook and the others authority and power is money from the Gulf states."

How does Israel attempt to solve the problem? I ask.

"The PA has an interest in keeping Palestinians in refugee camps. They remain in camps with no rights; they are not citizens. This is the best way to keep the problem going. This problem can only be solved with money. It can only be solved by providing the refugees with real housing and good jobs. Collective punishment and closures after attacks is bad because it hurts people economically. The solution is to build infrastructure, create an industrial zone, and give them the opportunity to work.

"In 1996, after the Ashqelon attacks, we asked permission to search Islamic schools in Ramallah. All the walls were covered with posters about the jihad. We found a videocassette of six-year-old children marching, saying, 'O my God, please take my life—I'm going to be a *shaheed.*' What should we do? Close the school? That would be a disaster. And we cannot give money to the PA to improve the schooling system because they are corrupt. We cannot close down the charitable organization because it would be counterproductive—it would only increase support for Hamas. Jordan is the only country that knows how to deal with terrorist organizations. It brought the Muslim Brotherhood into its government and coopted them."

How does Hamas attract its followers? I want to know.

"We have been researching the charitable and political wings of Hamas. In some places we found pornographic movies in their houses.

Some of them are not so religious. Some of them join Hamas because it's the best way to express themselves.

"One of the biggest issues is the gap between rich and poor. This gives Hamas two ways to attract followers. The first is economic. That's easy. Hamas tells families, 'We'll take your children to school. Then we'll take them to the club and we'll provide assistance with their homework. The children can join the Islamic sports club and it's free. Our bus will take the children. We'll pay their scholarship. And we'll find a job for you. Every Friday you should come to the mosque and we'll give you food and every month we'll give you fifty dinar [that's a lot—around one-quarter the average monthly salary].' After six to seven months of this, many families decide to join Hamas. Plus, everyone sees that the PA is corrupt, they are not providing essential services. Hamas looks very good in comparison.

"Second, the failures of modern, open societies make Hamas attractive. The failures of modern society are like a disease, like AIDS: people everywhere have the feeling that the only way to protect their families is to go back to tradition, to religion. Parents send their children to religious schools to protect them. Identity today is based in general on religion and culture, not nation-states. Hamas uses religion for political purposes. They use religion to achieve political objectives."

People all over the world feel wistful about an earlier, simpler time, and some of them turn to religious revivalism to help inoculate themselves and their children from some of the less appealing aspects of modernity and globalization. But I can't help but wonder whether this expert, like the soldier who asked Amira for a date, has any inkling of the role of Israeli policy in all this. It is not just the violence; it is the pernicious effect of repeated, small humiliations that add up to a feeling of nearly unbearable despair and frustration, and a willingness on the part of some to do anything—even commit atrocities—in the belief that attacking the oppressor will restore their sense of dignity.

At this point I had a pretty good understanding about the role of humiliation and alienation at both personal and national levels as risk factors for terrorism. In the next chapter we explore the impact of a government policy to deliberately shift an ethno-religious mix, granting a new ethno-religious group numerical dominance in a region.

THREE

Demographics

This chapter examines the violence that broke out in Maluku, Indonesia, in early 1999, and the evolution of a holy-war organization that formed in the wake of large-scale massacres of Muslims by Christians in several parts of Indonesia. The Soeharto regime had supported the migration of Muslims from overpopulated areas, such as Java, to more sparsely populated ones in a kind of internal colonization of far-flung regions. In some cases, including Maluku, the migration policy tipped the confessional balance to favor Muslim migrants, and indigenous groups lost the privileges they had traditionally enjoyed as a result of their demographic dominance. The Soeharto regime had held these ethno-religious tensions in check, but once Soeharto fell from power, violence erupted on both sides. Several new jihadi organizations emerged, and existing ones were revitalized in response to the massacres of Muslims by Christians in the late 1990s. Once these organizations formed, they instigated further violence, which spread to other parts of Indonesia.

The jihad in Maluku became a kind of second Afghanistan. Extremist clerics used the conflict to recruit small armies. Training camps sprang up, eclipsing Afghanistan and the Philippines as centers for instruction in jihad. The young men drawn to fight in Maluku developed military skills, combat experience, and a determination to wage jihad in defense of Muslims from a variety of perceived threats. Trainers and volunteers from Southwest Asia and the Middle East came to participate in this new holy war.

This chapter tells the story of an Indonesian cleric named Ja'far Umar Thalib, who fought in the Afghan jihad and returned to Indonesia to teach at a *pesantren,* or religious seminary. After the violence broke out in Maluku, Ja'far formed Laskar Jihad as a Muslim counterweight to the Christian militias that were active in the region. His goal, he says, was to ensure that fellow Muslims in Maluku would feel "safe in their own country."[1]

The Indonesian government's attitude toward the jihadi groups has been ambiguous. The army played a significant role in the creation of Laskar Jihad and other Indonesian jihadi groups. Off-duty military officers trained Ja'far's "soldiers." Vice President Hamzah Haz was a big supporter of Ja'far and other jihadis, denying the existence of terrorists or terrorism in Indonesia. But after the Bali bombing in October 2002, which killed two hundred Australian tourists and decimated Bali's tourism industry, it was no longer possible to deny the danger of Islamic militancy in Indonesia.[2] The government made an effort to shut Laskar Jihad down, and Ja'far recalled his troops from Maluku. But some of the graduates of that conflict are now seeking new "jihads," in some cases against Western targets. "Maybe in the future there will be another order" to mobilize again, a Laskar Jihad fighter said, shortly after his group was disbanded. "If so, we'll be ready."[3]

In the spring of 2001, I receveived a letter from a former student who was then working for the United Nations Children's Fund in Indonesia. He had recently arrived in Maluku. The entire city, he told me, was divided along religious lines, much like Belfast or Sarajevo. I was particularly intrigued with what he told me about militant organizations that had formed on both sides, especially by a new jihadi group that had formed only a year before. It reminded him, he said, of the Pakistani jihadi groups he had studied in my terrorism class. He urged me to come to Jakarta, where, he said, I would be able to meet Ja'far, the leader. In August 2001 I took him up on his offer.

On the streets of Jakarta you could feel tension, like the feeling just before a storm, a premonition of violence in late Summer of 2001. I would later discover that just before I had arrived, a suspected Al Qaeda cell from

Yemen had appeared in Jakarta with the intention of bombing the U.S. embassy, but fled when it became aware that Indonesian intelligence was monitoring it.[4] President Wahid had just been sacked by the Consultative Assembly in a vote of no confidence on July 23, 2001, and his successor, President Megawati Sukarnoputri, was just beginning to consolidate power. Wahid's moderate-Muslim supporters feared widespread violence.[5]

By the time I arrive in Indonesia, Ja'far was well-known inside Indonesia as a criminal religious radical. Earlier that year, he had presided over the execution by stoning of a follower who had confessed to adultery. Execution by stoning—or any form of privately administered punishment for "criminal activity"—is a violation of Indonesian law. The authorities placed Ja'far under house arrest for his crime. Ja'far claimed that an international Christian-Jewish conspiracy was responsible for his arrest.[6]

Although Ja'far had only recently been released from house arrest, he was on a business trip in Jakarta at the time I was there, meeting with the new vice president and other politicians. His assistant instructed me that if I wanted to meet with Ja'far, I would have to come to Yogyakarta, an hour away by plane, as his time in the capital was already spoken for. I flew to Yogyakarta the next day. A photograph of Ja'far hugging Vice President Hamzah Haz at an official reception appeared in the paper that morning.[7]

I book a room at the Hyatt Regency, a Disney-like resort on the outskirts of Yogyakarta. It has a pool with fantasy waterfalls, whirlpools, and slides. The golf course is decorated with copies of the temple at Borobudur, the world's largest Buddhist monument, which is located nearby.[8]

In my hotel, slim, young Javanese ply the mostly Western guests with drinks with flowers in them. At night there are noisy musical and dance performances on a grassy area between the restaurant and the pool. In my room is a basket of fruits—duku, passion fruit, water apples, and the divine mangosteen—thick towels, a robe. Complete luxury, some of which appeals, some of which repulses me. In any case, I know how Ja'far's followers would feel about this hotel with its scantily clad, alcohol-drinking Western tourists and its purported exploitation of Javanese cultural heritage. After a swim and an early-morning visit to the temple, I feel deeply relaxed, ready to try to see the world through Ja'far's eyes.

I hire a translator and a car to take me to Ja'far's house. My translator is a teacher at a local language school. She is Christian, pro-American, and wearing a miniskirt, not the ideal costume for meeting with jihadis. I wear a long skirt, long sleeves, and a scarf I had acquired in Lebanon. Our car is not air-conditioned, and the driver seems surprised by our request that he take us into a dirt-poor region on the outskirts of town. We drive past rice paddies and orchards. There is an active volcano nearby and the fields are intensely green, despite the heat. The volcano, the most active in all of Indonesia, is known as Mount Merapi (Fire Mountain).

We turn off the highway into Ja'far's neighborhood. Chickens squawk as our driver pulls up to a ramshackle house to ask directions. Farther on, an old man tells us, past those houses, near the end of the dirt road. At last we find the boarding school, called a *pesantren,* that Ja'far runs. Our meeting is to take place in his home, which the students tell us is located behind the school. I notice that the complex seems to include not only a school and Ja'far's house, but a number of additional buildings.

The house is surrounded by a high fence to protect the women who live in it from prying eyes. We ring the bell. A young woman in a Middle Eastern–style chador comes to look at us through a peephole in the gate. She is expecting us and lets us inside. The yard is filled with laundry lines, as though many people are living here. The young woman who has come to meet us speaks perfect English. She tells us she is a student of Ja'far's, and that her name is Fatima Ummu Yahya Lathifah el-Fadel. The name she gives me is Arabic, not Indonesian, and improbably long, as though she has added extra names to demonstrate her zeal.[9] She wears glasses and looks bookish. Fatima had been studying at Java's most prestigious university when she decided to drop out to study Islam with Ja'far. She had learned about Laskar Jihad at a rally. Ja'far is dissatisfied with the press coverage of Laskar Jihad, she tells us, and has asked her to sit in on the interview to make sure that his words are correctly translated.

Ja'far will not see me, I am told, but will talk to me from behind a curtain. The bookish young Fatima, my miniskirted translator, and I gather in front of the curtain. Ja'far has yet to make his presence known. Another young woman comes to join us. She is slim and elegant, with nutmeg-colored skin and glossy black eyes and hair. She is the most beautiful woman I have seen in Indonesia. She holds a baby boy whose features are as lovely as hers. She has the slightly apologetic air of a woman who knows that her beauty might sometimes be intimidating both to men and women, but I have the feeling that she could also be haughty if she weren't aiming to disarm. Fatima explains that this beauty who has joined us is Ja'far's third wife. She is twenty-five. Fatima treats the wife as though she were a goddess, completely in awe of her beauty and her authority.

An ancient lady servant offers us water and tea. Children run in and out hoping to catch a glimpse of me. Even young girls wear scarves to cover their hair. Ja'far's interpretation of Islamic dress is the strictest I've seen yet. Ja'far's followers are immediately obvious on the street. Their style of dress is "over-the-top extreme," in the words of Indonesia scholar Robert Hefner.[10] The children, the servant, and my hostesses are in a festive mood, as though a visitor from abroad is a rare treat. Later I realize that the women would not be allowed to meet male visitors, so their participation in such an interview was probably quite unusual.

After fifteen minutes, Ja'far suddenly informs us from behind the curtain that he is ready to begin. Is this exciting to him, I wonder, to think of all these women waiting to hear his voice, hanging on his every word, longing to see him?

I ask Ja'far whether he is promoting Wahhabism in Indonesia. Wahhabism, a branch of Salafism, is the puritanical form of Islam promoted by Saudi Arabia.[11] "No," he says. "Not exactly. Sheik Muhammad ibn-Abdul Wahhab emphasizes some hadith that are weak." The hadith are the collection of writings of followers of Muhammad that have become part of Muslim tradition. "Ahle Sunnah—the [moderate] Islamic teachings that we embrace—are more reliable, more fundamental. We are not taking Wahhab's point of view; we identify ourselves as people who follow Muhammad's friends. We criticize some of Wahhab's literature, as relying too heavily on weak hadith that may not be the word of Allah."[12]

Muhammad ibn-Abdul Wahhab founded the Wahhabi movement in the early eighteenth century, calling for a "return" to tradition, although the doctrine he promoted was new, according to Islamic scholars. Wahhab successfully converted the illiterate bedouins living in the desert of Nejd, including Muhammad Ibn Sa'ud, the leader of a gang of raiders, to his version of Islam. Wahhab and Sa'ud agreed to cooperate: Sa'ud would be the emir, the political leader, while Wahhab would be the sheik, the religious leader. Wahhab issued a religious decree, a fatwa, that all non-Wahhabis were infidels, giving Ibn Sa'ud "the cloak of religious legitimacy he needed to persecute innocent people," Professor Abdul Hadi Palazzi explains. "His gang was no longer a mob of traveling thugs and his victims were no longer innocent people. Now Ibn Sa'ud's goons were 'fighters for jihad,' authorized to murder 'unbelievers.' For the first time in history, jihad was proclaimed against Muslims and even against the Ottoman Empire, whose sultan was considered the heir of the prophet Muhammad and the highest Islamic authority." The Wahhabis' aim was to replace orthodox Islam with their puritanical doctrine, which would become the state religion of Saudi Arabia.[13] Wahhabis reject the tradition that war is the lesser form of jihad, while purification of the self is the greater form, a tradition widely accepted by mainstream clerics.[14]

Indonesia is the largest Muslim-majority country in the world. Most Muslims practice what is referred to in Indonesia as "traditional" Islam, which is influenced by Hinduism, Buddhism, and animism, the most common religions prior to the introduction of Islam.[15] Ja'far, in contrast, is a "modernist," a proponent of a Shari'a-based state. By rejecting Wahhabism as weak, he is proclaiming himself to be a more radical Islamist than bin Laden.[16]

Although the majority of Indonesian Muslims remain moderate, Islamist organizations and militia groups have become increasingly effective at mobilizing support for the cause of defending Indonesian Muslims from perceived or actual threats, often with the assistance of active or retired military personnel and with funding from Saudi Arabia. Christian extremism is also on the rise.

I was deeply curious about how Ja'far manages to attract youth to his version of Islamism in a country long known for its moderation in religious matters. I could feel a kind of anxiety in my hostess Fatima, a kind of urgent earnestness mixed with fear. But I wanted to know more about why youth join movements that require them to dress strangely and cut themselves off from society. Sociologists and economists have studied the appeal of "strict churches," whose followers embrace distinctive diet, dress, or speech, inviting ridicule, isolation, and persecution.[17] Rodney

Stark, Laurence R. Iannaccone, and others have found that in some cases it is precisely the demanding nature of such sects that make them attractive to certain segments of societies.[18] Sects break off from conventional religions with the aim of restoring religion to a higher level of tension with the environment.[19] People most likely to join strict sects are those with the least to lose in the outside world, either because they have limited economic or social prospects, feel deeply humiliated and confused about their future path, or are frustrated with the political regime in which they live.[20]

Another factor seems to be that Indonesia is undergoing rapid socioeconomic change. During the last quarter century, an extraordinarily rapid rise in the basic literacy rate and in the number of students who complete high school has resulted in the sudden development of an urban, Muslim middle class free to make its own choices regarding social and religious matters.[21] The new democratically elected government is striving to throw off the vestiges of military rule, including the human rights abuses that eventually led the United States and Australia to cut off all military ties, while at the same time trying to control centrifugal forces that threaten to tear the country apart. Strict religious communities simplify life by proclaiming an exclusive truth—a closed, comprehensive, and eternal doctrine that provides answers to life's most troubling questions. Ja'far offers rigid rules and severe punishment for transgressions, which for some youth is likely to be a comfort in a society under stress.

I have come to believe that modernity itself creates confusion and fear, in part because of a surfeit of choice. This is true not only for Indonesians, but for many people all over the world. Modernity introduces a world where the potential future paths are so varied, so unknown, and the lack of authority is so great that individuals seek assurance and comfort in the elimination of unsettling possibilities. Too much choice, especially regarding identity, can be overwhelming and even frightening.[22] Under these circumstances, some people crave closing off options; they crave discipline imposed from the outside.[23] The "strictness" of militant religious groups—and the clarity they offer about self and other—is part of their appeal.

Did you study Islam in Saudi Arabia? I ask. "No," Ja'far replies. Have you ever been to Saudi Arabia? I ask. "No, never," he says. I don't believe him, and later, when he feels more comfortable, he will confirm that he studied in Riyadh. He goes so far as to cite fatwas for a jihad in Maluku from a variety of Sunni ulema in the Arab world.[24]

Like many jihadi leaders around the world, Ja'far had spent some time fighting the Soviets in Afghanistan.

Did you like being a *mujaheed* in Afghanistan? I ask.

"Yes, very much," he says.

What did you do there?

"I fought the Soviets," he says, obviously not in an expansive mood.

Did you ever kill anyone?

"I don't know. We used long-range weapons."

What kind?

"RPGs. BM12s. Antiaircraft guns. M16s. AK-47s."

Did you ever learn to use Stingers?

"We had one, but I didn't use it."

I decide to try a different tack—forming questions based on research I had conducted prior to coming to Yogyakarta that will require him only to correct or amplify on my previous impressions.

I have read that Laskar Jihad was formed after a Christian massacre of Muslim villagers in Maluku in December 1999. Is this true? "Yes. The government was not fulfilling its role of protecting the Muslims in Maluku so we had to step in." So in essence you were replacing the government there? "Only in a military sense—to protect innocent Muslims."

In April 2000, Ja'far led a procession of fighters to meet with political leaders, including President Abdurrahmad Wahid, in Jakarta. Ja'far's followers were clothed in white robes, brandishing unsheathed swords and daggers. Ja'far threatened to launch a private jihad if the government didn't come to the defense of Muslims in Maluku, whom Christian gangs were then killing in great numbers. The president threatened to arrest Ja'far and his fighters if they went to Maluku. But despite the president's threats, within several weeks Ja'far dispatched an estimated three thousand white-robed warriors to Maluku, over a thousand miles from their base in Yogyakarta. They traveled by commercial ferry. No one stopped them. Observers reported that a container of military weapons had been sent separately to Maluku. The minister of defense complained that soldiers on the ground in Maluku did nothing to prevent the shipment of arms from reaching the militants.[25] Observers speculate that the military was using Laskar Jihad to discredit President Wahid, who had recently sacked General Wiranto, chief of the armed forces, for his role in the atrocities in East Timor. As we shall see, Ja'far's relationship with the military and with the government would remain ambiguous.[26]

I have read you are now training in Fakfak in Papua. Is this true? I ask.

"Yes," he says, reverting to one-word answers.

Where else?

"In Maluku, Sulawesi, Sorong on Papua, near Palangkaraya and elsewhere in Kalimantan."[27] His pride seems to have gotten the better of him here. When I share my notes with my colleagues upon my return, they are astonished that Ja'far would be so open about his plans.

Indonesia is made up of thirteen thousand islands spread over a three-thousand-mile long archipelago. The largest of these are Sumatra, Java, Sulawesi, Papua, and Kalimantan. Medium-sized islands include Bali, Lombok, Timor, and Flores. Half of the remaining islands are uninhabited. Seventy percent of the country is water. There are 731 languages, of which 726 are living, two are second languages without mother-tongue speakers, and three are extinct. Tensions between the center and the periphery have plagued the country since it achieved independence from the Netherlands in 1949. Tensions were largely held in check, however, by the brutal policies of the Soeharto regime and economic growth, until both came to an abrupt end in 1998.

Ambon, the capital of Maluku, is one of the cities hardest hit by violence. Christians had long dominated politics in Maluku, a legacy they inherited from the colonial period, when the Dutch favored them with educational advantages.[28] The migration of immigrants from South Sulawesi tipped the confessional balance in favor of Muslims, who now enjoy a slim majority in Maluku province.[29] Beginning in the 1990s, President Soeharto began appointing Muslims to the government bureaucracy as part of a broad campaign to woo Muslim groups. In 1992, Jakarta appointed a Muslim governor in Maluku.

Ambon was once the center of the lucrative spice trade. The Portuguese, Dutch, English, and Spanish vied to control trade in the spices grown in the

Moluccas, including cinnamon, cloves, and pepper, as well as the valuable nutmeg and mace, the carmine-colored filamentous arillode that covers the nutmeg seed. Several hundred years ago, nutmeg was the third most valuable commodity on earth after silver and gold. It was used to flavor and preserve food, freshen ladies' breath, promote virility, create hallucinations, and prevent plague. *Jeanne Rose's Herbal* (a book containing names and descriptions of herbs) describes nutmeg as a carminative, a stimulant, a narcotic, and an aromatic. It was believed to attract admirers if carried under the left armpit. When Christopher Columbus discovered the Americas, he had been looking for a new, western route to the Spice Islands.

Locals remain uncertain about exactly how the conflict in Ambon started, although the date of its inception is clear: January 19, 1999. It appears that a Bugi migrant from Sulawesi tried to mug an Ambonese bus driver, demanding money at knifepoint. Some say that the two involved in the incident were members of rival criminal syndicates. Whatever sparked the incident, machete-armed mobs were soon drawn into a brawl, and hundreds of people were killed.

Although the conflict arose in a climate of tension between indigenous Ambonese and migrants to the city, it was soon defined in religious terms. A vicious cycle of attack and counterattack took hold, with Christian gangs burning mosques and Muslim gangs burning churches in retaliation.[30] Christians became convinced that Jakarta intended to introduce Islamic law in the province, while Muslims were persuaded that Christians were conspiring with international backers to turn Maluku into a Christian state. Troops were widely observed siding with the Muslims in the clashes that followed, while police units sometimes sided with the Christians. In some cases security personnel stood by without intervening because, according to the provincial police chief, "rioters outnumber the troops and are armed with standard military weapons."[31] In other cases, according to a Muslim member of the national parliament, troops would demand money for protecting groups threatened with attack.[32]

Another factor increasing religious tensions in Maluku may be the erosion of interreligious social networks during the Soeharto period. Religious tensions in Maluku had long been held in check by a village alliance system known as *pela gandong.* The alliance system stressed mystical beliefs and ethnic similarities over religious differences. But the influx of Muslims from other islands reduced the power of these alliances because the newcomers did not participate in these interreligious social networks. They were weakened still further with the fall of Soeharto.[33] Political scientist Ashutosh Varshney has shown that interethnic or interreligious "networks of civic engagement" reduce the likelihood and severity of ethno-religious conflict in conflict-prone areas in India, and that, conversely, the lack of such networks makes conflict more likely.[34] Here, the erosion of such networks appears to have increased the potential for violence.

The violence soon spread beyond the capital city to other Moluccan islands. Militia groups, which formed on both sides, traveled to troubled areas with the purported aim of providing humanitarian assistance to their coreligionists. By early 2002, at least six thousand people had been killed in the conflict, and nearly a third of the population of 2.1 million in Maluku had become refugees.[35]

The governor attributes the increase in religious tensions to Christian prejudice against Muslim "newcomers" from Sulawesi and a widespread perception that Muslims are receiving unfair preference for jobs in the civil service, which are considered good jobs. Many believe that economic and political rivalries caused the conflict, not religion.[36]

Benny Doro, the head of a Christian militia group, says that God made him a military commander. He saw Jesus Christ hovering over his head like a bird while he was in the midst of battle, he says. With Jesus' help, he caught a bullet in his hand. He can no longer count the number of Muslims he has killed.[37] The Christian militia groups tend to be criminal gangs whose principal occupation is petty crime. They are nominally led by prominent religious figures, but their field commanders tend to be local toughs, recruited from criminal syndicates (known as *preman*).[38] They occasionally mobilize under a central structure, but they tend to be less organized and less effective than their Muslim counterparts.[39]

When I return to Cambridge, I want to investigate this further. How often does migration of an ethno-religious group into a region result in conflict? Does it occur only when the confessional balance is disturbed? I discover that this issue has largely been ignored by social scientists, who have focused primarily on the opposite problem: migration that occurs as a result of war.[40] Judging by what I saw in Indonesia, this strikes me as an issue that needs careful study.

One of Ja'far's "humanitarian missions" was to rescue the Christian village of Waai from what he called a "historical crime": the "illegal" conversion of the entire village to Christianity in 1670. Laskar Jihad destroyed the village. Ja'far and a group of local Muslim leaders decided to erect a new town in its place, to be known as Waai Islam.[41] According to local church officials, some four thousand Christians were forcibly converted to Islam on six islands in the Moluccas between January of 1999 and the end of 2000.[42] Similarly, Muslims in north Maluku were forcibly converted to Christianity in December 1999 when Christian forces attacked Tobelo and Galela.[43]

Forced conversions and returning villagers to their "true" religion is a common practice among religious extremists. I heard a lot about this issue when I visited Hindu extremists in India. Schools run by the Hindu nationalist group Rashtriya Swyamsevak Sangh (National Voluntary Service), which is usually referred to by its acronym RSS, focus on "returning" indigenous peoples and *dalits*—the lowest caste—to Hinduism. The RSS believes that all indigenous people, many of whom are animists, are actually Hindus. Through education, the RSS believes, they will come to realize their true religion. The Hindi word for indigenous peoples is *adivasi,* which means people of the soil, suggesting that they predate Hindu civilization. The RSS prefers to call them *vanvasi,* or people of the forest. "We believe that all tribals in India, they are originally Hindu only. Slowly they will feel like this. They will say themselves, 'We are also Hindus,' " Suresh Joshi, national coordinator of the education wing, told the *New York Times*.[44] The RSS clashed with indigenous peoples over Christian conversions in Gujarat, determined to bring them back to the fold. They were at least somewhat successful in this effort. Tribal groups are now taking the side of Muslims in Hindu-Muslim violence in that state.[45]

Christian missionaries have traditionally focused their efforts on converting the tribal peoples and *dalits* in India, in many cases luring them with social services otherwise unavailable to them. The RSS is creating its own social-service organizations to compete with the Christian ones, employing many of the same techniques, including dispatching doctors and drugs to treat the poor, and building orphanages and schools. An upper-caste teacher in one of the schools made clear that he chose his profession out of concerns about Christian conversions in the tribal belt. "Teachers control the mind," he explains.[46] In Indonesia, the indigenous peoples are called *pribumi,* sons of the soil, to distinguish them from Chinese immigrants, approximately half of whom are Christian.

I have read that you claim that Laskar Jihad has ten thousand members. Is this true? I ask Ja'afar.

"We have ten thousand fighters," Ja'far says. "There are a hundred thousand members of our organization."

I have read that you traveled to Pakistan in 1986 to study Islam. Is this true?

"Yes."

Where did you study?

"At a school run by Jamaat-i-Islami."

Did you go straight to Afghanistan from there?

"Yes."

What group did you fight with?

"I joined a group of Arabs under the command of Hekmatyar and later Rabbani." Hekmatyar is allegedly close to bin Laden and the Taliban.

How long were you there?

"I was in Pakistan until 1987, and Afghanistan, 1987 to 1989."

I have spent some time with Jamaat-i-Islami in Pakistan, and although they call themselves "fundamentalists," I can see that Ja'far has gone way beyond them in his desire to attain religious purity.[47]

Do you now consider Jamaat-i-Islami to be too liberal? I ask.

"Yes, much too liberal," Ja'far says. "They are adopting a version of Islam promoted by Western liberals."

Are you closer to the Taliban? I ask. "No. They rely on dreams and fantasies. They are also Tassawuf [Sufi]. We are not Deobandi [like the Taliban]—we are Ahle Sunnah."[48]

Where did you get the idea to promote this strict interpretation of Islam? At last I have posed a question he appears to want to answer, and he does so with a paragraph rather than a word or a phrase: "When I was in Peshawar, I met people from Syria. They introduced me to Sheik Muqbil, from Yemen. After that we went to Yemen to learn Islam there. He was my main teacher, but he introduced me to others in Saudi Arabia, Egypt, India, and Jordan. I went to all those places to study Islam."

How much time did you spend in these places?

"Between one and ten months—it varied. I studied with Siddik Hassan Khan in India, with Sheik bin Baaz in Riyadh. I brought these ideas back to Indonesia. I also sent my students abroad, and when they came back [from these same countries as well as Pakistan], they encouraged me to form Laskar Jihad."

Prior to coming to Yogyakarta, I had interviewed scholars, United Nations personnel, experts working for local nongovernment organizations, and local reporters, and I have come with a list of alleged financial sponsors of Laskar Jihad, including a foundation called Al-Irsyad and Saudis living in Indonesia. Ja'far denies that he receives funding from any of them.

Laskar Jihad has a highly sophisticated Web site, including detailed instructions for sending donations to the group. I see you are raising money on the Internet, I say. What fraction of the funds you raise comes from Internet versus fund-raising on the street?

"Most is from the Internet," he says, perhaps as a way to account for the money that purportedly does not come from foundations or Saudis.

What fraction of that comes from abroad?

"We don't know. It's always anonymous," he claims.

I have read that the Al-Irsyad foundation owns the seven hectares of land near Bogor, west Java, where several thousand of your operatives carried out their military training in advance of your mission to Maluku. Is this true?[49]

"Yes," he says, surprising me with his sudden openness. "We trained there, but I have nothing to do with Al-Irsyad foundation. My brother is a lawyer and he was involved in a case over that land. While the land was disputed, we were able to use it for training."

What is the most effective way to recruit followers? I ask. On the Internet, or in person—through social networks?

"In person, at universities and mosques. From one person to another. We used the Internet to correct the false impressions created by the media."

In February 2002, the two sides in Ambon agreed to end all violence, respect the rule of law, ban armed militias, and establish a national investigation team to examine the causes of the conflict. But in April, Ja'far gave a speech urging Muslims to "prepare our bombs and ready our guns." Two days later a dozen masked men entered a village outside Ambon with guns, grenades, and daggers. They killed fourteen people and torched about thirty homes. Some of the attackers were armed with M16s, a weapon that the Indonesian military purchases from the United States.[50] In May, the government ordered Laskar Jihad to withdraw from the region, but the group remained. Soon after this, Ja'far was arrested. Vice President Hamzah Haz visited Ja'far after his arrest, and a month later he visited Abu Bakr Ba'asyir in Solo, central Java. Abu Bakr Ba'asyir is the leader of Jamaah Islamiyah, which Singapore accuses of being closely affiliated with Al Qaeda. Mr. Hamzah Haz urged the government to arrest himself rather than the clerics, claiming, "There are no terrorists here," in Indonesia.[51]

I have read that the army supports you, I said to Ja'far. Is this true?

"No. We have no military support. The Christians—the RMS and the Laskar Christus—they get big, heavy weapons from outside Indonesia, even from U.S. Navy ships," Ja'far said.

President Soeharto initially saw radical Islam as a threat to his regime, and parties promoting a religious agenda were banned. But the armed forces used small radical Muslim groups to do their dirty work, often to fight communists[52] and in some cases to quell separatism, including in Maluku, Sulawesi (also known as Celebes, famous for its coffee), Aceh (the oil-rich western tip of Sumatra), Kalimantan (the Indonesian half of Borneo, and East Timor, which seceded from Indonesia in 1999.

Three years after the conflict began in Maluku, President Megawati ordered an investigation into its origin. Her army chief, appointed in spring 2002, said that army deserters were responsible for some of the bloodshed. Other government officials believe that generals loyal to Soeharto—who had been ousted by pro-democracy forces in 1998—had hoped to use the conflict in Maluku to undermine Soeharto's immediate successor, President B. J. Habibie and Abdurrahman Wahid.

Is it true that in east Java military backing for Muslim militias comes from Major General Sudi Silalahi, chief of the Brawijaya army command?

"No."

I have read in the Indonesian press that you admitted to having a hot line to armed forces commander Admiral Widodo A. S., through a sixty-year-old man called Bambang, who donated the land for your boarding school.[53] Is this true?

"No."

But you were quoted! I exclaim.

"Misquoted."

The Indonesian military has a long history of using semicriminal paramilitary groups *(preman)* for a variety of causes: to consolidate the new nation after it achieved independence from the Netherlands and to suppress secessionist movements; to fight Islamist opposition—including by promoting rival Islamist groups; to fight both communist and democratic opposition to Soeharto's autocratic regime; and to discredit political leaders.[54] Many of the Islamist groups active today were created by the military, including the group most closely affiliated with Al Qaeda—the Jamaah Islamiyah. Some of the groups created primarily as criminal syndicates occasionally take on a religious or political cause. But now the ideologically based groups are expanding their source of funding, and the military may be unable to control them. The government demanded that Laskar Jihad disband itself in 2002, but the military may not be able to control the genie it unleashed. This is exactly what has happened in Pakistan, as we shall see in chapter 5.

Is it true you have support from the navy? I ask.

"Completely false," Ja'far says.

Some observers believe that the security forces tolerate continuing low-level violence in various parts of Indonesia because it keeps them employed in the region. The conflicts also provide the military and the police with opportunities to engage in criminal activities, such as charging for their services, drug smuggling, illegal gambling, and "taxation" of trade.[55] Army, navy, and police personnel occasionally get involved in skirmishes among themselves, which the International Crisis Group attributes to competition for scare resources.[56] Soldiers reportedly rent or sell their weapons and trade ammunition for food. An Ambonese member of the national DPR (House of Representatives) explained, "The ammunition and guns are sold by soldiers who need money to live."[57]

Is it true the army gives you guns? I ask.

"No. We make our own," Ja'far says. "The Christians are getting guns from the army and from Australia, New Zealand, and the United States. But we have to use swords and knives and homemade guns. They supply the Christians, but not us." Observers report that Laskar operatives were equipped with military-style munitions.

Both sides in communal conflicts tend to create—and come to believe—myths about the enemy's capabilities and intentions. Each side is certain that the other is conspiring with outside forces to turn Indonesia into an uni-religious state. Like other organizations in the business of protecting groups from harm, militants have a strong incentive to exaggerate the enemies' crimes and capacity for violence because it helps them drum up business. "Those who supply protection are inclined to exaggerate and manipulate its desirability," Diego Gambetta argues with regard to the Sicilian Mafia. The same is true for governments, he argues, which often operate like racketeers in their exaggeration of threats to national sovereignty.[58] In other words, those in the business of providing protection have an incentive to persuade their customers that the adversary is unstoppable unless the protectors—in this case the militants—arm themselves and act.

I want to understand the effects of the Asian economic crisis, which hit Indonesia in late 1997, and from which, unlike Korea and other countries, it has yet to recover. It stands to reason that unemployed or underemployed urban youth would be susceptible to the lure of extremism for several reasons. The opportunity cost of their time is low. The groups provide structure and a social network. And the paramilitary organizations provide a variety of financial incentives.[59] Laskar Jihad acts as a kind of employment agency for underemployed students and university graduates. It actively recruits students and underemployed youth, providing stipends to fighters' families during their service.[60] But poverty, in and of itself, is unlikely to be the cause of violence.

Soon after September 11, a debate arose in the United States about the socioeconomic causes of terrorism, with one side arguing that socioeconomic deprivation is a root cause of terrorism, and the other arguing that bin Laden's wealth, and his use of educated operatives, proves this hypothesis wrong. A widely reported study undertaken by two economists for the World Bank purported to show that wealth and education are *positively* correlated with terrorism.[61] The study only served to intensify the debate, although the data it examined were limited.[62]

Measuring the effect of wealth and education on terrorism is difficult. Some terrorist groups, for example, left-wing extremists in the U.S. and Germany during the 1960s, attract well-educated and relatively wealthy youth, while others, for example, neo-Nazi groups, tend to attract less educated youth.[63] Moreover, there is often a difference between midlevel managers and low-level operatives. The latter are often less educated and trained, and therefore relatively expendable. One of the problems of the studies conducted so far is that they purport to demonstrate the role of socioeconomic factors on terrorism writ large, rather than for particular groups in a particular place under particular conditions at a particular time.

If we look at the Indonesian case, it appears that a combination of rising expectations followed by an economic downturn, leaving educated youth without jobs after the 1997–98 crisis, may have contributed to the appeal of extremist groups offering employment.[64] The same dynamic appears to have played a role in the first Palestinian Intifada and in the 1990s civil war in Algeria.[65] Another possible contributor to the extremist groups' appeal includes democratic reforms allowing Islamists to openly recruit militants. De Tocqueville's observation that the most dangerous moment for governments is when they begin to set about reform seems particularly apt with regard to Indonesia. It is at the moment life begins to improve that people are most prone to revolt.[66] Other factors include widespread corruption throughout the government and security services; the security services' support for paramilitary groups, in some cases in the hope of personal financial gain; and a weak state largely incapable of controlling communal violence.

Do you have any links with other jihadi groups? I ask Ja'far. "None," he says. I find this claim hard to believe, since I had already asked Pakistani jihadi leaders about Ja'far, and they knew all about him. I investigate further. Both American and Indonesian officials suspect that Laskar Jihad has had some association with Al Qaeda, but they lack evidence to make a conclusive link.[67]

The Indonesian intelligence chief, Lieutenant General Abdullah Hendropriyono, said that Al Qaeda was assisting Laskar Jihad in its battles with Christians in Poso and central Sulawesi.[68] He also said that Al Qaeda had used a Laskar Jihad camp in Poso, central Sulawesi. (Others claimed the camp was actually run by another group known as Laskar Jundullah, a group linked both to Jamaah Islamiyah and Al Qaeda.) The next day the minister of defense supported the intelligence chief's claims. Court documents in a trial of alleged Al Qaeda operatives in Spain indicated that camps in Indonesia had provided training to Al Qaeda operatives, and that an Al Qaeda leader in Indonesia, Parlindungan Siregar, was also a member of Laskar Jihad. At the Madrid home of one of the Al Qaeda suspects, police found weapons, travel documents to Indonesia, and photographs apparently taken at the Indonesian camp.[69]

Under pressure from conservative Muslims, the intelligence chief retracted his claim three days later, however. Other senior officials also denied that Al Qaeda was training in Indonesia.[70]

Ja'far admits that he met bin Laden, and that bin Laden had offered to fund Laskar Jihad. But Ja'far says he refused to accept the funding. Since September 11, he has distanced himself from Al Qaeda, perhaps out of concern that he will offend his backers in the military, who would not want to risk offending the United States.[71] Ja'far accused a rival Islamist group called Laskar Mujahidin, whose operatives are often called "ninjas" because they wear masks in battle, of accepting financial aid from bin Laden.[72]

Some experts on Southeast Asia argue that the threat posed by radical Islam in Southeast Asia is not nearly as great as American officials seemed ready to believe after September 11. John Gershman argues, for example, that there is no risk of state-sponsored terrorism against U.S. interests in Southeast Asia, and that ethnic and religious diversity (including in the practice of Islam) that characterizes the area militates against the establishment of a fundamentalist hegemony by any one group. He also observes that most governments in the region are democracies that tolerate dissent, making Islamist extremism relatively less attractive as a broader vehicle for opposition to the government.[73] Moreover, many of the groups do have predominantly local agendas, and none is formally a member of bin Laden's International Islamic Front against the Jews and Crusaders. They also compete among themselves for funding and support.

While all these points are true and important, under the influence of bin Laden, the Southeast Asian jihadi groups have found a common cause in opposition to the West.

Have you sent your students abroad to help in other jihads around the world?

"Yes," Ja'far says proudly. "They've been to Chechnya, Bosnia, Kashmir, Afghanistan. And they've been fighting the communists in Yemen."

Have any foreigners offered to join your jihad? I ask.

"We'd like them to, but until now we've had to tell them we can't afford their visas."

What is your level of secular education? I ask.

"Elementary school. After that religious education. Jihad is the highest possible form of Muslim worship. We have to face our enemies, who are clearly kafir [infidels]," he adds, changing the topic.

Who are the kafirs in Indonesia?

"Jews, Christians, communists, and atheists." The classification and required punishment of infidels interests Ja'far a great deal, and he continues, unbidden, "There are three classes of kafirs. The first is kafir *harbi.* This is someone who is physically attacking Muslims. This is an enemy we are obligated to fight. Then there is kafir *dzimmi*—these are kafirs that have a cooperative attitude toward Muslims. We must protect their blood and dignity and possessions. It is a big sin to bother them. Then there is the kafir *mu'ahhad.* They're the kafirs who are citizens of other countries. As long as they obey regulations, they have to be treated the same way as kafir *dzimmi.*" I am relieved to hear this, as I am obviously a member of that class of infidel.

What about Shia? I ask. Are they kafir?

"Yes. They are not Muslims."

You yourself are no longer participating in these battles? "No. Too much to do here."

How many wives do you have?

"Four." There are nine children. The women explain that there are four houses on the complex, that each wife has her own house.

Are you hoping for more children?

"Yes."

As I prepare to leave, Fatima tells me earnestly that she has written me a letter. She started drafting the letter as soon as she heard that a lady scholar was coming to visit the compound. She requests that I not open the letter until I return to my hotel, where I can meditate upon its meaning.

When I open it, I see that she has taken great care with the writing. Her grammar and spelling are surprisingly good, and there are no crossed-out portions. She must have written many drafts.

Here is an abstract:

> To Professor Jessica Stern,
>
> Professor, we don't know who you are, and it doesn't matter whether you're a spy or just somebody who's curious with our existence. And you've known that we're Moslems.
>
> But perhaps you haven't known how our religion is. We proudly say to you that, "There's no correct religion beside God but Islam." . . .
>
> We're not treasure and world adorers; we're not the authorities' flatterers. We're not dead people, cross nor statue worshippers either. We're the people who would bend and surrender their bodies and souls only to Allah, King of the kings. . . .
>
> Don't ever think that we're afraid of death in defending our religion. Even death is our goal to reach the true glory. Victory in this world is God's promise for us in our every war. . . . Allah has also told us that The Jews and The Christians would never be relief [*sic*] with our existence and us until they can get us out of our faith. We will not close our eyes from your ruses. . . .
>
> So we invite you, Professor Stern, to enjoy the noble of Allah's religion. . . .
>
> This is our call, Professor, and we seek no benefits from you. We're just seeking for bigger love from Him. . . . We hope that you will listen, and open up your heart. May Allah show you to the right way, to his bless and mercy.
>
> Yours sincerely,
> Fatima . . .

The case of Laskar Jihad teaches us several important things about religious terrorism. Once contestants claim to be fighting a "holy war," religious militants on both sides flock to the region. Embittered and traumatized refugees, living together in close quarters, often in harsh conditions with no jobs and little hope, become ready recruits. Religious passions are thus turned into a weapon in a war that is often actually about control of natural resources or political power. Ja'far excelled at fomenting and capitalizing on these passions.

When leaders express grievances in religious or spiritual terms, they give contestants the feeling they are fighting over eternal, spiritual values, rather than fleeting, material ones such as natural resources or territory. Religious language expands the pool of potential sympathizers, recruits, and funders beyond the contested region to wherever coreligionists are found. Religious charities and members of diaspora populations provide funding to the organizations for their social-welfare work, in some cases not realizing they are also funding militancy. We will see this pattern repeated in many parts of the world.

Once holy-war organizations form, purportedly religious conflicts spread as the groups seek new "humanitarian" missions. Indonesia is one of the best examples of this phenomenon. But in Indonesia, another element makes the situation especially volatile. Shifting demographics, in many cases brought about through deliberate government policies to increase the percentage of Muslims in restive regions, is often the spark that ignites violence between indigenous ethno-religious groups and immigrants.

The case of Laskar Jihad and the other Indonesian jihadi groups also makes clear that when governments promote religious militants to use them as mercenaries, they are playing a dangerous game.

The complex relationship between religious militants and the military is an issue that I felt I needed to explore more fully. Military commanders support the Islamist groups not necessarily because they favor a Shari'a-based state, but to use them, when needed, as mercenaries. As we will see in chapter 5, I would encounter a similar dynamic in Pakistan.

FOUR

History

Israel has always had "more history than geography," Isaiah Berlin once declared.[1] Jews see their history as an account of loss and longing. It is about destruction of their Temples, expulsion from their lands, endless pogroms, and attempted annihilation. This chapter tells the story of several Jewish extremists who tried to destroy the Muslim holy sites built where the Jewish Temples stood historically, or sought to expand the country's borders to encompass the entire biblical Eretz Israel (land of Israel). Ancient history, as we shall see, can be a powerful weapon in extremists' hands, including in their efforts to expand national boundaries and to seek redemption.

In 1990, a messianic group known as the Temple Mount Faithful announced its intention to lay a 4.5-ton cornerstone at the site where the ancient Jewish Temples once stood, now the site of the Muslim Noble Sanctuary, the Haram al-Sharif. The group's ultimate goal is to destroy the Muslim holy sites and build a Third Temple. Palestinians viewed the plan as the first step toward the destruction of the Haram-al-Sharif, and the group's presence near the site incited the deadliest riots in Jerusalem since the city was taken over by the Israeli army almost a quarter century earlier. Gershon Salomon, the leader of the Temple Mount Faithful, was satisfied with the results, boasting that the riots had sparked worldwide interest in the Temple Mount and his movement, particularly on the part of fundamentalist Christians.[2]

The Temple Mount, or Haram al-Sharif, is of profound spiritual importance to all three monotheistic religions. Jews, Christians, and Muslims all believe that when the Messiah appears at the end of days, he will reign from this small peak.

To Muslims, Al-Quds (Jerusalem) is the third most holy city after Mecca and Medina, and the Haram al-Sharif is at the core of Jerusalem's significance within Islam. According to Muslim tradition, the Al Aqsa Mosque, also located on the contested plaza, is the "furthermost place" to which the prophet Muhammad rode his horse on a nighttime journey from Mecca. From here, Muhammad ascended to heaven and met face-to-face with God. Muslims call the Western Wall "Al-Buraq Wall," after Muhammad's horse, which, they believe, Muhammad tied to the wall prior to his nocturnal ascent. Christians revere the Mount as the place where Jesus drove out the moneylenders and as the site of the Messiah's eventual return.

Although the Temple Mount Faithful is a fringe group on the Israeli landscape, it is not alone in its desire to build the Third Temple. The movement has attracted a number of messianic individuals and groups—both Christians and Jews. Among them is Yoel Lerner, an MIT-trained mathematician and linguist who was imprisoned for a variety of terrorist plots, including a plan to blow up the Dome of the Rock to make room for the new Temple.

In July 2000 I travel to Jerusalem to talk with some of these extremists. I call Yoel Lerner immediately upon my arrival in Jerusalem. He instructs me to meet him the following morning at the bus stop in the Jewish quarter of the Old City at 7:35 A.M. He will take me through the winding pedestrian streets to his home.

I drag myself out of bed early the next morning. I have a quick breakfast of Arabic coffee, dates, and cheese. I am staying on the Arab side of the city, so I hire a cab to take me to the bus stop. A crowd is waiting to meet the bus. Apparently I am easily identifiable as a visiting American researcher, for a plump, genial, middle-aged man comes straight toward me to introduce himself. It is already punishingly hot, but Lerner is wearing a long wool vest with fringes over wrinkled clothing. He is unkempt, with the look of someone too busy thinking to worry about combing his hair or ironing his clothing. I can see by the look in his eye that he is generally bemused by life, by its ups and downs, even the occasional prison sentence.

We walk on ancient white stone streets, past ancient houses, also of white stone. There is a calm here in the early morning, an inescapable feeling of spirit. This is a long way from Brooklyn, where Lerner was born. He now lives on one of the most hotly contested pieces of real estate on earth, as close as possible to the one remaining wall of the Second Temple known as the Western Wall, or Wailing Wall. I feel lucky to get to see this side of Jerusalem accompanied by a resident.

Lerner invites me into a house that is directly across from the Dome of the Rock, the extraordinarily beautiful Muslim shrine that he would like to destroy. From his roof, which Lerner shows me, you can see the glorious gold dome. Coming in from the blinding-bright light of the sun on white stone, Lerner's house feels dark and cavelike. Too much thinking and plotting going on here. Papers are piled everywhere. Newspapers, books, and legal briefs. He invites me to sit at the kitchen table, which is covered with a stained plastic tablecloth littered with sticky crumbs. He brings me coffee in a stained cup. I feel oppressed by the clutter and dirt. Once we start talking, I don't notice the disorder here. I am distracted by thought, just as Lerner is.

He tells me about his first run-in with the law, when he was involved in a plot to remove Christian "propaganda materials" from local churches. I ask about his later plot to blow up the Dome of the Rock. Jewish statehood requires that the Temple be rebuilt, Lerner says. To do that, the Dome of the Rock has to be removed. That was his aim back in the late 1970s; it was a necessary step in the establishment of a state based on Jewish law. Who are the people who would like to blow up the Dome today? "Today? I have no idea. And even if I did, I wouldn't tell you," he says. "I am now pursuing a different path." What path? I ask. "I have been trying to put together a political force, a political party we'll call it, over the past year or two. Essentially what I've been lacking is the funds to do it properly. I know the Israeli scene sufficiently well so as not to try to do it if I can't do it properly. The worst thing to do is start something like that and find yourself high and dry."

So, I say, you were once young and passionate, and you became a terrorist. And now that you're older, you pursue politics instead of terrorism. You no longer have the energy to sustain the necessary anger for violence. "You could look at it that way," Lerner concedes, "or you could say I have chosen other paths to achieve the same objective."

What is your party's agenda? I ask.

"It begins with the assumption that nobody—from the individual Jew up to the most powerful government you could imagine—has the moral right to give up any significant territory making up part of the land of Israel," he replies. "It would require bringing about a dissolution of the so-called Palestinian Authority. Moving tanks back into Gaza."

Lerner tells me he will explain more if I return in the afternoon. He has some business to attend to. I tell him I will.

When I come back, the house feels cheerier. Lerner's wife is in the kitchen, cooking. The house is permeated with sweet odors. A chocolate cake is cooling on a rack, and challah is baking in the oven.

We return to a discussion of Lerner's political movement. "The basic ideological premise," he explains, "is that we would like to make it possible for people to practice their religion. There are six hundred thirteen commandments in the Torah. The temple service accounts for about two hundred and forty of these. For nearly two millennia, since the destruction of the Temple, the Jewish people, contrary to their wishes, have been unable to maintain the temple service. They've been unable to comply with those commandments.

"The temple constituted a kind of telephone link to God," Lerner summarizes. "That link is now destroyed. We want to rebuild it."

The Jews' despair upon losing their Temple a second time was recorded by contemporary observers. The third century *Midrash Tanhuma,* a collection of rabbinical interpretations of the Torah, described the Temple Mount as the world's foundation. "Just as the navel is found at the center of a human being, so the land of Israel is found at the center of the world. Jerusalem is at the center of the land of Israel, and the Temple is at the center of Jerusalem, the Holy of Holies is at the center of the Temple, the Ark is at the center of the Holy of Holies, and the Foundation Stone is in front of the Ark, which is the point of Foundation of the world."[3] The fourth-century Christian leader Jerome observed Jews praying at the site of their lost Temple, destroyed by the Romans some two centuries earlier. "On the anniversary of the day when the city fell and was destroyed by the Romans, there are crowds who mourn," he wrote, "old women and old men dressed in tatters and rags, and from the top of the Mount of Olives this throng laments over the destruction of its Sanctuary. Still their eyes flow with tears, still their hands tremble and their hair is disheveled, but already the guards demand pay for their right to weep."[4] "In the sanctuary itself," a pilgrim wrote, "where the Temple stood which Solomon built, there is marble in front of the altar which has on it the blood of Zacharias—you would think it had only been shed today. All around you can see the marks of the hobnails of the soldiers who killed him, as plainly as if they had been pressed into wax. Two statues of Hadrian [the second-century Roman emperor who had been sent to Palestine to curb a Jewish revolt] stand there, and, not far from them, a pierced stone which the Jews come and anoint each year. They mourn and rend their garments, and then depart."[5]

Lerner's wife periodically interrupts us to provide her own opinions, often disagreeing with Yoel. I feel completely at home with this style of conversation, which is sometimes more like a fight. At one point she pronounces that she believes that within a few years I will make aliyah—the Hebrew term denoting immigration to the Holy Land, which literally means "ascent." In that moment I imagine myself living in Israel, hanging around in her kitchen helping her prepare for the Sabbath. Becoming a practicing Jew. Joining a community. The removal of skepticism and doubt. The image is remarkably comforting. Then I imagine what it would mean to want to rebuild the Temple. I see the beautiful Al Aqsa Mosque being crushed. I see Palestinians screaming. This guy is a terrorist, I tell myself. How can I have such a fantasy?

How else would life be different if your party took control? I ask.

"There is a prerequisite to Jewish statehood, which in our case was never fulfilled: the reestablishment of what Judaism used to call the High

Court, the Sanhedrin. It's a body of seventy-one judges who were entrusted with final rulings in connection with interpretation of Jewish law in everyday life. The Sanhedrin was also responsible for providing religious approval for the workings of the secular government. You can't have a seventy-one-man body run foreign policy in general. But a declaration of war, for example, would require their approval. That would make a very big difference. Today, every Jew picks his own final authority. If the Sanhedrin were active, every Jew would be bound by whatever it ruled."

But what about the murder of Rabin? I ask. Do you believe it was religiously acceptable, given that there exists today no ultimate authority to sanction such a step?

"You've got a ticklish point," he says. "Contrary to popular belief, the highest value for a Jew is not the preservation of human or even of Jewish life. The highest value is doing what God wants you to do. So in an attempt to put Jewish values in a hierarchy, human life in general, Jewish life in particular, is high on the list. But it's not the top."

But how, I wonder, does a Jew know what God wants him or her to do in any given instance? Why is it that the only people who seem to know with absolute certainty are the people who become terrorists?

"There are a number of circumstances under which the individual is enjoined to take a Jewish life if necessary without consulting a court," Lerner continues. "If you see a person preparing to commit a capital crime—rape or murder—it is your duty to stop him. You must stop him any way you can. It's similar in some respects to the right Jewish law accords the individual to restore his own property from a thief if it is stolen. You don't have to bring him to court. If you can catch up with him, you can take your property back by force. You don't have to bother the court with stuff like that. Rabin was stealing Jewish property, proposing to give it away."

So the death of Rabin was simply "collateral damage" in an effort to recover stolen property, according to Lerner's convoluted reasoning. His murder would not even have required a ruling by the Sanhedrin, if it had existed.

"I had been convinced for some time that Rabin's death was coming, that it had to come," Lerner continues. "I understand what motivated

Yigal Amir [Rabin's murderer]. I am convinced that he felt that Yitzhak Rabin was putting the survival of the Jewish people in danger by his policies. There was no other way of removing Rabin from the gun he was pointing at the Jewish people. I'm ninety-nine percent positive that that's what he thought. Honestly, I can't argue with it."

After his arrest, Amir proclaimed that the killing of Israeli prime minister Yitzhak Rabin was justified, even commanded, by the rulings of Din Mosser and Din Rodef, as described in the Jewish religious law, or *halakha*.

According to the halakah, the rulings of Din Mosser and Din Rodef apply to those Jews who have committed the most despicable crime imaginable—the betrayal of their fellow Jews. The punishment of the Mosser—a person who hands over sacred Jewish property to the gentile—as well as that of the Rodef—a person who murders or facilitates the murder of Jews—shall be death. Since the execution of the Mosser or the Rodef is aimed at saving the lives of other Jews, there is no need for a trial.

Amir admits that he was partly inspired by the book *Baruch the Man,* published to commemorate the death of Baruch Goldstein, the terrorist who massacred a group of twenty-nine Palestinians near Hebron in 1994. Rabbi Yitzhak Ginzburg, a mathematician and specialist in Jewish mysticism, contributed a chapter that discusses revenge and terrorism as purifying. Jews are encouraged to take revenge against those who harm them to recover their inner power after centuries of humiliation. Revenge, he argues, "is stressing my positive essence, the truth in my being. . . . It is like a law of nature. He who takes revenge joins the 'ecological currents of reality.' . . . Revenge is the return of the individual and the nation to believe in themselves, in their power and in the fact that they have a place under the sun and are no longer stepped on by everybody." Rabbi Cohen, another contributor, argues that seeking revenge not only helps the Jewish people, but also "provides the individual Jew with the satisfaction and consolation for the troubles the people of Israel suffered so long." Both rabbis were talking about revenge against gentiles, but Yigal Amir felt that Rabin was the Jews' worst enemy.[6]

"The land is a *sacred* thing," Lerner says, trying to explain Amir's decision to kill Rabin. He refers me to passages in Genesis and Deuteronomy for evidence that the land was given to the Jews in a sacred contract, and that no Israeli leader has the moral right to "give this sacred territory away."

"In that day the Lord made a covenant with Abram, saying: Unto thy seed have I given this land, from the river of Egypt unto the great river, the river Euphrates . . ." (Genesis 15:18). And also, "Every place whereon the sole of your foot shall tread shall be yours: from the wilderness, and Lebanon, from the river, the river Euphrates, even unto the hinder sea shall be your border. There shall no man be able to stand against you: the Lord your God shall lay the fear of you and the dread of you upon all the land that ye shall tread upon, as He hath spoken unto you" (Deuteronomy 11:24–25).[7] These areas include lands that today are in Jordan and Iraq. These are the words that inspired both Lerner and Yigal Amir. They also inspire many settlers. And they will probably continue to inspire terrorism for years to come.

The biblical promise of Eretz Israel, the land of Israel, to God's chosen people makes the concept of land a key component of religious doctrine to fundamentalist Jews, who believe that possession of the land of Israel is part and parcel of the Jews' Covenant with God. Accordingly, relinquishing or dividing any part of the land promised by God to the children of Israel would constitute a breach of the Covenant.

Israel's victory in the June 1967 war (the Six Day War) not only tripled Israel's territory, but the newly conquered areas included the biblical territories of Sinai, Judea, Samaria, and Gaza. For messianic Jews, the victory was a modern miracle of unimaginable proportions, a miracle that indicated the imminent arrival of the Messiah. The movement centered around the Merkaz ha-Rav Yeshiva in Jerusalem, whose students formed the nucleus of what became known as Gush Emunim (Bloc of the Faithful), a messianic group that became the vanguard of the Israeli settlement movement after the Yom Kippur War of 1973. The mantra of the students at Merkaz ha-Rav Yeshiva, and later on of Gush Emunim, was the settlement of the newly "reconquered" territory in the West Bank or, as they referred to it, Judea and Samaria. Fundamentalist rabbis swiftly declared every inch of the West Bank holy land and called on Jews to settle it. They declared that by settling on this land, Jews could accelerate the process of redemption, i.e., the coming of the Messiah, who is expected to remove the pain and agony from the Jews. The argument was that redemption would ultimately be achieved when the Jews controlled all of the biblical land of Israel.

Within hours of the arrival of Israeli troops on the Temple Mount, however, then Defense Minister Moshe Dayan decided to leave the compound in charge of the Muslim religious trust known as the Waqf. The belligerent parties agreed that Jews would be allowed to visit the Temple Mount only when Muslims were not conducting prayers there. Israelis would be treated as tourists, and Jews would not be allowed to pray on the compound at any time. In conceding control of the Temple Mount to Muslims, Dayan had made a pragmatic decision that enabled the Islamic world, however reluctantly, to tolerate Israeli sovereignty over East Jerusalem.

Dayan's pragmatism did not win him many friends among Israel's religious right. To them, he was relinquishing the biblical Mount Moriah, the site of the two ancient Jewish temples that were built to honor and worship God.

According to Jewish tradition, the first Jewish temple was erected by Solomon, King David's son, in around 1000 B.C. In 586 B.C., the First Temple was destroyed by the Babylonian king Nebuchadnezzar, who subsequently enslaved the Jews, sending them to the Babylonian exile. On their return from exile roughly seventy years later, the Jews rebuilt the Temple, where they worshiped for centuries. In 37 B.C., when Herod became vassal king of Judea, he embarked on a project to build a court around the second Jewish Temple. Art historians describe King Herod's enlarged Temple compound as one of the Roman Empire's most magnificent structures.[8] The Roman legion burned the Second Temple to the ground on the ninth of Ab by the Jewish calendar, in A.D. 70.

The Western Wall is the most well-known section of the remaining Herodian Temple Mount constructions and has stood exposed for almost two thousand years. The conquest of the Western Wall on June 7, 1967, was a national and emotional triumph to Israelis and many Jews worldwide and symbolized, more than any other conquest, a restored connection to the past. In contrast to the Temple Mount, Israel did not cede administrative control of the Western Wall to the Muslim Waqf during the 1967 war.[9]

In 688 to 691, the Muslim caliph Abd el-Malik built the Dome of the Rock, a spectacular Muslim shrine situated on the Temple Mount/Haram al-Sharif, believed to sit atop the earlier locations of the Jewish Temple. Among the most beautiful Muslim buildings that have remained intact and relatively unchanged, the shrine encloses a rock that, just like the mount itself, bears religious and historical significance for both Muslims and Jews. Jewish tradition holds that Abraham was preparing to sacrifice his son Isaac on this very rock when an angel of God called to him from heaven and ordered him to desist, just as Abraham was raising his knife. (In the Islamic version of the story, it was Ishmael, Abraham's first son, whom Abraham was about to slay.)[10] Muslims believe that it is the very rock from which Muhammad rose to heaven and that the Prophet's footprint and a handprint of the angel Gabriel are embedded in the stone. In early history, in many parts of the world, nonmonotheistic religions required occasional human sacrifice to propitiate the Gods for rain, abundant harvests, or other good fortune. Some scholars argue that the great significance of the story of Abraham on the Mount, which is common to the three monotheistic religions, lay in the new message that people need no longer engage in the wasteful sacrifice of human beings. The theory suggests that the sacrifice of animals or agricultural products was less costly to the community, and may help explain why the new religions eventually prevailed.

The main function of the Jewish Temple was to provide a dwelling place for the Shechinah, God's manifestation in the world. The main service done at the Temple was the offering of animal sacrifices. Regular public prayer could not be held in the Temple. However, the Temple served as the site for occasional communal gatherings to request divine favors. Anyone could go to the Temple to offer private prayers, even gentiles.[11]

Longing for the Temple's reconstruction on what is now a Muslim sanctuary is a central part of Jewish tradition. In the Amidah[12] prayer, observant Jews pray daily for the Temple to be rebuilt: "Lord our God, look with favor on Thy people Israel and their prayer. Restore worship to

Thy Temple in Zion, and with loving grace accept Israel's offering and prayer. May the worship of Thy people Israel find favor with Thee evermore. . . . May our eyes witness Thy loving return to Zion. Blessed are Thou, Lord who will restore His Divine Presence to Zion."[13]

Observant Jews commemorate the Temples' destruction yearly on the holiday Tishah-b'ab. The medieval philosopher Maimonides, in his Code of Jewish Law, urged every generation of Jews to rebuild the Temple if its site was retaken, if a leader descended from David could be found, and if the enemies of Jerusalem were destroyed. Of the 613 commandments of Judaism Maimonides listed in that code, about one-third refer to worship in the temple.

Millenarian Jews believe that at the End of Days, there will be a time of great troubles. Jerusalem will be taken in battle, but God will smite the enemies of the Jews. The wicked will act wickedly and not understand, while the knowledgeable will grow refined and radiant. The righteous among the dead will rise to eternal life, while others will be left to everlasting abhorrence. All three monotheistic traditions have a conception of an apocalypse, but each believes that its own group will prevail in the catastrophic events of the final days.[14] Some millenarians hope to bring on that very catastrophe, which they see as a necessary stage in the process of redemption. Evangelical Christians and Messianic Jews have developed a cooperative relationship, based on their common belief that rebuilding the Temple will facilitate the process of redemption, even though each believes its own group will ultimately triumph.

I want to learn more about the mystical aspects of the Temple. Lerner tells me he is too rational to understand or explain such things. He urges me to speak with two other people: Avigdor Eskin and Yehuda Etzion. Both are practitioners of Jewish mysticism. Eskin continues to run into trouble with the law, for a variety of activities the authorities call supportive of terrorism. Etzion, like Lerner, plotted years ago to blow up the Dome of the Rock and is now pursuing a more peaceful path.

At Yoel Lerner's recommendation, I take a bus to Avigdor Eskin's apartment, which is in an ultraconservative Orthodox neighborhood known as Kiryat Moshe. I am wearing a long skirt, long sleeves, and a scarf that covers my hair, neck, and shoulders completely. It is terribly hot. I begin to sweat as I wander through Eskin's neighborhood in search of his apartment building. I am ready to cry from the heat when I finally find the right address.

His wife meets me at the door. Smooth brownish skin. Coquettish, but obviously intelligent. Lithe, strong, efficient, I think. She is young, perhaps twenty. A long skirt and hair held under a scarf, just like mine, as is the custom for Orthodox Jewish women. But somehow she looks lively and bohemian in this getup, whereas the person I saw this morning in my hotel-room mirror looked stern, dour, and nervous. I hear children—laughing, shouting, crying. There are obviously lots of them; their cheerful detritus is everywhere in evidence. She speaks with a strong Russian accent. We switch to Russian, to her obvious relief. Avidgor will be with you shortly, she tells me.

Eskin is a mystic and a convicted criminal. On October 6, 1995, two days after Yom Kippur, Eskin organized a group of fellow mystics to intone a cabalistic curse called the Pulsa di Nura ("Lashes of Fire," in Aramaic). The ceremony was conducted outside the home of Yitzhak Rabin. They believed that chanting this evil prayer would result in Rabin's death within thirty days. Thirty-one days later, Rabin was dead.

"What Rabin was doing was the worst kind of betrayal to the existence of this country," Eskin tells me. "It had reached a certain historic point. People couldn't stand it anymore, what he was doing. His effort to make peace with the Palestinians. He wanted to change the course of Israeli history—in the wrong direction. We consider the PLO to be a terrorist organization, and Yitzhak Rabin was assisting people who want to kill us. We prayed that he'd change his behavior or be removed by heavenly forces."

Do you think the prayer worked? I ask.

"Yes," he says. "The prayer has been used only once before in the twentieth century. To kill Leon Trotsky. Rabbi Chafetz Chaim cursed him. It worked then, and it worked this time too."

I check on Rabbi Chafetz Chaim and find that his real name was Rabbi Israel Meir Ha-Cohen. He was known as Chafetz Chaim, which was the title of one of his most famous works. Ironically, *The Book Chafetz Haim* was an attempt to clarify the laws of "evil talk and gossip."[15]

Was it your idea to say the prayer? I ask.

"Yes," he says.

Where did you get the idea?

"I was thinking what to do against Rabin. How to express the deepest of the depths. I was ready to do anything against him. But not to shoot."

Why not?

"Simply because . . ." Words seem to fail him. "I was not ready to shoot."

Did you know Yigal Amir? I ask. Yigal Amir murdered Rabin thirty-one days after Eskin and the others chanted the prayer.

"No, I didn't know him then. But now I'm in touch with him."

Did he know of your prayer?

"I don't know—we never discussed it."

Weren't you curious whether he knew?

"No, not so much. I know he wasn't influenced by it—he would have told me. Whether he heard about it, I don't know."

How are you in touch with him?

"His parents, he calls them. Letters."

Could I hear the prayer?

"We don't talk about it. It was done once and I'm not going to be involved with this kind of thing ever again in my life."

The reason you're forbidden to talk about it is religious?

"It's a very strong magic thing. You don't go and give out this prayer to anyone."

What kind of rabbi knows about this kind of thing? I press him.

"You find out yourself. I'm not into this any longer. . . . We did it once and it worked and we don't do it again. Last year a woman called me. She says, Avigdor, I want to meet you alone. She refused to tell me in advance what she wanted from me. She wanted to meet me without my wife. It sounded pretty interesting. Why not? I told her to meet me at the airport; I was going to Moscow. She says to me, Avigdor, I have this mother-in-law, driving me crazy. She wanted me to use the Pulsa di Nura against her mother-in-law. I won't do it. You see? It's dangerous. It should only be used against public figures, like prime ministers. One is not allowed to use it unless one is certain that the person one is cursing is one that God would wish to be dead. If it's used incorrectly, or against a person whom God does not wish dead, it will backfire. It might kill the chanter instead."

"There are certain things you are not supposed to use the cabala for. Mystical powers are never used unless there is some extreme situation. You can't use it to make money. You can't use it to make your wife look younger," he says oddly, given how beautiful his wife is, and how much younger she is than he. "You can't use it to improve sexual performance. These things are wrong."

The phone rings. Eskin answers it. It's his lawyer. When he is finished, I ask him why he needs a lawyer. "Oh, there are many, reasons," he says. "There are four trials. One of them is about the Pulsa di Nura." How can you be arrested for a prayer? I ask. "They said it was support for terrorism. I was accused of supporting terrorism, and I didn't even know that Amir was going to do it. The GSS [General Security Service, Israel's internal security service known as Shin Bet] accused me also of collaborating with someone on a plan to build a catapult to throw a pig's head onto the Temple Mount while Arabs were praying. It was supposed to be a joke. And I didn't know anything about it. We were caught in a sting operation. Things happen here that couldn't happen in America. They torture us. They put sacks on our heads. They prevent us from sleeping. They do all kinds of things they do to Arab terrorists."

How do they keep you from falling asleep?

"Very simple, they keep interrogating you and in between interrogations they put you in a special chair so you can't sleep. These are the methods of the Israeli security service."

Were you observant when you were growing up?

"No. Absolutely not. My parents were not observant."

But you knew you were Jewish?

"Yes. I became a Zionist at age thirteen. I learned Hebrew at age fifteen, and then I started teaching it. I was expelled from secondary school. I was arrested. I was arrested for the second time at age fifteen or sixteen, for giving lectures. I have a long history of arrests."

Was your family afraid for you?

"Yes. My father was, my mother was, they all were. But back then I was the darling of the Jewish establishment. Now I'm the bad guy. Now I'm considered one of the most radical in the world."

We turn next to a discussion of America. "I would rather live in Moscow than America today," he says. "The whole idea of America being a country based on certain moral principles, values, is gone. The culture is absolutely destroyed by Afro-Americanism. The lowest of the low. All this pop music, rock and roll." I have heard that Eskin is an accomplished pianist. "All these homosexuals. Monica Lewinsky and Clinton. It was the only thing Americans were interested in. Their minds are disturbed. America wants to create in Israel the same situation it created in South Africa, meaning that Israel will cease to exist as a Jewish state. This is the ultimate goal of the United States of America."

What do you think of globalization?

"Americans think they've succeeded in using the Internet to occupy the whole world. They're mistaken. Some people use the Internet to promote the most extreme fringes of Islam, using the tools that America gave them. And other people use the Internet to promote the most vicious neo-Nazi ideology. The new strength given to neo-Nazis is primarily through the Internet. American Nazis can now communicate with Nazis in Europe. This is called globalization. America is also causing terrible damage to Israel and the West by exporting its culture. American culture should be treated the same as we treat drugs . . . as a poison. Pop music is created primarily by blacks who live on drugs. Homosexuals too. In America you have this homosexual revolution. It's a fashion—it's not an inborn thing. There's nothing left of what America used to be. And this is what America's trying to spread—to Israel, Russia. It's destroying the Western world and Islam is taking advantage of this. In America, people have no education, no religion. They play with computers, television, pop music. These people have no values to fight for. This is the way fascism is created psychologically. First by absence of values. Then by accepting violence. And American society is very violent."

Avigdor Eskin, like Yoel Lerner, is a follower of the teachings of Rabbi Kahane. Kahane's teachings remain a strong influence on Jewish extremists long after a Muslim extremist assassinated the rabbi in New York City in

1990.[16] To followers of Kahane, redemption is inevitable, now that God has helped create the modern state of Israel. But it is up to the Jews to reestablish a theocracy, and to remove any obstacles that stand in the way, including the Arabs.

The Kahanist ideology was institutionalized in the Jewish Defense League (JDL), which Kahane founded in 1968 under the slogan "Never again!" The group's activities have included fighting "black and white anti-Semitism," supporting the emigration of Jews from the former Soviet Union, hunting for Nazi war criminals, and practicing "Jewish self-defense."[17] Under the motto "Every Jew a .22," the vigilante JDL urged American Jews to arm themselves.

In September 1968, Kahane moved to Israel, where he founded the organization Kach (Thus) in 1971. An offshoot of Kach, an organization called Kahane Chai (Kahane Lives), was founded by Kahane's son Binyamin[18] following his father's assassination. Both groups have a violently anti-Arab outlook and call for the expulsion of Arabs from Israel.

In 1973 and 1977, Kach participated in the Israeli parliamentary elections, but failed to gather a sufficient number of votes to elect anyone. The movement ran for elections again in 1984, and Kahane was elected to the Knesset, the Israeli parliament, with twenty-six thousand votes. In the following year, the Knesset banned Kach from participating in the next elections based on its incitement to racism. In his appeal to the Supreme Court's decision to ban Kach, Kahane claimed, to no avail, that "security needs justify severe measures of discrimination against Arabs."[19]

Kach and Kahane Chai were declared terrorist organizations in 1994 by the Israeli cabinet. The banning of the two groups followed one of the most well-known incidents of Jewish extremism, namely the massacre of twenty-nine Muslims in Hebron by Dr. Baruch Goldstein on February 25, 1994. Goldstein, a thirty-seven-year-old doctor and father of seven at the time of the shooting, was a prominent member of Kach. The group had issued statements supporting Goldstein's attack.

Both Kach and Kahane Chai organize protests against the Israeli government and harass and threaten Palestinians in Hebron and the West Bank. Groups affiliated with them have threatened to attack Arabs, Palestinians, and Israeli government officials. They claimed responsibility for several attacks of West Bank Palestinians in which four persons were killed and two were wounded in 1993. In April 2002, the current leader of Kach, Baruch Marzel, was arrested by Israeli police in connection with a plot to leave a trailer laden with two barrels of gasoline and two gas balloons outside a Palestinian girls' school in East Jerusalem.[20] The West Bank settlements of Tapuah and Kiryat Arba are strongholds of the Kahnist movement. According to the International Policy Institute for Counter-Terrorism, both organizations receive support from American and European sympathizers.[21]

"America is the most antisexual society," Eskin continues. "American women try to be like men. They turn men off with feminism. This is ruining the white people. While Western society is dying, all the savages around the world, they are not doing this. Their women aren't feminists. They have real culture, real strength. The Muslims, they are ready to fight. They are ready to die for something. They're ready to die for their ideas. Democratic liberal societies are getting globalized, they are rotten to the core, while the enemies of the free world are becoming stronger and ideologically firmer. More determined to fight and win.

"Today you see two very strong trends in the world. One you can define as a masculine trend—presented by Islam, which has no mercy. Only power, force, and violence, and no place for the individual. On the other hand you see America getting rotten with liberalism, which is more feminine—being submissive and being poisoned by no will to live."

Do you associate this feminine liberalism with too much emphasis on the individual?

"Yes. America emphasizes the individual to the extent that individual becomes nobody. An individual grows when he gets higher and higher through education, by being with God. But without education and prayer, you don't develop. You lead an animal life, you have no freedom of choice, you become the simple and despicable animal of the earth. When you have desires stronger than the animal, this is the psychological basis for fascism. . . . Judaism has the special task to be in between these two extremes, to preserve certain values, family values. To make sure that men are men and women are women. Homosexuals should be somewhere else, in Africa maybe, with those that prefer that lifestyle."

Eskin tells me that he and Yoel Lerner are raising money to start a political party to create a truly Jewish state in Israel, based on Jewish law. He wants to know whether I would like to contribute. I politely decline.

It turns out to be easy to find the killing prayer on the Internet. It reads, "I deliver to you, the angels of wrath and ire, Yitzhak, the son of Rosa Rabin, that you may smother him and the specter of him, and cast him into bed, and dry up his wealth, and plague his thoughts, and scatter his mind that he may be steadily diminished until he reaches his death. Put to death the cursed Yitzhak. May [he] be damned, damned, damned!"[22]

The next day I hire a car to take me to Yehuda Etzion's house. He lives in a West Bank settlement, in vigilante territory, where Jews and Arabs regularly shoot one another in "self-defense." We drive through arid countryside, which is dotted by farmland and settlements. It is here you notice how the Israelis have made the desert bloom, as my grandmother told me breathlessly many years ago. What extraordinary ingenuity. What impressive energy. But there is one part of the story my grandmother didn't tell me: the settlements get a disproportionate share of the region's scarce water supply. Palestinians will readily admit, at least in private, that they envy the Jews' drive. But it is hard to build farms in the desert without water, no matter how strong one's energy and determination.

In the summer of 1967, the Israeli cabinet deliberated on what to do with the land they had won in the war. The crucial session, Israeli historian Amos Elon tells us, began on a Sunday in mid-June and lasted, with brief interruptions for food and sleep, until the following Wednesday. In the end, the decision taken was not to decide, Elon explains. As a result, Defense Minister Dayan, by then a national hero, Foreign Minister Allon, and "assorted right-wing and religious fundamentalist militants and squatters" successfully established "dubious facts on the ground." The settlements and outposts, which multiplied, were lavishly subsidized and eventually legalized through a patchwork of formal and semiformal agreements. "It was said of the British Empire that it was born in a fit of absentmindedness. The Israeli colonial intrusion into the West Bank came into being under similar shadowy circumstances. Few people took it seriously at first. Some deluded themselves that it was bound to be temporary. Those responsible for it pursued it consistently," Elon recounts.

Michael Ben Yair, Israel's attorney general to the Rabin government, describes the occupation thus: "The Six Day War was forced on us; but the war's Seventh day, which began on June 12, 1967—continues to this day and is the product of our choice. We enthusiastically chose to become a colonialist society, ignoring international treaties, expropriating lands, transferring settlers from Israel to the occupied territories, engaging in theft and finding justifications for all this."[23]

As of this writing, the total settler population of the West Bank and the Gaza Strip is roughly two hundred thousand, about double what it was at the time of the 1993 Oslo Agreement. Another two hundred thousand live on former Jordanian territory in East Jerusalem and in the Golan Heights. The boundaries of the approximately 190 settlements (130 in the West Bank alone) enclose some 10 percent of West Bank territory.[24]

Israeli law does not prohibit the creation of new settlements, but since the signing of the Oslo accords in 1993, official Israeli government policy has been that no new settlements will be built.[25] But the law is unclear and therefore problematic, explains Dror Etkes, who heads the settlement watch section of the Israeli peace organization Peace Now.[26] The government allows existing settlements to expand, sometimes into new neighborhoods located miles away from the original settlement site. And construction of outposts, or fledgling settlements, is continuing. It has become a "twilight zone," he says, "a Wild East."[27]

Opponents of the settlements say that all Israeli settlements violate United Nations resolutions and international laws, not just the new outposts. The Fourth Geneva Convention, for example, prohibits an occupying power from settling its own citizens on the territory it is occupying. But past Israeli administrations have asserted that the territories are "disputed" rather than "occupied," and that these laws do not apply.[28]

Some settlement residents have become frightened of the violence and are moving back to Israel proper. But a new generation, significantly more radical than their predecessors, is coming to take their place. The "Hilltop People" are armed settlers who have taken it upon themselves to police the new outposts. They have engaged in violent confrontations not only with local Arabs, but also with the Israeli military. When the military tried to destroy Gilad Farm, an outpost manned by twenty-seven-year-old Itay

Zar and ten Hilltop People, hundreds of young people from surrounding settlements came to Zar's rescue, throwing rocks and rotten vegetables at the soldiers. A hundred protestors lay down in front of a bulldozer dispatched to destroy the settlement outpost. After his house was destroyed, Zar and his family set up tents. "Until all the land of Israel belongs to the people of Israel according to what has been promised in the Bible, there will be no peace," he told the *Washington Post*.[29]

We turn into a dirt road off the main highway, leading into a cluster of small homes. Etzion's house is easy to find, as everyone in the neighborhood knows where he lives. He invites us to wait in his living room while he gets some oranges for us to eat.

I ask Etzion to explain his feeling of urgency about rebuilding the Temple. "If you seek the kernel of meaning in the Temple," he says, "it is akin to the meeting of love between the Jewish people and God, or the attraction between men and women. The Jewish people are the female aspect, and they are missing their other, an other which can only be recovered when the Temple is rebuilt. The view of God is symbolized by the man, and the Jewish people as a woman.

"It is something so wonderful you can hardly imagine it. None of us has ever seen or touched anything like it. It is not just the stones it's built of. That's just the framework, like the peel of an orange. The Temple is the collective spirit of the people." Etzion is clever, like Lerner. But he is also poetic. Listening to him, I start to feel the loss of this mystical place. I feel the longing. For the Temple, and for this sensual union between God and man that he describes. Fundamentalism is always about longing, I remind myself, often for something that never existed.

In 1984, Israeli authorities uncovered a plot by Yehuda Etzion and coconspirators to destroy the Dome of the Rock, which the group called "the abomination." The group, an offshoot of Gush Emunim, was known as the Jewish Underground, or Makhteret. Until that point, the Gush Emunim settlers had eschewed violence, despite their messianic and fundamentalist outlook. Beginning in the 1980s, in the wake of the Camp David peace accords, the group began to despair of achieving its goals peacefully. Some members of the group, among them Etzion, turned increasingly violent, prepared, in the end, to risk a world war in pursuit of religious redemption for the Jewish people.[30]

Etzion subscribed to the teachings of Rabbi Shabtai Ben Dov, who promoted the idea of active redemption as the best strategy to achieve the total transformation of Israel into a sacred state run according to Jewish law. The group had stockpiled weapons to use against the Dome as soon as they had received rabbinical authority, which, fortunately, did not occur prior to their discovery by law-enforcement authorities.

Carmi Gillon was chief of the Shabak, the Israeli general security service, when Yigal Amir assassinated Prime Minister Rabin. He was head of the department that uncovered the Jewish Underground. Gillon describes Etzion as an unusually clever terrorist. On May 2, 1980, Fatah threw a grenade into a group of Jews who were praying in Hebron, six of whom were killed. The Jews in Hebron wanted to take revenge. Most of them wanted to go to a market and blow up as many Arabs as they could or do the same in a mosque. But Etzion persuaded his colleagues that wounding, not killing, several Palestinian leaders was a better strategy. Etzion felt that killing them would only make them heroes. This was a clever strategy. The group managed to wound several Palestinian mayors.

In subsequent attacks, Etzion failed to prevail over his more violent colleagues. On July 17, 1983, an Israeli yeshiva student was killed in Hebron. Etzion's colleagues entered the Islamic College in Hebron, determined to kill as many Arabs as they could. They killed three and injured over thirty. Yehuda Etzion objected to this. He does not believe in random violence or acts of pure vengeance, Gillon explains.[31] The group was shaped initially by Etzion's desire for redemption, but over time its goals shifted to sheer revenge. The group had the potential to become a professional organization of killers, but was stopped before it got that far.[32]

Although Etzion appears to have given up violent struggle, at least for now, he has not given up his efforts to prepare Israelis to rebuild the Third Temple when the time is ripe. Yehuda Etzion, Yoel Lerner, and Avigdor Eskin are all members of the Temple Mount Treasury, a group that continues to raise funds to rebuild the Temple.[33]

Gillon believes that the radical right continues to pose a grave threat to

Israeli national security, perhaps even more than Hamas. "Here in Israel we don't like to say this very loudly, but the radical-right Jewish groups have a lot in common with Hamas," he told me. Hamas and the radical-right groups have twin objectives: one religious, the other political, Gillon explains. Both use selective readings of history and of religious texts to justify violence over territory.

Etzion tells me sadly that he has learned the Jewish people are not ready for redemption. He serves as the leader of the group Chai Vekayam (Alive and Existing), which regards itself as "the catalyst for a Jewish renaissance."[34] The group focuses on encouraging Jews to prepare themselves for the imminent redemption through prayer.

The Temple Mount is the only holy place for the Jews, Etzion explains. "The one thing I am sure of," he says, "is that the Dome of the Rock is a temporary building. It must come to an end. Exactly when and exactly how I cannot say. But as a principle, I am sure its end is near."[35]

Selective reading of history is a powerful tool for mobilizing terrorists seeking to settle conflicting claims to the same territory. Both sides in the Kashmir dispute use the same tool, as we shall see in the next chapter.

FIVE

Territory

It was still possible, before September 11, to persuade yourself that talking to jihadis was safe. This was before a group of Pakistani militants murdered *Wall Street Journal* reporter Daniel Pearl, leaving his pregnant wife and future son to fend for themselves in a much changed world. Nevertheless, the first time I traveled to Pakistan I was nervous. I spent months preparing, soliciting advice from diplomats, reporters, academics, and businesspeople who knew South Asia well.

This chapter begins with a conversation with a group of managers and leaders of a Pakistani jihadi group called Lashkar e Taiba (Army of the Pure), the first group of jihadis that I met. Few Americans had heard of the group at the time I first met them, but it has since become known because of its alleged connections with Al Qaeda and its involvement in a brazen attack on the Indian parliament in New Delhi in December 2001, resulting in the deaths of fifteen people.[1] The group is a member of the International Islamic Front, bin Laden's umbrella organization. From our first meeting, members told me proudly about the financial contributions they received from Saudi Arabia; and how those donations enabled them to build factories, residences, farms, and schools, in addition to what amounts to a privately held town outside Lahore. The group claims to train forty thousand youth per year at its schools and military training camps.[2] The group is still referred to by the name Lashkar e Taiba (LET), as well as Jamaat-ud-Dawa, although it changed its formal name to Pasban-e-Ahle Hadith after it was banned in Pakistan and America.[3]

Through conversations with these militants and with others discussed later in this chapter, we learn that the half-century-long dispute over the territory of Kashmir has become the raison d'être for a wide variety of jihadi organizations and businesses on both sides of the border. These groups represent jihad in an advanced stage, in which the original, purported motivation—to help the Kashmiri people—has become less important than the organizations themselves and the political or financial interests of their leaders. The groups have often functioned as mercenaries for the Pakistani government, serving as a kind of second army, but one that also receives outside funding from Islamist charities abroad. That the groups are no longer beholden to a single sponsor has emboldened them to the degree that they are prepared publicly to threaten Pakistan's leadership. Pakistan's decision to join the U.S. side in the "war on terrorism," and its halfhearted efforts to shut the groups down, infuriated the jihadi leaders. "We are angry," a Lashkar e Taiba leader told the *New York Times*. "No Pakistani leader has ever betrayed Kashmir and survived," he said, implicitly threatening General Musharraf.[4] The last Pakistani leader to "betray" Kashmir was Prime Minister Nawaz Sharif, who was deposed by General (now President) Musharraf in a 1999 military coup.

I embark on my first trip to Lahore in February 1999. With the help of local reporters, I make some phone calls. I find a local guide who is well-known to the jihadi groups. With his help, I contact Lashkar e Taiba. The group agrees to talk with me, but insists that two of its members come to my hotel to meet me first. One is a senior manager named Yahya. The other is a new recruit who gives me a nom de guerre, Ahmed. The purpose of our first meeting is for them to look me over to determine whether I am fit to be taken to their headquarters and to meet with their emir, a former professor who runs the Wahhabi educational organization, LET's parent organization, then known as Markaz-Dawa-Wal-Irshad (MDI), now called Jamaat-ud-Dawa.

This is the first time I have met any mujahideen. Yahya, the senior manager, looks like a *mujaheed* from a movie set, but one that has gone somewhat soft. He is massive, like a football player, with a protruding belly. He has flashing black eyes and a long black beard flecked with gray. He wears an elephant-gray *shalwar kameez*, which wrinkles humidly, like slept-in pajamas, over his colossal frame. He wears traditional Pakistani slippers, with turned-up, pointy toes and mirrors, giving his feet an incongruously elfin look. He walks heavily, ostentatiously relaxed, but his latent power communicates itself clearly to an observer. I notice that his hands, which are soft and brown, look big enough to crush me with a single swat. He projects a kind of chivalrous serenity.

Ahmed, the young new recruit, is beautiful. He is slim but strong-looking, with luminous skin and clear, intelligent eyes. He has the obligatory beard of a fundamentalist, but it is neatly trimmed. His hand-embroidered *kameez,* the long shirt worn over the pajama-like pants known as *shalwar*, is made of fine white cotton. I notice that his clothing is perfectly pressed. The young man's English is refined. He is trained as an engineer. The emir was his professor.

We settle down on plush sofas in the hotel foyer, just outside the twenty-four-hour restaurant, which is crowded with busy businessmen at almost any hour. My visitors do not look like businessmen, and I wonder what the hotel management thinks of us. (Later I will discover that I am "taken care of" at all times, even when I am staying in a hotel.) Yahya's voice is soft, and he appears to be somewhat shy around me. His English is not great, but Ahmed's is perfect.

They have some questions, which Ahmed puts to me politely. Are you Muslim? Where are you from? For whom do you work? What sort of project is this? I tell them I am an academic studying religious militancy around the world. That the project is funded by various foundations. That I used to work for the U.S. government, but no longer do. I avoid answering the question about my religion, but it is put to me again. I am Jewish, I admit. Both militants study my face intently throughout the interview.

They apparently find my answers satisfactory. Yahya instructs me to hire a car for the next day. We will all travel together in a hired car out to Muridke, a suburb around twenty-five miles from Lahore where Lashkar owns a two-hundred-acre compound. They have one demand: I must wear Pakistani clothing, and I will have to wear an extra sheet on top of my head when I meet with Hafez Sayeed, their emir.

My guide, Muzamal Suherwardy, then a reporter for the Urdu paper *Nawa e Wakt,* had told them in advance that he felt it was safe to bring me to Muridke, but the two militants wanted to look me over for themselves. After they leave, Muzamal tells me that they were trying to determine if I came to Lahore on a mission to kill their leader. They wanted to know if I was there on behalf of RAW (Research and Analysis Wing), India's intelligence agency. Or working for the Mossad.

"As a result of their inspection, they have determined that you work for the CIA," Muzamal informs me, seemingly bemused. I cannot tell whether he concurs with the militants' assessment.

Well, I'm not, I tell him hotly, foolishly. How can you persuade someone you're not working for the CIA? It is clear in any case that speaking with an American woman is a novelty for all three of my new acquaintances. My guide admits that I am the first American he has ever met, and the first Jew he has ever seen.

"Anyway, it's okay; they are flattered if the CIA is interested in them," he says.

"How did they decide that I'm not planning to assassinate their leader?" I ask.

"It is obvious. You can tell a person's character by looking into her eyes. You have innocence in your eyes."

I often recall this conversation. I no longer retain the belief that character can be discerned by visual inspection, and this feels like a terrible loss. I also wonder whether I still have innocence in my eyes, after talking with so many killers, after observing so much pain.

My guide Muzamal takes me to buy a *shalwar kameez.* He helps me choose one made of silky silver fabric flecked with gold. The shopkeeper stitches the slacks and scarf while we wait. I could almost be Pakistani, I think, when I see myself in the mirror. "You look like Benazir in that outfit," Muzamal's friend tells me later, referring to Benazir Bhutto, the former prime minister who is in self-exile because of various corruption-related cases pending in Pakistani courts. But the first time I wash it, the sleeves fray into wispy silvery shards. The dress still hangs in my closet, looking like the cast-off clothing of an exotic princess.

Muzamal takes me to his parents' home, where he lives with his wife, their son, and two young girls who work as servants. They live in a working-class neighborhood. The barefoot servant girls look about twelve years old. So far I have visited only upper-class Pakistanis, who speak perfect English, send their children to elite Western universities, and are not shocked by a woman like me. Here, I feel slightly awkward. I feel too independent, too tough, a kind of strange species halfway between man and woman. The idea of a woman traveling on her own to a faraway land is utterly alien to a family like this. Still, Muzamal's family immediately takes me under its wing. His mother brings us a large meal to eat sitting on the sofa bed in the living room. Coca-Cola. Cucumber salad. Curried chicken. Rice. Also whole-wheat roti. I am embarrassed that I can't drink the cola or eat the salad; Westerners' digestive systems are too weak to manage the water here. But the curried chicken is delicious.

Pakistanis dine late, and it is midnight by the time Muzamal drops me back at my hotel. I give my *shalwar kameez* to the laundry, instructing them both verbally and on the laundry tag to return the pressed clothing at 8 A.M. I soon fall into a deep, jet-lagged sleep. Two hours later, fierce pounding awakens me on my door.

What do you want? I shout from my bed.

"Room service" is the answer.

Now I am petrified. I know for certain that the terrorists have come to get me. I phone the room service operator. Why are you sending room service to me, I ask, when I didn't order anything?

The operator tells me he didn't send room service.

I phone hotel security. A guard promises to come to my room immediately.

I am afraid. I cannot control my legs, which are shaking uncontrollably. By the time the guard arrives, I feel empty, as though I've lost five pounds in the several minutes I had to wait.

"Madam," the security officer tells me, "your laundry is ready."

I am irritated, but also embarrassed. This was the most frightening ten minutes I spent on this book project. My husband is fond of telling me that I am afraid when I shouldn't be, and not afraid when I should be.

When I awaken, it is already sunny and bright. Lime-colored parrots fly past my window. In the café, cheerful piano music plays on an endless tape. I order extra strong tea with hot milk. I add extra pepper and sliced ginger to my *channa dal,* a breakfast curry made of chickpeas and potatoes.

Yahya comes to retrieve me at the appointed hour. He is on time to the minute. "You don't need to hire a car," he tells me. "My driver will take us." He informs me that Muzamal will not be joining us because he is busy. I consider the idea that they intend to kidnap or kill me, but I quickly put it out of my mind. They are too chivalrous, I think. I am too much like their sisters in my new outfit, I tell myself. My night terrors have not survived the parrots or the sweet tea.

We drive on the famed Grand Trunk Road, originally conceived by a sixteenth-century ruler of the Indian subcontinent named Sher Shah Suri. Kipling set much of *Kim* along this road. He described it as a "rutted and worn country road that wound across the flat between the great dark-green mango-groves, the line of the snow-capped Himalayas faint to the eastward, . . . such a river of life as exists nowhere else in the world."[5] During partition in 1947, millions of refugees used the road to escape into safety, Muslims into Pakistan and Hindus into India. The road traverses much of the subcontinent, passing through the Indian cities Calcutta, Varanasi, Delhi, and Amritsar, and from there westward to Lahore and north-west into Afghanistan. We pass grazing oxen and rice paddies, deep into farm country.

I talk with Ahmed, the young recruit, in the car.[6] He tells me he has a master's degree in engineering and that he is doing computer work for Lashkar e Taiba, which uses its sophisticated Web site for fund-raising and recruitment. One of its Web sites solicits funding for additional equipment. Visitors to one of the group's sites are requested "to provide as much money as you can . . . or otherwise provide us a fast enough computer and accessories," noting that "jihad work on Internet is very far behind the Non-Believer's advancement in information technology field."[7]

Ahmed is intense, neat, polite. He tells me he is from a wealthy family, which I can in any case see. Later I will learn that most militants are from poor families, but Ahmed is not the only upper-class *mujaheed* I will meet.

Still, I have the strong impression that Ahmed was chosen to meet with me because he is so presentable.

I asked how he came to join LET. "I came to Islam intellectually. I read a lot. I realized that the Islamic way of life is the best way of life. A single event can turn your life around. That is what happened to me," he adds, without specifying what that event was.

He looks me straight in the eye now and says, "People who have a pen in their hands have a serious responsibility. I hope you will take this responsibility seriously. Writers can kill people.

"Many who report on Islam and Islamic movements distort the facts. You must tell the truth, not distort the truth. America accuses others of being terrorists, but America is a terrorist state. The sword is not the only weapon. The pen, ideology, financial systems—these are the weapons used against Islam. It is important to realize that capitalism is also an ideology just like Islam. The West is trying to force its capitalist ideology onto Islamic states."

We talk about Pakistan's poverty. "People are efficient in the West because you fear losing your job. You are not working for the welfare of society, but to enrich yourself. In Islam you work for yourself but also to improve society, to serve Allah. There is less poverty in America [than here] because Americans benefit from the poverty of the developing world. America cares only about its own wealth—improving its own financial situation. Maximizing its own wealth, to the detriment of others, is its only ideology. I don't believe in capitalism," Ahmed tells me, making clear that he is rejecting the lifestyle of the Lahore high society he comes from. "In Africa, America provides weapons to one tribe at war with another. They fight. Then America provides weapons to the other side. This is absolutely a fact," he adds, for emphasis.

Are you going to be fighting in Kashmir? I ask.

"The emir decides who goes to fight. He decides each person's role in the struggle. He has not selected me to fight." Tension flits across Ahmed's jaw. He does not want me to see his disappointment. He is a good boy, determined to follow his superiors' orders, not only in action, but also in spirit. He tells me he plans to pursue a doctorate: "With a doctorate in mechanical engineering I can help LET with technology."

Is there anything else you would like me to know about your views? I ask.

He thinks for a minute, then says, "Westerners have an image in their mind of the mujahideen as antisocial, crude people. We would like you to make clear in your book that mujahideen are smiling and joking sometimes and very sociable. The emir says there is a one-sided view of us in the Western media. We would like to come to America; we would like to talk to the media. We want people to understand: Islam is not terrorism. You have a stronger weapon in your hand than we do: the pen is stronger than the sword. You must be careful to portray us accurately."

After half an hour of driving, we turn right off the main highway. We pass fish farms, cotton fields, and a brick factory.

"All this belongs to us," Yahya says. "We make our own bricks here and grow our own vegetables."

A mile in from the highway we can see the main compound. The first thing you notice is the construction. An enormous brick structure appears half-complete. "This is our new university," Yahya says, a gift from Saudi donors. Some say that Osama bin Laden himself was the donor, but Yahya denies that his organization has any links with bin Laden or Al Qaeda. There are grammar and secondary schools on the compound. There are trucks and tractors, cattle and horses. "These are our housing complexes," Yahya says, pointing to a group of apartment buildings. "Those are the apartments for the families of the martyrs." He points to a separate block. "We have our own mosque and our own clinic. Over a thousand students are housed here."

To the right of the road is an obstacle course and tightropes, where students practice carrying one another on the high wire. There are no nets. This is how they learn to cross Kashmir's steep ravines. Most of LET's military training takes place elsewhere, however, at mobile training camps in Pakistan-held Kashmir.

The students are required to relinquish all forms of entertainment. No music. No dancing. Boys who want to become fighters have to smash their television sets to demonstrate their willingness to live a pure Islamic life, Yahya tells me. Only then are they considered worthy of becoming mujahideen.

Yahya brings me to a living room in the main building. He instructs me to sit on one of two matching beds, both covered with a green cloth, which face one another across a low table. He gives me a heavy white cloth the size of a crib sheet and tells me to hold it over my scarf to ensure that my hair and part of my face are hidden before the emir enters the room.

Four men enter the room and seat themselves across from me on the second sofa: Emir Hafez Sayeed and three "elders." I notice that the emir has a long beard, dyed red with henna. Before long, I realize that my interlocutors have an advantage over me visually. Sunlight pours through the windows behind their backs. My eyes quickly tire and the four men blur into black lumps with no features. But I fear they can see me clearly—and I wonder whether this is deliberate.

What are the main objectives of your organization? I ask.

"Our mission is to invite all of humanity to Islam, to persuade the whole world to worship only Allah. In Pakistan our goal is to help people get a better understanding of Islam. Islam is not just a religion. It regulates every aspect of life, including politics. We would like to see implementation of divine laws here. Fortunately, many people are beginning to agree with us. In Kashmir our goal is to end the butchering of Muslims; there we are practicing jihad. We do not provide physical training for mujahideen here; we train closer to Kashmir. But we train them in Islamic thought; we prepare their minds for the jihad. Once the mind is prepared for jihad, very little additional training is required."

The war between India and Pakistan over the fate of Kashmir is as old as both states. When Pakistan was formally created in 1947, the rulers of Muslim-majority states that had existed within British India were given the option of joining India or Pakistan. The Hindu monarch of the predominantly Muslim states of Jammu and Kashmir chose India, prompted partly by a tribal rebellion in the states. Pakistan responded by sending in troops. The resultant fighting ended with a 1949 cease-fire, but the Pakistani government continued covertly to support volunteer guerrilla fighters in Kashmir. Islamabad argued then, as it does now, that it could not control the "volunteers," who as individuals were not bound by the cease-fire agreement. (On the other hand, Maulana Abul A'la Maududi, the late founder of the Islamist party Jamaat-i-Islami, argued that as individuals, these "volunteers" could not legitimately declare jihad, either.)

The two countries again went to war over Kashmir in 1965, but the status quo division remains unchanged. The cease-fire line was converted into the Line of Control by the Simla Agreement, signed by India and Pakistan at the end of their 1971 war, in which Pakistan lost the noncontiguous eastern part of its country, which became Bangladesh. The agreement stipulates that the Line of Control "shall be respected by both sides," and that "neither side shall seek to alter it unilaterally, irrespective of mutual differences and legal interpretations." It also called for both sides to "refrain from the threat or the use of force."[8]

India has interpreted the agreement as reaffirming its rights over Kashmir. Pakistan, which negotiated the agreement at a time of weakness, argues that the pact does not preclude holding a plebiscite called for in UN resolutions of the late 1940s which would allow Kashmiris to choose their own fate. Pakistan's diplomatic strategy since 1972 has been to get India to admit that Kashmir has remained a "disputed territory," and to involve the United Nations in the dispute's resolution. India has done everything in its power to prevent precisely that.[9]

Pakistani officials admit to having repeatedly tried to foment separatism in Kashmir in the decades following the cease-fire. But the attempts were largely unsuccessful. An independence movement started to gain strength in the late 1960s, but was suppressed by the Indian military.[10]

When separatist violence broke out in the late 1980s, the movement was largely indigenous. Indian officials I have interviewed admit their own culpability in creating an intolerable situation in the region.

Indian-administered Kashmir was run essentially as a kleptocracy. One longtime observer of the region noted that Delhi had ruled the region through compliant Muslim politicians who monopolized illegal businesses and franchised out areas of criminal and legal commercial enterprises. Corruption, extortion, and bribery were common. "India's Muslim henchmen made themselves very rich and deeply hated," he writes.[11] India ignored Kashmir's significant economic troubles, rampant corruption, and rigged elections and intervened in Kashmiri politics in ways that contradicted

India's own constitution. The rigged 1987 state-assembly elections were the final straw in a series of insults, igniting, by 1989, widespread violent opposition, this time by the Kashmir Muslims themselves.[12]

The idea of jihad as a multinational armed struggle of Muslim believers, according to the late Pakistani scholar Eqbal Ahmad, had all but died out by the beginning of the twentieth century. The coalition of forces that fought the Soviets in Afghanistan, sponsored by United States, Saudi Arabia, and Pakistan, turned the Afghan resistance movement into a modern multinational conglomerate—a "Jihad International, Inc.," in Eqbal Ahmad's now famous words.[13] Jihad, along with guns and drugs, became the most important business in the region.

The United States and Saudi Arabia together funneled some $3.5 billion into Afghanistan and Pakistan during the Afghan war, according to Milt Bearden, CIA station chief in Pakistan from 1986 to 1989. By financing and training the mujahideen, the United States helped to create a multinational jihadi organization, which eventually evolved into the biggest threat to U.S. national security.

After the Afghans defeated the Soviets, graduates of the war were looking for new jihads to fight. The so-called Arab-Afghans were discouraged from returning home because of the threat they might pose to their own governments. Followers of bin Laden moved their headquarters to Sudan and from there to Afghanistan. Other veterans moved to Somalia. In chapter 3 we discussed the warriors that returned to Indonesia. Pakistani nationals and other "foreign mercenaries," as the Indian government refers to the mujahideen, got involved in a new jihad to wrest control of Kashmir from India. "After two or three years, our movement was hijacked by religious extremists," many of them veterans of the Afghan war, "and we were crushed by both India and Pakistan," complains Amanullah Khan, founder of the secular Jammu and Kashmir Liberation Force.[14] Jihadis involved in the Kashmir dispute have told me that they are certain they will prevail. After all, they managed to rout a superpower in Afghanistan, they say, neglecting the role of outside powers in their victory.

Pakistan's official denials of any involvement in supporting the jihadi groups, who were closely aligned with the Taliban regime in Afghanistan, became more difficult to maintain after the Kargil crisis of 1999. In the winter of that year, jihadis occupied the remote and sparsely populated area of Kargil, which is located in the mountainous border area of Indian-held Kashmir. While Pakistan initially insisted that the incursion was carried out by volunteer Kashmiri Muslims acting on their own initiative, the Indian government released intercepts and other intelligence findings making clear that the Pakistani military had planned and executed the operation. General Musharraf, Chief of Army Staff, had reportedly not informed then Prime Minister Sharif of his troops' involvement in the operation until it was well under way. India responded to the Pakistani army's movements with full military force, and by midsummer 1999 heavy fighting along a one-hundred-mile stretch of mountainous terrain ensued, bringing the two nuclear powers to the brink of a nuclear confrontation.[15] After an emergency meeting with President Clinton, who had agreed to help resolve the crisis, Sharif ordered his troops to retreat. The troops complied, and all-out war was averted, but Sharif was eventually deposed by the military for his "treachery." The Kargil crisis resulted in the deaths of between thirteen hundred and seventeen hundred people. It is generally regarded as the prelude to the October 1999 military coup that brought Musharraf to power.[16]

Indian forces and paramilitary personnel, often disobeying orders, have responded to the militants' violence with atrocities of their own. Extrajudicial killings have become commonplace. Gang rape is practiced by both sides. There are many stories of raped women becoming mentally ill or committing suicide. Indian forces arm irregulars to work with the security forces, but without any official accountability. Human Rights Watch reports that custodial killings—the summary execution of detainees—are a "central component of the Indian government's counterinsurgency strategy," and that "disappearances" of detainees also "remain a serious problem." Indian security forces engage in brutal forms of torture, claiming that it is the only way to obtain information from suspects, but torture is also routinely used to punish suspected militants and their alleged supporters, and to extort money from their families. Methods of torture include severe beatings with truncheons, rolling a heavy log on the legs, hanging the detainee upside down, and use of electric shocks.[17] Indian security personnel in Indian Kashmir regularly target Muslims suspected of supporting the jihadi groups, employing arbitrary arrests, torture, and staged "encounter killings."[18] Human Rights Watch also reports that Indian security personnel have opened fire on demonstrators on a number of occasions, claiming to be acting in self-defense. They have also beaten journalists covering attacks against their facilities.[19] Hundreds of innocent civilians, at the hands of the Indian military and Pakistani-sponsored jihadis, die every year in this conflict.[20] Tens of thousands have died.

The U.S. State Department Human Rights Report for 2001 reports that Indian security forces "committed numerous significant human rights abuses, particularly in Jammu and Kashmir and in the northeastern states," including extrajudicial killings, excessive use of force by security forces combating active insurgencies, torture and rape by police and other agents of the government, and arbitrary arrests.[21]

The jihadi groups operating in the Kashmir valley also target civilians. They routinely assassinate suspected informers, in addition to political leaders and civil servants who are believed to be pro-India. They are increasingly involved in massacres of Hindu families. Kashmiri pandits, an upper-class Hindu minority based in the Jammu area, have been migrating from the valley, claiming that the jihadi groups have threatened to kill them if they don't leave.[22] For example, LET was suspected in connection with an attack in Jammu in July 2002 in which twenty-seven civilians were killed and about thirty were injured.[23]

Religious extremists on both sides make it significantly more difficult for the leaders of India and Pakistan to reach any kind of agreement. In India, Hindu extremist groups oppose any peace initiative with Pakistan, whereas, in Pakistan, the religious parties strongly object to giving any ground to India in the dispute. In Pakistan, Hafez Sayeed has played a significant role in efforts to stop and hamper the peace process.[24]

Sayeed is skeptical not only of Pakistan's rulers, but also of rulers of all Muslim countries. "Rulers of Muslim countries do what they are told to do by the West," he tells me. "They are not true leaders. They are puppets of the West. The West enslaves the Muslim countries through debts to the IMF, the World Bank, foreign aid, and loans. We want our leaders to raise their voices on behalf of all those who are trapped by the West." Pakistan is some $40 billion in debt, much of which it owes to international banks.[25]

Would it be better if the Muslim countries stopped accepting Western loans or aid? I ask.

"Yes," Sayeed says. "The Muslim countries are rich in natural resources. We do not need money or assistance; we should utilize our own resources."[26] This is followed by a long, rambling complaint about the evils of globalization and of international institutions, some aspects of which Western antiglobalization activists would find familiar and persuasive.

"Globalization is similar to what the British did when they established the East India Company," he says. "They established a company as a pretext for occupying the land. That is what globalization is all about—a pretext and a prelude to occupation. When the United Nations was established, its purpose was to protect basic human values. Now it is a puppet of America. UN workers are just spies for the American government. The UN works against Muslims. We have asked all Muslim nations to drop out of the UN and form their own organization."

Where were you trained? I ask. He tells me he received his higher education in Saudi Arabia.

A servant arrives with a large bowl of apples, bananas, and oranges, enough for twenty people, even though only four of us are in the room. Another follows with a tray of soda. My hosts insist that I eat. None takes fruit onto his plate until the emir is satisfied that I am eating. I eat two bananas and an orange quickly, to show my appreciation, hoping to deflect attention from my refusal to drink.

This is the first time I have encountered the jihadis' hospitality. I feel the briefest moment of concern that my hosts may try to poison me, but quickly reject the thought.

Here is the unspoken bargain between us. I make myself vulnerable and they will not harm me. I must strive not to reveal fear, and to trust that they won't hurt me, despite their machismo and manufactured rage. And they, in turn, will consider telling me the truth, but only half-truths. That is our bargain.

By this time I have heard enough about the evils of the West and I decide to take a slightly different tack.

What if I decided that I wanted to teach here at the university? I ask.

How would it work? Would you pay me? I explain this is a hypothetical question, that I am not actually applying for a job.

"You would have to be married or have a father, brother, or son prepared to protect you. If you were not married, we would have to find you a husband. You would be taken care of," the emir reassures me.

Later, in 2002, the then sixty-two-year-old emir marries a second wife, half his age, a widow of a Kashmiri "marytr."

I ask the emir about his views of nuclear weapons. He supported Pakistan's 1998 nuclear tests and opposes Pakistan's signing of the Comprehensive Test Ban Treaty, he says. He tells me he would also favor exporting nuclear technology to other Islamic countries "so that they can resist Western oppression."[27]

What about Iran? I ask. Do you support exporting nuclear technologies even to [Shia] Iran?

"Yes. We have more important things to worry about than the Sunni-Shia conflict right now. We should export nuclear technology to Iran and other Muslim countries. If Iraq had nuclear weapons, it would solve all their problems."

What do you think of Osama bin Laden?

"He is a *mujaheed.* He is very well respected here. America is pushing him into a corner, making his life miserable. He raised his voice against oppression. Should he be punished for that? He is not an oppressor."

I ask whether any of them have any questions for me. There is a murmur of momentary uncertainty, then one of the elders decides to make some comments.

"Jews are brutalizing Muslims all over the world," a gray-bearded elder says. "All senior executives around the world are Jews. All members of the U.S. Treasury are Jews. How do you explain that?" I see that my admission to being a Jew has been transmitted to the emir and the elders.

I mumble something about this being an exaggeration. Another elder jumps in: "Anyone who goes against America is labeled a terrorist. We believe that the [1993] World Trade Center bombing was actually carried out by the CIA. How could a terrorist organization be so stupid as to go back to get the money from his rental car? Obviously it was a CIA operation."

The first elder adds, "Take Algeria. The FIS won. The people selected the FIS. What happened to Westerners' supposed respect for democracy then?" FIS, an Islamist political party of Algeria, had won 188 out of 231 seats in the national legislature in 1991 before the second round of elections were canceled by the Algerian military. It is widely believed in Algeria, as well as by many Muslim states, that the United States backed the Algerian military and also sponsored this move to keep the Islamist party out of the corridors of power.[28]

This is a very good question indeed, so I am relieved when I realize that a response is not expected. The views of an American woman are of little interest to them. She should listen, she will learn.

After our trip to Muridke, Ahmed insists on reading over my notes to make sure I am not distorting his words. He approves. Two years later I discover that Ahmed's emir granted his wish to fight in Kashmir. He was killed soon afterward.

Later that week, I met my first operative from the Pakistani militant group known as Harkat-ul-Mujahideen (HUM). Like LET, HUM is a member of Osama bin Laden's International Islamic Front for Jihad against the Jews and Crusaders. The U.S. government blames it for having hijacked an Indian airliner in December 1999 and killing a number of Americans.[29]

The operative agrees to come to my hotel, the safest place to meet. He wears a camouflage-colored frontier cap, a *mujaheed*'s vest, and hiking boots. He smells strongly of sweat and dirty clothing. He is thin—almost dainty—but obviously strong. He has a haunted look in his eye. He is a local boy from Lahore. I ask to photograph him. Muslims don't allow that, he says.

I take him to the business center at the hotel. My guest stands out among the businessmen in the center, even more than I do. He seems high on something—probably adrenaline, but maybe drugs. I order green tea and pastries. The waiter brings a large plate of cookies and other sweets. My guest appears famished. He eats the pastries quickly. I order another serving. Each plate seems enough for four or five people, but he finishes the second one too, like any other teenage boy hungry for sweets. Every time I return to the hotel, which is called the Pearl Continental, the waiter tells me that he remembers serving us tea and pastries and asks me when this book will be done.

Aren't you afraid of fighting? I ask the operative.

"What is there to be afraid of," he responds. "I pray for death every day. During my studies, reading the Koran, I decided to sacrifice my life for jihad. If I die in the jihad, I go to paradise. Allah will reward me. This is my dream. The Taliban have the best understanding of Islam and the jihad. You must cover them in your book too." A minute later he realizes this might be difficult, since "they don't like talking to women."

How do you feel about meeting an American?

"I respect you. You are American but you are not snob. I respect everyone who is a man of Allah. Indians in Kashmir—they are men of the devil."

How do you recognize a man of the devil when you see one?

"I can tell by what people look like. Any person who does not obey Allah is a man of the devil. He doesn't have innocence on his face," he says, echoing my earlier conversation with Muzamal.

Are you sure you can always tell?

"This doesn't work one hundred percent of the time. There might be men of the devil that I don't recognize. A man of the devil drinks whiskey, gambles, he kills women and children."

Is there a difference between drinking whiskey and killing children?

Yes, he concedes. "If he drinks whiskey, he is a man of the devil but at a lower level. He is sinning and Allah will punish him. We would try to put him on a straight path.

"In Kashmir everyone loves us. Also in Lahore—everyone knows I am a *mujaheed* and they love me. Muslims all over the world love the mujahideen." He pauses to reconsider. "Secular people don't love mujahideen.

"My emir has just sent me back here to Lahore on a temporary assignment. We rotate. I remained in Kashmir for six months. One year for training and trying to get into the valley. We have to climb fifteen thousand meters [sic]. Seven days of constant walking. You have nothing to eat for seven days—just biscuits. See these pockets in my vest? That's where we carry our biscuits." (This is higher than Mt. Everest!)

What kind of mental training did you receive at the *madrassah* (seminary school)?

"Practical training for war. The rules, the disciplines, and advantages of jihad. You must never humiliate a woman. You must never kill a child or an innocent person. We hike a lot, we learn to live hungry."

How were you trained at the camp?

"We have a daily routine. We do target practice, but we don't have a lot of bullets. We have to save them for the enemy. We emphasize running in our training. I can run five kilometers without stopping. We also play a lot of football [soccer]."

I ask him to describe his day when he was at the training camp.

"We get up at five A.M. for the morning prayer. Then we recite the holy Koran. Then we take exercise. Then we have breakfast."

What did you eat for breakfast? I ask, trying to visualize the details.

"Honey, cream, mangoes, dates. We eat only things you can find in Kashmir. We were taught to use guns, hand grenades, and also how to make time bombs and guns. Then we slept for around one and a half hours. Then we had lunch. Then afternoon prayers. Then religious education, and education about jihad. We were taught about the Islamic way of life. We had evening prayers at four-thirty P.M. Then we played football or volleyball. In the evening we read books or heard lectures on Islamic history or about atrocities against Muslims around the world."

Did you ever listen to music?

"Sometimes we sang songs."

How about rock music?

No rock, he says. "Too much sex, too much drinking. Rock songs are all about sex and drinking. Singing brings nudeness and the devil to the spirit. We sang songs of jihad. It's kind of a chant."

How about literature that is not related to religion?

"No, we don't read it. It's a waste of time."

No poetry?

"No. Muslims are made in order to build the world, not to waste their time. We come into this world to do the will of Allah. Jihad is the will of Allah. Our lives are only for Allah, not for others. That's why Muslims are brave—because you know you will meet Allah when you die.

"Jews are the most cowardly nation. We were three hundred and thirteen, and we defeated an army of one thousand non-Muslims. When we were three thousand, we defeated an army of two million non-Muslims.[30] In the Afghan jihad, three mujahideen destroyed three hundred Russians. Non-Muslims love their life too much, they can't fight, and they are cowards. They don't understand that there will be life after death. You cannot live forever, you will die. Life after death is forever. If life after death were an ocean, the life you live is only a drop in that ocean. So it's very important that you live your life for Allah, so you are rewarded after death.

"God looks at those who sacrifice their lives in the jihad with love. The love is seventy times stronger. I feel this love. Had God not been good to me, I would have lived a luxurious life. The devil would have overcome me. But the devil has not overcome me—God is with me. Two people might say the same prayer—one understands the prayer, the other does not. I'm very lucky—I'm one who understands the point of the prayer. Both hard work and luck.

"I like Saudi Arabia, Pakistan, and Afghanistan, but not America," he says, shifting to a new topic.

Why don't you like America? I ask.

"Americans are not free to practice Islam so I don't like it. In European countries and in America there is too much sex.

"We respect women. She must always be in full purdah. In America women are treated like sheep and goats. The American president raped a girl the age of his daughter," he says, referring to President Clinton.

"In Afghanistan everyone is punished for sin openly in front of the public. That's why crime is stopped. If you punish a rapist in front of the whole city, then you have no more rapes. That's the kind of system we want to see here. Also in America! We would preach to everyone to practice Islam. We would punish every sin.

"Why don't you become a Muslim? In your system there is no respect for women. So many rapes. Men look at you. In your system rapists are arrested, but there are rapes still going on every day. Rape is not stopped. This is due to bad policy."

Are you married? I ask.

"No. I can't afford to be married right now, my business is not going very well."

I ask him what kind of business he runs.

"I am importing dry food from Afghanistan—nuts, almonds, dried fruit. But I am too busy in the jihad to run the business."

Toward the end of the interview I ask him why he was willing to meet with me.

"I want to explain to the West that we are not in favor of violence for its own sake. We are for peace. People have a misconception of mujahideen in the West; you see us as fundamentalists. We are fundamentalists—but we are for peace. We believe in the teachings of the Prophet—we will not attack anyone unless we have to."

I was beginning to get some sense of the variety of persons who become mujahideen, but there was still a lot more I wanted to learn. When my former colleague Michael Sheehan was named counterterrorism coordinator at the State Department in 1999, he had a premonition that the world of terrorism was about to change and that the locus was about to shift from the Middle East to South Asia. He was determined to energize the office in advance of the changes he suspected were coming. Mike and I decided to try something unprecedented—to run a conference, jointly sponsored by the State Department and the Council on Foreign Relations (my then employer) that would include both academic experts and government officials from around the world. In June 1999, we invited counterterrorism officials—Ambassador Sheehan's counterparts—from seventeen countries in the Middle East, South Asia, and Central Asia. The conference was followed by a series of meetings and counterterrorism "games," in which participants practiced responding to international terrorism crises. Lieutenant General Ghulam Ahmed Khan (known in Pakistan as GA) was one of the attendees. I got to know him during the period he spent with us in Washington. At the time, he was the head of the domestic wing of the ISI, Pakistan's intelligence agency. At the end of our conference, GA urged me to return to Pakistan as his guest, offering to arrange for me to travel to Kashmir and the Line of Control.

I had mixed feelings about GA's offer. On the one hand, I was anxious to travel to Kashmir, and it is not possible for foreigners to travel to the

Line of Control except under the auspices of the military. GA struck me as unusually straightforward—possibly even a good man. But he worked for an agency that specializes in dirty tricks. Traveling under the ISI's auspices could mean that I would continue to excite that agency's interest, forever; and that any Pakistanis I talked to would be vulnerable to being visited and even harassed by their government. I decided, in the end, that my contact with the jihadi groups made me interesting to the ISI either way, and that the relative anonymity I might once have enjoyed in South Asia was no longer available to me.

A year after the conference, in June 2000, I decide to take GA up on his offer. By then General Musharraf had seized control of Pakistan in a bloodless coup and had asked GA to serve as his chief of staff. I wrote to him asking whether his offer still held, and he wrote back immediately, urging me to come. His colleagues would meet me at the airport. They would provide for my safety and take care of me in Kashmir.

Judy Miller, a prominent reporter for the *New York Times* and the coauthor of *Germs: Biological Weapons and America's Secret War,* asked to accompany me. I called the defense attaché at the Pakistani embassy. I would like to take Ms. Miller with me, I tell him, but I would have no control over what she writes, I explain. My contact checks with my hosts and later informs me that Ms. Miller will be allowed to accompany me. Judy wants to go to Afghanistan first, and we agree to meet in Islamabad.

My plane arrives at 10:30 P.M. on June 5, 2000. I am hopelessly rumpled, in a "wrinkle-free" traveling skirt and shirt. I have bought a new computer for this trip, knowing that the ISI will quickly copy my hard drive if I leave my computer within their reach. I carry my new laptop and little else in my hand luggage. Pakistan International Airlines (PIA) flies direct from New York to Islamabad, so the risk of losing my luggage seemed minimal. I pack an external A drive and disc in my checked luggage.

At the luggage carousel I notice a middle-aged man, dressed like a driver, scanning the crowd. Finally he allows himself to consider the possibility that the small, wrinkled-looking person might be the female "VVIP" he seeks. Are you Dr. Stern? he asks uncertainly, clearly not expecting a "very very important person" or even a professor to look like me. When reassured that I am indeed she, he takes me to a small truck, which is parked on the landing strip, and drives me to a building marked VVIP. Three military men are waiting for me in an enormous, private lounge. Other than the servants, we are the only people using the room, which is decorated for receiving diplomats. Sofas are arranged in small groups for private conversation, and the room is lit with crystal chandeliers.

A servant brings me mineral water in a crystal glass. My hosts explain that they will retrieve my luggage for me. We make small talk. An hour passes. At last someone from PIA comes to inform me that my luggage is lost. My hosts are deeply apologetic: You know how it is with airlines these days. They take me, in two cars, to the guesthouse where I'm to meet Judy. The guesthouse prides itself on not allowing ISI "thugs" to spy on its guests, in contrast to the main hotel in the center of town. The guesthouse is popular with journalists, in part for that reason.

By now it is nearly one in the morning, and I am surprised to discover Judy still awake. She is talking to Brigadier X, who, we soon discover, is the boss of my interlocutors at the airport. He is gracious, if a bit officious. I mention that, since my luggage is lost, I have no clothing other than what I've been wearing on the plane for the last day and a half. He promises to send one of our guides early the next morning, to take me shopping, before we leave for the Line of Control.

I sleep for a few hours, until the guide comes to retrieve me. He takes me to a dress shop. He helps me choose a peach-colored *shalwar kameez,* which he insists on paying for. The scarf is too long, so we staple it; we don't have time for alterations. I will wash out my underwear nightly, I think. I am too embarrassed to mention underwear to a government guide. There is toothpaste in the hotel, but no sunblock, no deodorant, no comb, and no shampoo. I will look a bit wilder than normal, but at least I have one clean outfit.

Soon after this, we embark on our journey. Judy and I are to travel with our main minder, whom I will call Zahir, and a driver. Our minder is an academic working at a military think-tank, he tells us, although later he will tell us he works for Pakistan's intelligence agency, the ISI. A police car drives in front of us, lights flashing. Behind us is another jeep, filled with half a dozen soldiers, just in case. Just in case of what, I'm not sure, but I suspect it has more to do with preventing Judy and me from running off to visit terrorist training camps than with protecting us from attackers.

The police car forces other traffic to the side of the road to ensure that we make good time. The road out of Islamabad is Pakistan's autobahn, a newly paved, four-lane highway. It is not long before we are on a more typical two-lane road, which winds steeply toward Muree, a British colonial hill station now popular among Islamabad's nouveaux riches. Enormous mansions dot the side of the road. Drug kingpins, politicians, industrialists, and military men grown rich off the Afghan or Kashmiri jihads are said to summer here, which, at seventy-five hundred feet above sea level, provides a respite from Islamabad's punishing heat. After Muree, the road narrows to follow the Jhelum River. It is nauseatingly winding and steep. We look down over steep gorges to the banks of the river below, where cars that couldn't manage the climb lie rusty and abandoned. We share the road with pedestrians, goats, mules, carts, trucks, and the public buses known as flying coaches, which strain, pollutingly, up the narrow road. Luggage is strapped to the top of the buses, making them top-heavy. They seem in danger of toppling over into the Jhelum.

Eric Margolis, a reporter accustomed to traveling in the region, describes the six-hour trip to Muzaffarabad as "unrelenting torture," in which one cannot escape "Pakistani road terror," a condition "that combines utter helplessness with panic, flashes of fatalism, and nervous frenzy. It's impossible to close your eyes for a second, even during a twenty-hour ride, lest you fail to see your last moments on earth. Screaming at the drivers, who keep their pedals to the floor and horns blaring, is pointless. They shrug, laugh at the foolish *farangi,* mutter curses in Baluchi or some obscure dialect, and go even faster. You cling desperately to the backseat, or straps, brace your battered body and pray to Allah, Vishnu, Buddha, and the Holy Virgin of Santiago that you will survive the next mile."[31] Mr. Margolis captures the feeling well.

The air grows cooler as we climb. I am getting excited. I have wanted to visit Kashmir since childhood. My fantasies about traveling in Kashmir included friendly wild horses, moonlit rides, and fording emerald streams, not traveling by jeep with a military escort, but our images of exotic locale often turn out to be slightly different from reality.

It is difficult to understand the dispute over Kashmir without seeing its beauty. It lies between two mountain ranges: the Himalayas and the Karakorams. It is fertile, lush, and green, with waterfalls and lakes, whereas the flatlands of north India and Punjab are hot and dry. The snowcapped peaks and clean air are reminiscent of the Alps. The Kashmiri people are tall, slim, and fair-skinned, the coveted attributes of upper-caste Indians.

Kashmir has been described as the jewel of India and the abode of the gods. Indian and Pakistani poetry is filled with references to the beauty of Kashmir. But violence now dominates life there, and tales of bloodshed have replaced the romance and beauty of the earlier literature. Here is a particularly sad example:

A DREAM OF GLASS BANGLES

BY AGHA SHAHID ALI (1987)

Those autumns my parents slept
warm in a quilt studded
with pieces of mirrors
On my mother's arms were bangles
like waves of frozen rivers
and at night
after the prayers
as she went down to her room
I heard the faint sound of ice
breaking on the staircase
breaking years later
into winter
our house surrounded by men
pulling icicles for torches
off the roofs
rubbing them on the walls
till the cement's darkening red
set the tips of water on fire
the air a quicksand of snow
as my father stepped out

and my mother
inside the burning house
a widow smashing the rivers
on her arms

Halfway to our destination, our jeep suddenly stops. Our guide invites us to a tailgate picnic of mineral water, Coca-Cola, biscuits, and cigarettes. Our military escorts join us. Although the Jhelum rushes below and the vegetation is greener at this altitude, it is not entirely pleasant to picnic here. Trucks and buses pass in rapid succession, expectorating exhaust from poor-quality gas.

After nearly five hours on the road we reach Muzaffarabad, the capital city of what the Pakistanis call Azad (free) Jammu and Kashmir (AJK), and the Indians call Pakistan-occupied Kashmir (POK).[32] (I will refer to it as Pakistan-held Kashmir, and the Indian side as Indian-held Kashmir.) The region covers 13,292 square kilometers. It is technically an independent state with its own prime minister, legislative assembly, Supreme Court, High Court, auditor general, chief election commissioner, and chief secretary. But all matters related to defense, foreign affairs, foreign trade, security, and currency are managed by the Pakistani federal government. The Kashmir Council oversees the provincial government. It is composed of six elected members, three ex officio members, and five members nominated by the Pakistani National Assembly. The prime minister of Pakistan serves as the council's chairman.

The legislative assembly, which is composed of forty elected members, has the authority to appoint the prime minister. But the appointment is actually subject to Pakistan's approval. Members of the legislative assembly are not allowed to call for independence for Kashmir, as their oath of office requires commitment to the accession of Jammu and Kashmir to Pakistan.

Ninety-one percent of Pakistan-held Kashmir's 2.8 million people live in rural areas, where they are dependent on forests and agricultural land for their livelihood. There is little industry other than farming. Annual per capita income is U.S.$184—half that of the average Pakistani.[33]

Upon our arrival in Muzaffarabad, we are immediately taken to meet the prime minister of Pakistan-held Kashmir, Barrister Sultan Mehmood, who receives us in his private office. Our next stop is the state guesthouse, where Judy and I will be staying. We are directed to two rooms, one marked VVIP, and the other VIP. Both are enormous—fit for visiting dignitaries, although they are somewhat deteriorated. Lizards and insects vie for dominance in our rooms.

The next day we rise early for our trip to Chokothi and the Line of Control (LOC). We stop at a refugee camp, where Muslim refugees from the Indian side are housed in tents. We receive a briefing by the manager and a tour of the site. Since 1990, the manager tells us, the Indian government has killed 71,204 persons in Jammu and Kashmir, wounding an additional 29,561. Indian attacks resulted in fires at schools in which 553 schoolchildren were burnt alive; 7,613 women between the ages of seven and seventy were raped by the Indian military, and another 16,607 were sexually assaulted, according to the manager's figures; 6,726 persons were sexually incapacitated through torture; 617 dead bodies have been recovered from the Jhelum River; 41,760 people have been disabled for life.[34]

The camp has no running water. Whole extended families live in a single tent. The ground is muddy. The children are wide-eyed, beautiful, barefoot, and covered with mud. Each looks like a poster for Save the Children. Electricity is carried by extension cord from tent to tent.

We are taken to see a young man who was shot in the back. He is paralyzed and lying on a cot. Judy asks what kind of medical treatment he is receiving. None, the manager tells her. Nothing can be done for him, he says. We are skeptical. In 1999, 1,382 hospitals beds were available in Pakistan-held Kashmir, an average of .46 beds per thousand population, compared with .67 beds per thousand in Pakistan.[35]

A mother preparing lunch on a hot plate invites us to join her and her family. The thought of taking food from this woman and her emaciated family is nauseating. Our hosts tell us there are ten refugee camps in Muzaffarabad alone, housing over nine thousand people. Nearly sixteen thousand refugees are housed in camps throughout Pakistan-held Kashmir.

Judy and I are horrified by the squalor and poverty of the camp, as our hosts have hoped. We have read about Indian human-rights violations in Kashmir, including figures published by the State Department, NGOs, and Indian sources. As usual, seeing the victims is more persuasive than reading figures. But the Pakistani government's complicity in the refugees' plight is perhaps even more horrifying. The Indian government has an equally horrifying story to tell. An estimated quarter of a million displaced Kashmiris live in or around Jammu, many of them in the nine refugee camps located in the Jammu District. An additional estimated one hundred thousand Kashmiris are displaced in other parts of India, primarily in the New Delhi area.[36] Keeping refugees in camps like this to intensify the desire for revenge and to demonstrate the enemy's misdeeds is a standard ploy all over the world. At about this time I begin to realize, in a visceral way, how truly sad the Kashmiris' situation is. No government cares about them. Although they claimed to care about the Kashmiris' plight, the jihadis I met in Lahore seemed focused on Kashmir as a symbol. They seemed more interested in jihad for its own sake, or on their conception of Pakistan's interests, than on the people who live in Kashmir. The refugees are shunted to camps on both sides of the border to save money and to manipulate public opinion.

Our next stop is the Line of Control. Green hills jut straight into the sky here; it is stunningly beautiful and the air is drier and cooler. A soldier guards the gate where the road continues into Indian territory. A sign warns passersby that if they attempt to cross, they will be shot. Our hosts concentrate on showing us evidence of further Indian atrocities, the places where bullets have strafed schools and shops.

We have been traveling with a large party of officials and soldiers, one of whom introduced himself as a reporter. Soon after our hosts show us the border, the reporter takes a television camera out of the jeep, telling us he would like to interview us for the evening news. He wants to know what we think of what the Indians have done here. What do you think of the refugee camps? he asks us. Of the Indian military's shooting at schoolchildren? Judy immediately realizes what is going on. The aim is to capture us while the emotion is still fresh, in the hope that we will condemn the Indians on the evening news. We say something innocuous, about the situation being difficult for all parties.

On our way back to Islamabad our minder takes us to tea in a lovely hotel in Muree, overlooking the valley and the Karakoram Mountains. From our table on a covered porch we can see rain clouds moving toward us, and soon there is a downpour. The rain ends quickly, and two rainbows appear. We all get up from the table to watch the changing light and look more closely at the rainbows; I have never seen two at once. This is like a fairy tale, I think to myself. I tell Judy that I'm going to the ladies' room. Suddenly our minder notices that I'm not there. He became extremely agitated, Judy told me later, demanding to know where I had gone. At that point I began to take more seriously the possibility that my luggage had been "lost" for a reason, and that our hosts were taking good care of us for a particular reason—that they thought I might be a spy. It seemed quite unlikely that GA thought I was a spy—he seemed far too sophisticated and had invited me to come. But perhaps his former colleagues weren't sure. I had thought they would consider this possibility but then reject it. If I were undercover, would I willingly submit myself to a tour under military escort? And if I were overt, wouldn't I be talking with them about areas of mutual concern? Why would I pretend to be an academic? The Pakistani government's uncertainties about who I am—and how they might be able to use me—would puzzle me for years to come.

When we return from Pakistan-held Kashmir, our hosts present us with green velvet albums filled with photographs they took of us on our trip. On the cover is a gold-covered plaque with the words "Inter Services Intelligence Pakistan." Inside is a note from Lieutenant General Mehmood, who was then director of the ISI.

Judy and I are anxious to remove ourselves from the clutches of our hospitable hosts, and we both leave Islamabad soon afterward.

Later I meet with a "retired" Kashmiri militant. We meet in a hotel bar in Delhi. Firdous Syed, a slender thirty-six year old, orders a Coke. I order green tea with lime. Firdous has light skin and almost blond hair. My eye is immediately drawn to a gem set in silver on his finger. It is dark gray, like the sea in a storm, with a kind of watery stripe down the center. Later he will tell me that a philosopher, killed by the militants, gave him the stone. His green eyes, too, have a watery quality.

He wants to know about me. I tell him I teach a course on terrorism. What do you teach your students, he asks me? I teach them that there are many reasons why people become terrorists, some political, some personal, some spiritual, and some material; and that the reasons are likely to change over time. I urge them to put themselves briefly in the shoes of the terrorist—to imagine themselves fighting on behalf of some beleaguered group, to see themselves as saint-like killers. I always warn my students that the course will be painful and confusing. Firdous listens quietly. I feel as though I'm the one being examined, on a spiritual level.

Firdous tells me he founded the Muslim Janbaz Force, a Kashmir-based group. The group no longer exists, although another small group based on the Pakistani side has taken its name. I ask him why he quit militancy. "It was a gradual process," he tells me. "I realized the futility of violence."

Why did you start the group, I ask. "I was a fundamentalist," he says. "I thought a lot about the golden period of Islam. I wanted to re-create this golden period, to recover what we lost. We wanted to recover the glory of the *Al-Khilafat er-Rashida,* the Rashida Caliphate, the golden period of Islam." He thinks for a minute and adds, "Muslims have been overpowered by the West. Our ego hurts. We are not able to live up to our own standards for ourselves."

The four caliphs that ruled Muslims after the death of Mohammed in 632 are referred to as the "rightly guided" ones. This period—considered the golden age of "pure Islam"—lasted thirty years. It was the period of expansionist battles. Syria, Jordan, Palestine, and Iraq were conquered during the first decade after Mohammed's death. Egypt was taken from Byzantine control in 645. There were frequent raids into Armenia, North Africa, and Persia in the succeeding decades.

How old were you when you started the group? He tells me he was eighteen. Why would a teenage boy be so moved by ancient history that he would start a militant group? I ask. "We were influenced by what happened in Afghanistan—the victory over the Soviet Union, the Iranian revolution, the situation in Palestine. This was the first generation of militancy. We were euphoric. Our group varied in size—sometimes 500 militants, sometimes 2000. Our immediate target was to throw India out of Kashmir. Our next goal was to join Pakistan. Our ultimate goal was the revival of the caliphate.

"I was able to engage with the best Pakistani minds. I met Hamid Gul [the director of Pakistan's intelligence agency during the Soviet-Afghan war]. He was our hero. I discovered that he was just an ordinary person. I could see in his face that he was a cruel man, that he didn't care about the people of Kashmir. This was a spiritual project for me, but he was running a conflict enterprise. He was just playing with our lives.

"Soon after this I realized that everybody was making money," he continued. "When we started out, life was miserable for us in Muzaffarabad, even for the militants that Pakistan was sponsoring." He names a leader of one of the biggest Kashmiri militant groups. "He started out an ordinary fellow, a teacher. But when I saw him in Pakistan he was driving around in a big jeep. I realized that the jeep was influencing him—influencing how he sees the world. I saw that the emir of Jamaat-i-Islami—a man who had earlier lived very simply—had acquired a fleet of big cars. This realization that people were profiting from the jihad, and that people like Hamid Gul, who had been my hero, were unworthy of admiration—that was only part of what turned me away from jihad. The most important factor was my realization that our fight on behalf of the umma—the Muslim people—had been transformed into a fight on behalf of a nation-state. We were being used to serve the interests of Pakistan.

"Intellectuals glorified our movement. They called it a freedom movement. India's repression and occupation of Kashmir were factors for us; part of our motivation was nationalism. But nationalism is only part of what causes Islamic militancy. This wasn't a fight for freedom, at least for me. It was a civilizational battle," he says, weirdly echoing the title of Samuel Huntington's book, *The Clash of Civilizations.* "You have to understand: resolving the nationalist conflict between Israel and Palestine won't end the violence. The goal is much bigger—to recover our lost civilization, to recover the golden age of Islam. Nationalism is only part of the picture.

"When I see young Kashmiris donating their lives to what they think of as a jihad I feel a deep sense of regret. I feel that we initiated this violence. We initiated this destruction. I regret my decision to put people onto that course. With each generation Islamic fundamentalism becomes uglier and uglier. When I look at fundamentalists today, I see a bleak future for them. The first generation of fundamentalists—Qutb and Maududi—was focused on Dawa—education. We focused on freedom. This generation is much more rigid, stricter, than my generation. They are focused on hate. It is a painful journey. Bitter and sour, like eating a lemon. To hate is venom. When you hate, you poison yourself. This is the typical mentality of the fundamentalist movement today. Hate begets hate. You cannot create freedom out of hatred. Today's jihadis are confused—they are trying to revive old structures. We shouldn't be seeking structures, but something more spiritual."

I tell him that I have come to think of religion as having two sides—one that is spiritual and universal, the other a marker of identity, often in opposition to others. It is relatively easy to be pluralist in the political scientist's sense—to provide different religious and political groups equal opportunity, and to allow power to be shared by several parties. But to be pluralist in a philosophical sense—to recognize more than one ultimate principle or pathway to God—and to see these as equally valid—this is harder to achieve within monotheism.

He agrees. "I now realize that my desire to help the umma—the Muslim community—was in itself a spiritual error. It is wrong to focus on your own people's suffering, to imagine that the suffering of your people is greater than others'. Faith that is not able to make you understand the suffering of all peoples—not just your own—is unworthy of the name."

He continues: "I now realize that we become prisoners of our rituals. Our rituals help us pray, but they also divide people."

We talk about how confusing and difficult it is to maintain faith while avoiding ethnic identification. It is something he struggles with every day, he tells me. "I feel isolated and alone," he says. "I was accused of being afraid of death. I cannot escape this accusation. But people judge you according to their own weaknesses."

The bottom line, I now understood, is that purifying the world through holy war is addictive. Holy war intensifies the boundaries between Us and Them, satisfying the inherently human longing for a clear identity and a definite purpose in life, creating a seductive state of bliss. I was now ready to begin investigating how leaders and organizations bring about this state of bliss, the topic we will explore in part II.

Part II

Holy War Organizations

Part 2 addresses the question: How do leaders run successful holy war organizations? It looks at several organizational types. The first type is a virtual network, which right-wing extremists call "leaderless resistance." In a leaderless-resistance network, leaders inspire operatives to take action on their own, without communicating their plans to others. Leaders avoid participating or planning the attacks themselves. We also look at individuals, "lone-wolf avengers," who act entirely on their own. They are often influenced both by terrorist ideologies and personal grievances. The Internet has greatly increased the impact of virtual networks.

At the opposite end of the spectrum from lone-wolf avengers are terrorist armies or commander-cadre organizations. In this kind of organization, the leader may provide inspiration, but he also commands his followers. He trains them, provides housing and/or salaries, provides for their families in the event that they die as "martyrs," and, in many cases, punishes them if they disobey his orders. Unlike a virtual network, this type of organization is capable of carrying out massive attacks. It tends to be found only in states that are poorly governed, or where state agencies or their agents promote terrorism. We look first at the leaders and managers of some of the Pakistani jihadi organizations—how they live, and how they see their role; and then move on to assess the cadres and how they are recruited. We then turn to assess hybrid organizations, which combine virtual networks, cadre organizations, franchises, and freelancers.

The most important aspect of the organization is the mission. The mission is the story about Us versus Them. It distinguishes the pure from the impure and creates group identity. The organization's mission statement—the story about its raison d'être—is the glue that holds even the most tenuous organizations together. Without this mission statement, the organization is little different from an organized criminal ring.

The mission is not static. It can change over time. Its function is not only to attract recruits but also to raise cash, a critically important requirement for running a commander-cadre organization. Leaders change their mission at will sometimes, because the orginal mission was achieved and the operatives wanted to keep their jobs as holy warriors (as was the case for the organization that became Al Qaeda); and sometimes because there is no longer funding to support the original mission (as was the case for the Egyptian Islamic Jihad). When there is money for Islamist causes but not communist ones, Islamist terrorist organizations will rise, and communist ones will begin to fail. Some terrorist groups frequently change their mission, while others have "sticky missions." They stick with their original objective, even when the cause is no longer fundable, impossible to achieve, or already attained. Organizations with sticky missions are unlikely to last as viable organizations; those with more flexible missions have the potential to persist.

We will find in the pages that follow that the requirements for running terrorist organizations are similar to those of running a firm or a nongovernment organization (NGO). Today's multinational terrorist leader is an entrepreneur who brings together mission, money, and market share. He hires skilled and unskilled labor and often pays competitive rates. Money is more important for commander and cadre-style organizations that carry out large-scale attacks than for virtual networks, in which participants are expected to fund themselves or raise money on their own initiative. We will study cases in which leaders of commander-cadre organizations abandoned not only their original mission, but even the population they were "serving," because the organization was cash-starved.

We will learn that commander-cadre organizations raise money on the Internet, through NGOs, from diasporas, and through licit and illicit businesses. We will see some terrorist groups joining forces with organized criminal rings, with both entities benefiting; while other terrorist groups get so involved in making money through kidnapping for ransom or drug running, for example, that they become essentially organized criminal rings in their own right. Many also run licit businesses, including farming, manufacturing, and shipping.

We will see terrorist groups competing for market share in the same way firms or humitarian organizations do. They advertise their mission and accomplishments. They meet with high-level donors. Just like humanitarian NGOs, they may begin to view their donors as the most important entity to please, rather than their clients, as the appearance of accomplishment becomes more important than actually achieving social or religious justice.

One of the requirements for a maximally effective network is the free flow of information between nodes (a node could be an individual operative, or it could be a cell or network), but there is a balance of strong and weak ties between the nodes because too much communication (as in an all-channel network, with each node in communication with every other node) bogs down the network rapidly. I will define effectiveness as the ability to optimize the terrorist "product" (garnering attention, usually through attacks) while minimizing costs in terms of the number of personnel required. The network is held together by a common mission, but power and decision-making are distributed.[1] But terrorists and criminals require secrecy, so it is not possible to achieve the ideal of constant communication.

We will learn that terrorist organizations face a trade-off between resilience and capacity. I will use the term resilience to refer to an organization's ability to withstand the loss of part of its workforce, and capacity to refer to its ability to optimize the scale of attack. Effectiveness is a function of these two attributes. Resilience is improved by secrecy, weak links among operatives and between operatives and management, and redundancy in the organization (perhaps three CFOs rather than one). Maximizing capacity requires recruiting personnel with special skills, including in fundraising, acquiring and using weapons, collecting intelligence, and planning operations. It requires managers, cadres, public-affairs officers, recruiters, and diplomatic personnel responsible for coordinating with government agents, as necessary.[2] Maximizing capacity also requires careful coordination, making the group vulnerable to law-enforcement penetration. The group's effectiveness is a function of its resilience *and* its capacity. When groups operate in law-enforcement-rich environments, their resilience will make a big difference to their effectiveness.

Most high-capacity, hierarchical terrorist organizations are not resilient. Hierarchies can be penetrated and unraveled. For example, the 1956 Battle of Algiers was essentially won as soon as the French learned that every terrorist-member of the Front de Libération Nationale (FLN) organization knew two other members—the person who recruited him and the person that he in turn recruited.

Electronic communication is perhaps the biggest vulnerability. Unless successfully encrypted, law-enforcement authorities can intercept electronic communication. In a June 2002 interview with al Jazeera, Ramzi bin al-Shibh, an Al Qaeda manager, called communications "the dangerous security gap through which the enemy could infiltrate and attempt to foil any operation. Therefore it is imperative to determine the most secure means of communication and determine an alternative means in case this becomes necessary." Bin al-Shibh was captured by a Pakistani SWAT team on the one-year anniversary of September 11 in his Karachi apartment.[3] By the time he was captured, management of the Al Qaeda network was sufficiently dispersed that the loss of a single leader will make minimal long-term difference.

The maximally resilient style of organization is a network with widely distributed leadership and minimal (or successfully encrypted) communication among nodes. Capturing an operative or cell will not help law-enforcement authorities find other cells. In a virtual network (or "leaderless resistance" network, which is not really leaderless), the cells do not communicate with one another or with the leadership. And the leader cannot be captured because he never breaks the law. The organization with the greatest capacity is likely to be a commander-cadre organization.

As we will see in chapter 9, the best way to balance the requirements for capacity and resilience is to develop a hybrid organization that is a network of networks of various types. It will include leaderless resisters, lone-wolf avengers, commanders, cadres, freelancers, and franchises. The leader will be partly inspiring and partly commanding. Like an inspirational leader, he will aim to transform many of his followers into leaders. He will inspire some through appeals to spirit and emotion, but he will also provide tangible rewards, punishment, and coercion. The International Islamic Front is the best example of this type of organization.

SIX

Inspirational Leaders and Their Followers

This chapter contains a series of conversations with members of the save-the-babies movement, the name I use to refer to the part of the pro-life movement that supports murdering doctors and attacking abortion clinics. The chapter opens at a fund-raiser for the movement, where ex-convicts are celebrated for their antiabortion crimes and attendees bid on handmade items produced by currently incarcerated activists.

After attending the movement's annual banquet, I visit two major leaders: Michael Bray, the movement's leading intellectual, and Paul Hill, now on death row for murdering a doctor and his escort. In addition to exploring the save-the-babies movement, the chapter assesses the nature of leadership in a virtual organization, referred to by American right-wing extremists as "leaderless resistance."

In the winter of 1999 I decide to attend the White Rose Banquet, a charity dinner held every year to honor the "saints of Christ," violent antiabortion activists who are now in prison. The banquet has two purposes. The first is to promote esprit de corps among proponents of "defensive action" at "baby butcheries," which means, in plain English, killing doctors and their staff and bombing clinics where abortions are provided. The second is to raise money to help support the families of the "martyrs." You have to send your forty dollars in advance to the Reverend Michael Bray, the organizer of the banquet, so he can decide whether you are worthy of attending.

The Reverend Michael Bray is the intellectual father of the extreme radical fringe of the antiabortion movement, which engages in terrorism rather than nonviolent protest. Bray spent four years in prison for conspiring to bomb ten clinics near Washington, D.C., but now sees his role as inspiring others through his writings, sermons, and events like this.

Bray is a postindustrial-style leader of a virtual organization, which has no headquarters, no established hierarchy, and no regular planning meetings. Members are likely to learn about the organization and its mission over the Internet. Some establish friendships by e-mail long before they meet. Bray and other leaders of the save-the-babies movement mobilize rather than supervise their followers. They do not get involved in day-to-day management issues (in this case, planning attacks at clinics, the product this virtual organization produces) or providing tangible rewards such as salaries to their followers. This style of leadership involves what James MacGregor Burns calls a "transforming relationship," in which leaders and followers influence each other's thinking and actions, with the purported aim to make the world a better place. In the process, followers become leaders, and leaders may become moral agents.[1]

Bray sees himself as a moral agent, and he aims to convert many of his followers into leaders in their own right. To this end, he and other inspirational leaders in the save-the-babies movement motivate their followers through their writing, their Web sites, their sermons, and most importantly, the White Rose Banquet. The banquet has been held nearly every year since it was established in 1996.

The fourth annual banquet, which the Reverend Michael Bray permits me to attend, is held at a Holiday Inn in suburban Washington, the evening before the twenty-sixth anniversary of *Roe vs. Wade.* It's raining hard, and I get lost. Eventually my research assistant and I make our way to a nondescript hotel, where we drive around to a parking lot for the ballrooms in the far back corner. I park my white Volvo among a profusion of pickup trucks. People have driven long distances to get here—including from Alabama, Ohio, Florida.

At the entrance to Ballroom B you have to pick up your ticket. A woman in a black, sequined gown checks me off: Jessica Stern, fellow, Harvard University. I am obviously somewhat exotic. There are people milling all around—in the ballroom, and in the rooms outside it. Refreshments are served. I look through the literature on a table. In the White Rose Banquet program I read:

> Some of them are dead by having their small, soft bodies literally wrenched apart and pulled through suction tubing; others are neatly cut here and there by a knife-like instrument—an arm brought out first or maybe a leg with other appendages and organs to follow. . . . Among the pieces is a heart that was warm and beating only moments before. Tiny fingers and thumbs that once sought the comfort of this baby's mouth lay gently curled and discarded next to what was a liver and a foot. The face of this infant has been nearly shorn from the rest of his head, the eyes open and dark with sudden terror . . .
>
> Joseph Grace, quoting Cathy Ramey
> Somewhere in a Virginia jail since 1983.

I tell myself I must not be shy, I must talk to these people. Many seem to know one another from past actions at "abortuaries" or from banquets held in earlier years. None of the people I approach seems hesitant to talk to me, even though I tell them I am writing a book about religious militancy and terrorism.

The first person I meet is Katherine Horsley, the seventeen-year-old daughter of Neal Horsley, the man who is best known for the "Nuremberg files," a Web site that lists physicians and clinic staffers who allegedly provide abortions, including, in some cases, their office and home addresses. The names of doctors and personnel who have been killed are crossed out, and those that are wounded are grayed out. When Dr. Barnett Slepian's name was crossed out within hours of his death, Planned Parenthood and a group of doctors filed suit, and Horsley's Internet service provider took the site down.[2] But Horsley relaunched the site on his own server and has expanded it since then.[3]

Katherine is a daughter of royalty in this setting. She is beautiful and sweet and obviously excited to be here, among the most important people in the violent antiabortion movement. She is talking to Jonathan O'Toole, a fresh-faced nineteen-year-old who works for her father.

I ask him how a nineteen-year-old man would get involved in the movement. "To begin with," he says, "I am a Christian, and therefore opposed to abortion. Unborn babies are dying by the millions, and I feel compelled to help." He shows me a grisly picture of an aborted fetus. It is horrifying, and I understand why it moves him. He informs me proudly that he is a member of the Army of God, a shadowy organization that advocates killing abortion providers as "justifiable homicide."[4] A number of attacks on clinics and personnel have been carried out in its name, but it is best described as a virtual network, or in the language of the movement, a "leaderless resistance" network, rather than an actual organization.[5]

The Army of God manual explains that it is "not really an army, humanly speaking. . . . God is the General and Commander-in-Chief. The soldiers, however, do not usually communicate with one another. Very few have ever met each other. And when they do, each is usually unaware of the other soldier's status. That is why the Feds will never stop this Army. Never. And we have not yet even begun to fight."[6]

A leaderless resistance network—with no central office and no known leaders involved in planning operations—is almost impossible for law-enforcement authorities to penetrate and stop. It has been adopted by a number of Christian and Jewish extremist groups in America.[7] Despite the name, the network is not actually leaderless.

The doctrine of leaderless resistance was developed by Louis Beam, who calls himself ambassador at large, staff propagandist, and "Computer Terrorist to the Chosen" of Aryan Nations, a neo-Nazi group.[8] Beam writes that hierarchical organizations are extremely dangerous for insurgents. This is especially so in "technologically advanced societies where electronic surveillance can often penetrate the structure revealing its chain of command," such as the United States. Those who oppose state repression must be prepared to adopt a new organizational style, he argues. Success will depend on the following factors: "avoidance of conspiracy plots, rejection of feeble-minded malcontents, insistence upon quality of the participants," and "camouflage," which Beam defines as the ability to blend in the public's eye with mainstream associations that are generally viewed as harmless. In the leaderless form of organization, "individuals and groups operate independently of each other, and never report to a central headquarters or single leader for direction or instruction, as would those who belong to a typical pyramid organization. Organs of information distribution such as newspapers, leaflets, computers, etc., which are widely available to all, keep each person informed of events, allowing for a planned response that will take many variations. No one need issue an order to anyone." Beam's goal was to develop a more effective means to resist the "tyrany" of the U.S. government.[9]

David Ronfeldt and John Arquilla argue that groups organized this way need an operational doctrine, and the most potent one, they argue, is what they call "swarming." Swarming involves widely dispersed but networked units converging on their targets from multiple directions. Networks must be able to coalesce rapidly and stealthily on the target and then disperse. But they must be ready to recombine for a new pulse almost immediately. "The Chechen resistance to the Russian army, the rush of NGOs into Mexico to support the Zapatista movement, and the Direct Action Network's operations in the 'Battle of Seattle' against the World Trade Organization (WTO) all provide excellent examples of swarming behavior," they explain.[10]

For virtual networks promoting terrorism in technologically advanced countries such as the United States, swarming operations may be an attractive fantasy, but not one easily carried out in the near term. The planners of the Battle of Seattle were not promoting assassinations, bombings, or mass-casualty attacks, and if they were, law-enforcement authorities would have stopped them. Until they get access to impenetrable mass communication systems (and they might), they will not be able to coalesce rapidly and stealthily on the target, and then disperse, as Ronfeldt and Arquilla envisage.[11] Henry Felisone, a Florida-based minister involved in the save-the-babies movement, describes his vision of a swarming operation to end abortions: "So we have the Army of God, which in the future will organize and coalesce like those of Europe who had centuries of underground work, and there will be skilled assassins and skilled saboteurs after the abortion industry, which is not only the abortionists but also the people on top of them, including Supreme Court judges. Now Paul Hill has called for the Supreme Court judges to be killed and also for chemical and biological weapons, and we support this call, at least I do."[12] The prospect of swarming is not the most frightening aspect of virtual terrorist networks, in my view. It is, instead, the increasing availability of more and more powerful weapons usable by smaller and smaller groups, leading to the potential for mass-casualty attacks that require minimal coordination and communication.

Jonathan O'Toole and I move together toward the ballroom, where there is a cheerful din. The feeling of bingo night. He suggests that I talk to the Reverend Donald Spitz, another leader in the movement. Spitz is the head of Pro-Life Virginia. He seems determined to communicate something important to me. He tells me that the world system is becoming worse and worse. "Evil is increasing, iniquity abounds," he says. "Just look in Matthew 24." When I do, I find it refers to an apocalyptic period preceding an imminent return of Jesus Christ:

> "And ye shall hear of wars and rumors of wars: see that ye be not troubled: for all these things must come to pass, but the end is not yet. For nation shall rise against nation, and kingdom against kingdom: and there shall be famines, and pestilences, and earthquakes, in divers places. All these are the beginning of sorrows. Then shall they deliver you up to be afflicted, and shall kill you: and ye shall be hated of all nations for my name's sake. . . . And because iniquity shall abound, the love of many shall wax cold."

Spitz continues, "Peter says evil means shall become worse and worse. Pontius Pilate slaughtered the unborn children."

I ask him how he ended up focusing his life on stopping abortion.

He tells me, "I was taught from a very early age that abortion is the worst thing a woman can do."

Do you support violence against doctors or abortion clinics? I ask.

"That's not how I would put it," he says. "I support defensive action. If a born person were being murdered right here, it would be our duty to defend him. It would be wrong to allow him to be murdered in front of our eyes. It is also our duty to defend the unborn. An unborn person is no less a person than you and me."

Defensive action sometimes influences doctors' decisions. The movement specializes in applying psychological pressure on doctors and clinics and has in many cases intimidated doctors and other abortion providers to the extent that medical practices have been shut down. In late 1999, for example, Steven Dixon, a forty-year-old obstetrician-gynecologist, shut down his practice, telling patients that he was "terrorized by antiabortion activists."[13] In a letter to his patients explaining his move, Dixon wrote that "the ongoing threat to my life and my concern for the safety of my loved ones has exacted a heavy toll on me, making it necessary that I discontinue practicing OB-GYN."[14] According to the *Washington Post,* antiabortionists mailed threatening letters to his office and home and distributed "Wanted" posters with his photograph. In addition, his name was added to Neal Horsley's Nuremberg files Web site.[15] Other studies have shown that between 1992 and 1996, a time when antiabortion violence was particularly common, the number of abortion providers in the United States decreased by 14 percent. It is unclear how much of that drop is due to antiabortion violence or to other factors such as reduced incidences of unintended pregnancies.[16]

Waiters have been carrying steam trays to a buffet table at the back of the room and are now ready to serve us. The food is what you might expect in an elementary school cafeteria—overcooked, but comforting.

Jonathan suggests that we sit with a middle-aged friend of his sitting at a table in the middle of the room. The older man introduces himself as Bob Lokey. I notice two things right away: his muscled arms are decorated with tattoos, and he has the strangely bright eyes of a person who meditates a lot. Lokey tells me he spent twenty years in San Quentin for first-degree murder. I ask him how he spends his time now. He is a long-haul trucker. But his great passion is "saving the babies" and his Web site, which was linked at the time to Neal Horsley's site.[17]

"Everything I found on the Internet looked wimpish to me," he says. "I wanted to establish my own site that makes the evil of baby murder more clear. I indicted the Supreme Court for its support of baby murder. I explain it all on my Web site."[18]

How did you get interested in this work? I ask Lokey.

"In 1973 I had a powerful vision," he says. "I was in a forest. A great power came to me and instructed me to paint an image of Uncle Sam, dragging a baby by its neck with handcuffs. You can see the image on my Web site now."[19]

He tells me he has two children. A look of pain comes briefly into his eyes.

Lokey is a vegetarian. Jonathan is obviously very much in awe of him. Jonathan asks him, "How do you maintain your bulk on a vegetarian diet?" Lokey describes his weight-lifting regimen in some detail. "I'm still bench-pressing at fifty-eight," he says. He tells Jonathan, who looks frail sitting next to Lokey, meat is bad for you, you shouldn't eat it. Jonathan smiles sheepishly. I don't want to hurt anyone, Lokey says. Not even animals. Lokey does yoga and meditates every day. He learned in San Quentin, he says, from a teacher who was allowed to visit them in prison. He tells me he can make an *om* sound on his guitar. He often plays the guitar with a young girl in the neighborhood. When they play, birds start acting oddly, he says.

I ask Jonathan how he ended up working for Neal Horsley.

"My dad is a Southern Baptist preacher," he says. "I'm the oldest of five children. My whole family was involved in the antiabortion movement in Kansas. Most of the elders from our church got involved in rescues, and I got involved too. Participating in rescue missions was my Sunday school."

In the late 1980s Operation Rescue began staging "rescue operations" outside abortion clinics. The rescuers would surround clinics and attempt to prevent patients from gaining entry. They would terrorize the women with photographs of bloody aborted fetuses, with the goal of making them reconsider their decision to have an abortion. In a competition of horrors, pro-choice activists, also likely to be on the scene, would carry placards illustrating abortions performed with coat hangers to emphasize the dangers women would face if abortion was made illegal. During the late 1980s and early 1990s, hundreds of demonstrators were arrested as they tried to block entrances to abortion clinics. In 1994, Congress passed the Freedom of Access to Clinic Entrances Law, which made it a crime to block access to abortion clinics and mandated stiff penalties for harming anyone during demonstrations. After the law was passed, blockades dropped off precipitously, from a high of 201 with 12,358 arrests in 1989, to 2 blockades and 16 arrests in 1998. But the number of violent incidents rose. Since 1993, antiabortion activists have shot and killed seven people and attempted seventeen other murders since 1991. They have set fires at clinics and exploded bombs, sometimes with lethal results. They have placed a noxious chemical, butyric acid, inside clinic doors, hoping to nauseate or burn the skin and eyes of building occupants. They have sent letters to clinics purporting to contain anthrax, temporarily halting activities at the clinics and terrorizing their patients. Some of the clinics and the people within them have had to be decontaminated, at great expense to city governments.[20]

"I saw his Web site and assumed that he was somebody closely connected to the terroristic kind of, real radical antiabortion crowd," Jonathan says, explaining what initially attracted him to Horsley.[21]

I ask what he does for Horsley. "I help him on the Web site," Jonathan says. "I didn't have a lot of computer experience, and I wanted to learn. Also, I sort through hundreds of letters and e-mails daily."

Horsley started the Nuremberg files Web site in 1995.[22] The site provides information about personnel at abortion clinics all over the country, and about judges and politicians "who pass or uphold laws authorizing child-killing." Visitors are urged to send names and birth dates of abortion providers' family members and friends; social security numbers; license plate numbers; photographs and videos; affidavits of former employees, former patients, or former spouses; or "anything else you believe will help identify the abortionist in a future court of law."[23] Horsley added President George W. Bush to the Nuremberg files in August 2001, after the president announced that he would allow federal funding for fetal stem-cell research under certain conditions.[24]

I ask Jonathan about his schooling. He tells me he was homeschooled until about sixth grade. After that, he says, he attended a small Christian school, then public high school. "I've been to three colleges," he says, "William Jewell in Liberty, Missouri, Ross Hill in Aiken, South Carolina, and Maple Woods in Kansas City. But now I'm taking a sabbatical leave from school."

Later I learn that homeschooling had left Jonathan feeling ill equipped for public school, especially gym class. He had never played team sports.

He had a dream in his sophomore year in high school about killing his classmates with an automatic rifle. He felt no guilt about it, he said. He just mowed them all down. When Jonathan went off to college, he brought a pistol. His classmates grew so alarmed hearing Jonathan talk about his desire to attack abortion clinics that thirty of them held a meeting to decide what to do. Jonathan decided to leave and returned to his parents' home.[25] Jonathan learned about Neal Horsley from his Web site, just as he learned about Bob Lokey from his. After Jonathan dropped out of school, he showed up on Neal Horsley's doorstep, offering his services for free. Now that he has met Bob Lokey in the flesh, Jonathan seems uncertain about which man to admire more: the muscle-bound ex-con sitting with us now, or the Internet agitator he works for.

It is hard to explain certain kinds of leader-follower relationships without considering the possibility that, in Barbara Kellerman's words, "some persons, under some circumstances, experience the need or wish to look *up*."[26] In many cases the leader may be responding to subconscious needs. Freud saw leaders as "great men" who tapped into the majority's "strong need for authority which they can admire, to which they can submit, and which dominates and sometimes even ill-treats them."[27] This need to look up seems to apply to Jonathan. The leaders of the save-the-babies movement don't mistreat their followers in an overt sense, but they may try to dominate them spiritually and emotionally.

According to Abraham Maslow's famous work on human motivation, human needs can be arrayed in a hierarchy, from the most fundamental ones such as the need for food and shelter, to more abstract ones, such as the need to feel part of a community or to be esteemed for one's work. The urge to belong to a group comes immediately after the need to satisfy basic physical requirements for life and security.[28] But today, people are largely free to choose their identities. Just as nations are largely imagined communities, to use Benedict Anderson's well-known phrase, so too is individual identity.[29] Identity and identification are thus two separate concepts.[30] Individuals must identify their identities. As we discussed in chapter 3, too much choice regarding identity can be overwhelming.

For a person living in a closed, ethnically homogenous, rural society, there is a limit to the number of identities he might choose. But today, individuals are exposed to many possible identities. Cities are multiethnic. Businesses are globalized. We are flooded with images and ideas from around the world on television and the Internet. One job of a leader like Bray is to present a narrative about the mission of the movement that helps a person like Jonathan find an identity and a sense of purpose in life. Part of the appeal of militant religious groups, as we have discussed, is the clarity they offer about self and other.

Transforming leaders, as Burns defines them, tend to focus on followers' psychological and spiritual needs rather than their physiological ones. In a relationship based on transforming leadership, "leaders and followers raise one another to higher levels of motivation and morality. . . . Transforming leadership ultimately becomes moral in that it raises the level of human conduct and ethical aspiration of both leaders and led, and thus it has a transforming effect on both," he argues. This kind of leadership is to be distinguished from what Burns calls "transactional leadership," which involves an exchange of things: jobs for votes or subsidies for campaign contributions or money for work.[31] Both leaders and followers get something out of the relationship. Many of the leaders of the save-the-babies movement are clergy with tiny congregations. Bray's congregation consists of seven families, for example. At the White Rose Banquet and on their Web sites, these leaders become important. They get the satisfaction of feeling that they are doing the right thing, and they also get the satisfaction of being leaders.

Many scholars now accept the idea that leadership, as distinct from management, involves a transforming relationship as Burns defined it, in which influence runs both ways, but reject the notion that it always involves the promotion of moral action.[32] Most people would consider Hitler and Stalin leaders, for example, although they promoted evil rather than good. It is also important to recognize that one leader's conception of moral action may be another's conception of evil action.[33] The moral dilemma, and the possibility of just terrorism, is probably most obvious in precisely the case examined here. If we take the view that abortion is not murder, then those who kill abortion providers are guilty of murder themselves—in a moral as well as a legal sense. But if we accept the prolifers' premise that "ensoulment" (or potential viability) begins at conception, attempting to force doctors to stop practicing abortion could be justified as the defense of innocent persons who are under attack. Still, as we discussed in the introduction, even if we accept the doctor killer's ends, if only for the sake of argument, his means are morally indefensible because, among other reasons, murder is not a last resort.[34]

Lokey wants to tell me more about his visions: "In 1968 I awakened from sleep. I heard a lot of clanking noises. Suddenly I realized those clanking noises were actually in my head. My head expanded, then my whole body. My covers seemed to be inside my skin. As I moved my arm, I realized there were billions of stars between my arm and my body. I was at one with God at that moment. There were stars all around outside me."

Were you on drugs? I ask.

"Awareness is greater than any LSD high," he scoffs. "I felt power—it was God's power. It's primitive—you know it when you feel it. Time did not exist. I said, 'I want to go home.' I saw a tiny dot. I began rushing toward that tiny dot. Then I realized that that tiny dot was the earth.

"I also had the experience of remembering my birth. I'm in the joint, in the typewriter repair shop. I felt a tight band around my head. My friend the foreman called me, 'Lokey!' I turned sideways. The band was proceeding down my body. I fell out into space. I tried to breathe—I knew I had to fall down and cry. I felt someone spanking me—the pain was incredible. I heard a baby crying—I heard the words of the doctor, I heard my mother travailing. Satan embodied himself in the man I had been talking to—my friend. I knew I was looking straight into the eyes of the devil himself. I heard the voice of an angel. Satan said, 'This child is mine.' The angel said, 'Deliver him from evil.' I was being delivered from evil and into life. This happened in 1968."

Lokey tells me excitedly that all doctors that provide abortions will be killed, and all the women who have had abortions will be killed. He then modifies his statement to make clear that women who had malice aforethought have committed first-degree murder, but if they only have malice in their hearts, it's second-degree murder.

How do you know which ones have malice aforethought versus malice in their hearts? I ask. Lokey seems not to like this question. I sensed his muscles tensing. Perhaps he hadn't thought this through. He says something unintelligible, then tells me, "If we don't bring this to a conclusion soon, everyone on earth is going to die.

"You know I'm celibate," he adds, as if that were something that everyone knows. "I've been celibate since 1984." He tells me he was "vaginally defeated," but now he's "free." Later he explains this concept of vaginal defeat in somewhat more detail. "I've been vaginally defeated all my life," he told Neal Horsley while the two were being filmed for a television documentary. "Finally, God said to me 'Son . . . you have got to leave this thing alone.' I was so attracted to women, at one time I thought women were gods. And he made me quit women then and there. . . . I quit smoking, quit drinking, quit meat; I even circumcised myself. All those things that I've had to do. . . . women were the toughest. . . . I have to be the only grown man to have circumcised myself."

Sociologists argue that the first requirement for mobilizing a group that feels oppressed is the identification of a common enemy. "Without the identification of an adversary, or another social actor in conflict with the group for control of certain resources or values, discontent and protest will not engender a movement," sociologist Alberto Melucci argues.[35] Religion is thus the ideal mobilization tool for violence because the Other is often inherent. "Whatever universalist goals they may have, religions give people identity by positing a basic distinction between believers and nonbelievers, between a superior in-group and a different and inferior out-group," Samuel Huntington observes.[36] Defining "us" automatically entails defining "them." In this case, the "abortionists" and their clients are "them."

Leaders of the save-the-babies movement identify "us" versus "them" through stories of heroes and villains, martyrs and saints. The right story, Ronfeldt and Arquilla explain, can help keep people connected even in a network whose looseness makes it difficult to prevent defection.[37] Bran Ferren, a former Disney executive, argues that the ability to tell stories, to articulate a vision and communicate it, is a "core component of leadership."[38] The stories strengthen group identity.

But here, language is as important as narrative. The words *abortuaries, butchertoriums,* or *baby butcheries* are used for abortion clinics; *defensive actions* or *justified homicide* for violence against clinics or personnel; *saints, martyrs,* or *prisoners of Christ* for those who break the law on behalf of the movement. No crimes will be committed here at this meeting, but the language itself is thrillingly illicit. The words are like a drug. Uttering them aloud heightens the mood of self-righteous insurgency. The story is also told through images on the movement's Web sites, through pictures of bloody, mutilated, aborted fetuses and streaming-video close-up shots of women's genitalia and abortions actually taking place. The Internet is a critically important part of the network's strength.

Lokey goes back to describing his visions: "I saw a vision of a perfect woman. Next thing I knew I was looking through a hole and the woman was old and wrinkled. And then she was dead. Her eyes were open. They were filled with vengeance, loathing, disgust. She said over and over again, 'Kill him! Kill him! Kill him!' "

Whom did she want you to kill? I ask.

"I don't know," Lokey says sadly. "After that, when God woke me up that time in Georgia, when I wrote 'Holocaust II,' I took it to the print shop. There was a woman there at the print shop, her back was to me, she backed into me. She was collating my document. I told her, 'I only paid you to copy that material, not collate it.' She said to me, 'I do stuff like this for my kids all the time. I'm happy to do this for you.' I knew then and there that she was the very one—I had met the perfect woman of my vision—my vision had come true."

What is a perfect woman? I ask.

"A perfect woman is a good mother. Most women are vile," he says politely.

At 8 P.M. the official program begins. First we sing a hymn, "Rise Up, O Men of God." Michael Bray comes to the podium and tells us, "We know that government agents are in this room. But we know who you are." He peers around the room. I also look around the room, wondering which of the activists is actually working for the FBI. "Some members of the media are also in this room. You don't have to talk to them. You make up your own minds. But we're not in here to hide. Remember that Cheryl Richardson spent some time in jail, and the media helped get her out. Their goal is to use us. But our goal is to use them. It's a contest of wills."

Bray provides updates on all the prisoners of Christ now serving time for violence against abortion clinics. The prisoners have written short letters for the banquet, and the aforementioned Cheryl Richardson, now released from prison, reads them to us.

The Reverend Donald Spitz now comes to the podium to tell us excitedly that the auction of relics will now begin. The saints in bonds have donated the items that will now be put up for auction. Proceeds of the auction are for the benefit of the families of the prisoners. Well-known activist Shelley Shannon has knit a number of items to be auctioned off, all in camouflage: a pair of mittens, a pair of gloves, a hat, a scarf, and a pair of baby bootees. Shannon is serving time for attempted first-degree murder. She shot and wounded Dr. George Tiller in Wichita, Kansas, in August 1993. Spitz jokes that if you wear these items, the feds won't find you. "Eric could use one of these things in those woods," he says, referring to Eric Rudolph, wanted by federal officials for a series of bombings at abortion clinics, a gay nightclub, and the Atlanta Olympics. Rudolph, who was captured in 2003 after six years on the run, is one of the movement's major heroes. The mention of his name brings a frisson of excitement to the room and titillated giggles. The knit items go for $75 among them.

David Lane, serving an eighteen-year sentence for vandalizing an abortion clinic and a doctor's office on March 18, 1995, has donated a video of gospel messages dictated in prison, various original artworks, a cross and pin worn by himself, and a prison ID card. These go for a total of $245.

The last item is a hooded sweatshirt with a prison ID number, donated by John Brockhoeft, who is in the room with us. He is introduced as an "ex-con for life." Brockhoeft is a forty-seven-year-old truck driver. His face is gaunt and glowering. My assistant points him out to me, telling me she finds him "the scariest guy in the room. After Lokey, of course." He is wearing black fatigues and a black beret and is known here as the Colonel. He served seven years for firebombing a Cincinnati clinic and attempting to blow one up in Florida. During his probation he had to wear an electronic surveillance bracelet and was forbidden from talking to anyone affiliated with the antiabortion movement. Tonight he is among his fellow believers, who treat him like a homecoming hero. The sweatshirt fetches an impressive $125.

His first wife divorced him while he was in prison. But he married a young antiabortion activist who had been writing to him while he was incarcerated. She is here by his side. She has long hair and a sweet, wholesome-looking face. She is watching over their little girl, a pretty blond toddler. She and John are great supporters of Paul Hill, perhaps the most important visionary leader in the movement, who was then on death row in Florida.

Bray now introduces two other ex-cons for life, John Arena and Joshua Graff. Graff is a twenty-four-year-old who was incarcerated for bombing the West Loop "abortuary" in Houston. Arena is a seventy-seven-year-old who spent four years in prison for his involvement in blockades and "covert rescue techniques," including butyric acid attacks.

The evening ends with another hymn, "Jesus Shall Reign," and another benediction.

Later, I visit Michael Bray at his home in Bowie, Maryland, to learn more about his philosophy.[39] Bray lives with his wife and their ten children in a small tract home and runs the Reformation Lutheran Church in a nearby town. He invites my assistant and me into his office. He is utterly charming. He is handsome, intelligent, and intensely charismatic. I understand why people fall under his influence, now that I am seeing him up close.

I ask Bray why he believes that violence is justified, given Jesus' Sermon on the Mount and the admonition to turn the other cheek.

"Christians tend to be opposed to violence," Bray says. "Some oppose capital punishment. But there is nothing in the Scripture to support this view. Violence is amoral—its moral content is determined on the purpose of the violent act."[40]

Some Christians argue that the New Testament reflects a progression in human understanding of God and His intentions, I say. The God of the Old Testament is harsh and violent, that of the New Testament kinder and gentler. Why do you focus so much on the Old Testament? I ask.

"There has been a progression of understanding, but still there is judgment of sin," Bray says. "The grace of God was manifested in his sending His son to earth. But God did not change His standards. Take a look at Pascal, at John Wesley, at Jonathan Edwards's encounter with God, at Saint Thomas Aquinas. They all make clear that God still judges, even in the New Testament."

You refer to obedience to God's calling; that when Joan of Arc heard God's call, nobody accused her of being psychotic. But if you as a pastor are going to encourage your parish to listen uncritically to the voice of God, how do you know that won't be encouraging the mentally ill to listen to the voices they hear, possibly instructing them to murder innocent people? What if a serial killer hears a voice telling him to kill, and he believes he hears the voice of God? What if you encourage Islamists or Hindus to kill Christians? They are equally convinced they are killing in the name of God, as are antiabortion activists, I argue.

"You should only listen to the voice of God if the action called for is morally justified," Bray says unsatisfactorily.

How should a pastor react to a modern Abraham who claims that God has instructed him to kill his son? I ask. Bray seems unprepared to answer this question.

Later I will learn that Paul Hill, one of Bray's protégés, felt he was following Abraham's example when he killed Dr. Britton. Hill knew he would be leaving his three children essentially fatherless. He felt that God would be pleased by his willingness to sacrifice the well-being of his own children for the good of countless children still unborn. "It occurred to me that I was making a sacrifice—thinking about the promise made to Abraham that if he was willing to sacrifice his son, that God would bless Abraham and grant to him descendants as numerous as the sands in the seashore and the stars in the sky."[41]

We move on.

In your book *A Time to Kill* you draw a distinction between vengeance and protective force; that the activist who engages in "defensive action" does so not out of vengeance, but to save the unborn. In that case why would it be necessary ever to kill? Why not wound doctors like Shannon did—shooting them in the arms?

"Shannon *intended* to shoot Tiller in the arms," he says. "She did that on purpose. But Tiller went back to work right away, so it shows that wounding doctors doesn't necessarily work."

We spend some time talking about Christian Reconstructionism, a movement to turn America into a fundamentalist Christian state with laws in accordance with the Old Testament.

I ask Bray, do you foresee a Christian revolution anytime soon?

"The necessary structures are not really in place at this point," he says.

"You have to consider the difference between a legitimate revolution and anarchy. The problem is that people like the status quo—they're not ready for any kind of revolution at this point. There would have to be a war, an economic crisis. A plague. You can't have a revolution when the president says I'll give everyone free medical care and grow the economy. People are generally happy right now—the economy is doing well.

"Not everyone is called to be a missionary. The work entails sacrifice. Similarly, we wouldn't expect everyone to become a prisoner of Christ—not everyone wants to, feels called to, or can afford the sacrifice. You must count the cost. But the truth is, I feel more fear of being charged with not encouraging activists more."

It occurs to me that Michael Bray is managing to encourage contributions to the public good of "saving the babies" without providing typical selective incentives. Hafez Sayeed, for example, whom we discussed in chapter 5, provides his followers with food, housing, schooling, weapons, and in many cases, salaries. His organization punishes and, in some cases, kills disobedient operatives. But Bray persuades his followers to take action without threatening them and without giving them material things. Although he raises money for the families of the "martyrs," the proceeds of the yearly auction do not amount to much—$615. It certainly does not rival the $10,000 to $25,000 that Saddam Hussein offers the families of Palestinian "martyrs."[42]

Individual operatives can have their own reasons for turning to terrorist violence unrelated to the group's purported goals. "Individuals are drawn to terrorism in order to commit terrorist violence," Jerrold Post argues. They feel "psychologically compelled" to commit violent acts, and the political objectives they espouse are only a rationalization.[43] Some of the people attracted to the save-the-babies movement may be more attracted to violence, for example, than they are interested in "saving babies." (Dave Grossman estimates that 2 percent of soldiers actually take pleasure from killing.[44]) For these individuals, terrorism is an end in itself rather than a means to an end: it is consummatory or expressive rather than instrumental. Men who take pleasure from killing people may be relatively easy to recruit to terrorist movements, but they are also likely to be harder to control than those who are committed to the cause.

Although some of the people involved in the movement may fall into this category, more is going on here. Michael Bray is persuading people to take action by appealing to their values and needs without offering material incentives. He is displaying a kind of leadership that is different from Hafez Sayeed's style.

When I began reading about leadership in an effort to understand Bray's approach, I discovered that there is no general agreement about what leadership entails, let alone how it's practiced. Leadership has become a popular area of inquiry in a wide variety of academic disciplines, but there is no consensus about what the word leadership means, even within disciplines.[45] But none of the definitions I found in the literature accurately describes what Bray is doing.[46]

I will call the kind of leadership that Bray and his colleagues employ "inspirational leadership," to distinguish it from other forms. Inspirational leadership involves a relationship between leaders and followers in which each influences the other to pursue common objectives, with the aim of transforming followers into leaders in their own right. But, unlike Burns's transformational leadership, inspirational leadership may promote immoral action.[47]

Inspirational terrorist leaders are different from commanders, whom we will discuss in chapter 8. Bray and his colleagues do not punish or threaten wayward followers. They use moral suasion rather than cash to influence their followers, appealing to higher-order deficiency needs in the Maslow hierarchy, including the desire to be part of a community and to gain recognition for one's achievements. Some of the leaders are charismatic, but not all. Commanders, as we will see in chapter 8, appeal to their cadre's most immediate needs for food, shelter, and safety (although they also appeal to their higher-order needs).[48]

Inspirational terrorist leaders work best in postindustrial, virtually networked organizations. They inspire "leaderless resisters" and lone-wolf avengers rather than cadres. They run networks or virtual networks rather than bureaucracies, and they encourage franchises. Inspirational leaders rarely if ever get involved in breaking the law themselves. That is why this style of leadership can persist even in states where the law is generally respected. If Michael Bray started paying his "saints," he would soon be incarcerated.

A few weeks after visiting Bray, I decide to call Bob Lokey to see if there is anything else he wants to tell me.[49]

What fraction of the antiabortion movement supports killing abortion providers, what you call defensive action? I ask.

"A small core would actually carry it out in my view," he says. "But one hundred percent of the people I talk to believe the things I say about it. I sometimes ask people, 'Do you believe America needs a civil war?' and everybody I talk to about that says yes. And I talk to a lot of people. A civil war would be pretty violent. Most people that I know and that I talk with agree with me on this—it's just that they're not as vocal as I am.

"People don't tell you the truth in polls. But I have a knack for talking with people—I have a knack for getting them to tell the truth. The major part of America thinks there should be civil war . . . there will be a civil war. When would it begin? I don't know about that. Everybody asks me that. Probably soon—within a few years. People are getting more upset and angry not just about abortion—it's about all manners of things. Justice. There are racial overtones. People are fed up with affirmative action, immigration. The white male is being pressed real hard.

"When I'm out there with other truck drivers, I say we should have a shooting war and people say, 'Yeah, we should.' Most white people I talk to feel they're being discriminated against. There's a lot more of that on the Internet as well. I didn't realize the movement was as large as there [*sic*] appears to be. When I was in prison, there were Nazi groups, but they seemed to be tiny splinter groups and the members were poor. Hardly ever did you see anyone who was middle-class. But now I see normal middle class people—married people—who are white supremacists or whatever you would call it. They are defensive about their race. They're opposed to discrimination against the white male. People feel attacked. Everybody that I know feels attacked.

"I've been picking up on the Internet; people are talking about anthrax, poisons. I think that very soon that will be perceived as one of the modes of getting at Them. . . . When you are at the bottom, when you have no power at all . . . There was a time when I was thinking about these things.

"Paul Hill is a good example for people to follow . . . he's a real martyr, no doubt about it. The people at a distance, they're conservative. They say they oppose violence when you ask them in a poll. If you took a poll right now, 'Do you believe in violence to stop violence?'—they'd say no. That's hypocrisy speaking. Polls lie because people lie to the polls. I'm expecting it—I'm expecting civil war soon, and hoping for it. I've had everything in my life that I wanted. The tragedy is that I didn't know what I wanted until it's too late. I'm going to get it. Civil war will come."

Paul Hill is a former Presbyterian minister. On July 29, 1994, he shot and killed John Britton, a doctor who provided abortions in Pensacola, Florida, and the doctor's security escort, a seventy-four-year-old retired air force lieutenant colonel named James Barrett.

Hill argues that "the abortionist's knife" is the "cutting edge of Satan's current attack" on the world.[50] He believes that anyone who opposes abortion on moral grounds is obligated to defend the "innocent unborn."[51] As citizens, we must always distinguish between what is legal and what is right, he says. "It is self-evident that a government may declare an act legal that is actually unjust according to God's law. A slave owner prior to the Civil War may have abused his slave in a way that was legal, but ultimately unjust. The present abortion laws legalize the killing of unborn children, they are unjust in God's eyes," he asserts.[52]

"The Bible clearly teaches that we may protect our own lives from unjust harm with deadly force if necessary," he argues, quoting Exodus 22:2, which says, "If the thief is caught while breaking in, and is struck so that he dies, there will be no blood guiltiness on his account." "The Scriptures also clearly teach that as we should defend our lives with force, we should also do so for our neighbor," he argues. "When the state or any other authority requires one to do what is contrary to God's law, the child of God 'must obey God rather than men.' This was clearly the opinion and practice of Peter and the Apostles." Prayer and fasting are not enough, moreover, because true faith "shows itself by good works."[53]

Killing fetuses is the moral equivalent of Hitler's killing of Jews in gas chambers, Hill argues, and those who don't take action in the face of such atrocity are the moral equivalent to the acquiescent church leaders in Hitler's Germany, who "also shrank from resisting the evils of an unjust, oppressive government. . . . Dietrich Bonhoeffer is an example of a church leader who, as an individual, sought to protect innocent life by plotting the death of Hitler. . . . We are certain that the counsel of restraint today will be regretted by those who look back on it in the future," he says.[54]

Hill has become an inspirational leader in his own right. He admonishes his followers not to remain at home, leaving others to respond to the "call from the womb." "Death opens her cavernous mouth before you," he says. "Thousands upon thousands of children are consumed by her every day. You have the ability to save some from being tossed into her gaping mouth. As hundreds are being rushed into eternity, other questions shrink in comparison to the weighty question 'Should we defend our born and unborn children with force?' Take defensive action!"[55]

It is difficult to visit Paul Hill. You need permission from the prison authority in Florida where Hill is incarcerated. And you need Hill to request that you come. Both the Reverend Donald Spitz and Michael Bray agree to write to Paul Hill, encouraging him to meet with me. I also write to Hill myself. Eventually I receive a letter from Hill. It is written in carefully lettered calligraphy. No errors, no smudges. Like a wedding invitation.

> Dear Ms. Stern,
>
> Thank you for your letter of April 15. I was glad to approve your request for an interview. . . . I am glad to hear you are interested in the Christian Reconstruction movement. My worldview is based on Reconstruction principles. I will, thus, be happy to answer your questions on the subject. Since I will, hopefully, get to meet you soon, and be able to discuss these matters in depth, I will not now descend into particulars. I am, however, looking forward to meeting you.
>
> Sincerely, Paul J. Hill[56]

Florida State Prison is located in Starke, fifty miles southwest of Jacksonville. The prison is surrounded by rows of chain-link fences, razor wire, and guard towers. A colleague who accompanies me and I pick up passes from the guardhouse, and a prison official directs us to the building where death row prisoners meet with visitors. We have to pass through a metal detector so sensitive that the steel in my heels triggers the alarm. I walk through in my stocking feet while a security officer inspects my shoes.

A guard walks us to the room where we are to meet with Hill, inmate number 459364. Hill is the only one of the fifty-four prisoners on death row who remains entirely unrepentant for his crime, the guard tells us, with a look of irritation and perhaps puzzlement.

Hill is waiting for us in a room next to the meeting room. He is wearing a neon-orange prison shirt that looks as though it would be visible in the dark, blue athletic pants, and sneakers. When he comes into our room, his hands are cuffed behind his back, but the guard recuffs them in front of his body. I expect to see shame or resistance or pain in Hill's features when the guard locks the cuffs in our presence, but I see something like pride, or maybe glee, instead. Hill appears entirely at ease. I sense that his submission to the cuffs gives him the feeling that he has the moral advantage. Perhaps he finds life easier not having to worry about what to do with his hands.

I have come to the prison having read Hill's manifestos, and I ask him about an apparent inconsistency in his work. If you favor defensive action rather than vengeance, why didn't you try to incapacitate the doctor? I ask.

"If I wounded him, just shot him in the leg or shoulder, I knew there was an excellent probability that he would return to killing innocent children." He pauses, then adds, "In my thinking, it just became: I had to kill him. . . . I was totally justified in shooting the abortionist, because he was actually the one perpetrating the violence." Moreover, he says, it was an act of defense, not an act of violence. "I would not characterize force being used to defend the unborn as violence."

We ask him whether the antiabortion movement will be successful. He says yes. "Christ's kingdom and principles will ultimately prevail. God is in control—he will bring about victory—we must obey him. Sooner or later America will become a Christian nation. Only Christians will be elected to public office. No false worship allowed."

Do you advocate killing Supreme Court justices? I ask.

"Killing Supreme Court justices, considering the majority of them favor mass murder . . . It's hard for me to escape the conclusion it would be just for someone to kill them. But I'm not altogether certain it would be wise," he says primly.

The Army of God manual promotes the use of chemical and biological weapons, I point out. Would you support their use?

"Yes, yes, I wouldn't want to rule those out. I'd want, of course, to use them wisely to try to minimize unnecessary harm. . . . If you sent [the weapons] to an abortion clinic, I would think your chances of harming an innocent person would be greatly reduced." It is only much later that I understand why he referred to *sending* weapons to a clinic; violent antiabortion activists have specialized in sending chemicals and biological-agent simulants and hoaxes to abortion clinics through the mail. According to the National Abortion Federation, 2001 saw a sharp increase in the number of anthrax hoaxes. In October and November 2001 alone, 550 anthrax-threat letters were received by abortion clinics and other women's health organizations.[57]

I ask Hill whether he sees himself as a martyr. "Yes," he says, "I would be willing to die to promote the truth. I am glad to do so, standing for principles for which I stand." His excitement causes him to speak in a slightly officious style. "I'm not resisting their efforts to kill me," he explains, by which he means he is not appealing his death sentence. "The heightened threat, the more difficulties forced on a Christian, the more joy I experience if I respond appropriately."

He tells us that he has sacrificed his life in the service of promoting good, and this knowledge has left him "experiencing more joy and inner peace and satisfaction" than ever before in his life. "I think it's because of the increased adversity." Knowing that what I do is "for Christ's sake makes it an experience I can rejoice in. I can rejoice and give thanks for the privilege of suffering." He claims to feel no remorse, professing he would do the same thing again. "I wouldn't advise them to give me my shotgun back . . . unless they wanted a similar outcome. I feel what I did was right.[58]

"There are many things I go through that can legitimately be compared to what Jesus went through," he says, summarizing his views on the topic of martyrdom.

You once predicted the emergence of a kind of pro-life IRA. Do you believe this terrorist organization has emerged? I ask.

"I would hope that a few people making symbolic acts such as the one I made would cause people to come to grips with the issue [of abortion]. And the thing could be resolved without causing undue chaos. . . . But as time goes on, there will be more and more need of war.

"There is absolutely no question that an example is one of the best teachers, and there is also no question that I hope others will act in ways similar to the way I acted. So, yeah, I hope to encourage others to defend the unborn as much as I did," he elaborates in a subsequent interview. Hill hopes, through his example, to inspire "justifiable homicide at a butchertorium," what most Americans would call murder.[59]

Jonathan O'Toole seems to be getting the intended message. Every time he thinks of Hill, he feels he's not doing enough to stop baby murder. "It really puts me to shame," he says.[60]

After spending time with members and leaders of the save-the-babies movement, I had a pretty good sense of how leaders inspire followers to take violent action, even when they cannot offer material rewards in return for participation. As we have seen, inspirational leaders create a narrative and a secret language, which they use to create a community of like-minded believers, very much like a "normal" religion or church community. But unlike most churches, the aim is to inspire followers to take *violent* action on behalf of the in-group in opposition to an out-group.

Some terrorists are even more "leaderless" than the individuals discussed in this chapter. Lone-wolf avengers are often inspired by strains of anomie expressed on the street or on the Internet, in addition to personal grievances. They may sympathize with the grievances expressed by particular terrorist movements, or they may choose complaints and goals from several movements, creating a kind of patchwork movement of their own. After studying the doctor killers and virtual networks, I turned to assess the problem of lone wolves, which we discuss in the chapter that follows.

SEVEN

Lone-Wolf Avengers

This chapter tells the story of two lone-wolf avengers—Mir Aimal Kansi, a Pakistani immigrant to the United States who shot several CIA employees in 1993; and James Dalton Bell, whose various schemes for ridding the world of his purported enemies are at the cutting edge of the virtual-network organizational style. Although only two people died in Mir Aimal Kansi's attack, it significantly affected how CIA employees view the safety of their workplace. Analysts became aware, in a visceral way, that they are vulnerable to lone-wolf shooters; that even in their cars driving to work, they may suddenly find themselves combatants in some terrorist's war.

Lone wolves often come up with their own ideologies that combine personal vendettas with religious or political grievances. For example, John Allen Muhammad, who, together with a seventeen-year-old protégé who called him Dad, carried out a series of sniper shootings in suburban Washington, D.C., in the fall of 2002, appears to have been motivated by a mixture of personal and political grievances. He told a friend that he endorsed the September 11 attack and expressed admiration of the small group that had managed to cause more damage to the United States than an army could have done. He said that he disapproved of U.S. policy abroad, especially in regard to Muslim states. But he appears to have been motivated principally by anger at his ex-wife for keeping him from seeing their children, and some of his victims were personal enemies.[1]

There is a limit to the damage a lone-wolf avenger can cause. An individual can terrorize a city, as the sniper case makes clear. But he could not carry out a September 11–type attack, which required coordination among a large number of operatives and supporters. Lone wolves are especially difficult for law-enforcement authorities to stop, however. As military technology continues to improve and spread, enabling what political scientist Joseph Nye calls the "privatization of war,"[2] virtual networks and even lone-wolf avengers could become a major threat.

On the morning of January 25, 1993, a Pakistani immigrant named Mir Aimal Kansi walked into rush-hour traffic and fired a Chinese-made AK-47 at commuters waiting to enter CIA headquarters in Langley, Virginia. Lansing Bennett, sixty-six, a physician and intelligence analyst, and Frank Darling, twenty-eight, a communications officer in the covert operations branch, died. Three other people were wounded. Although Kansi seemed at first to be shooting randomly at drivers, he went back to Frank Darling's car and shot him many times, making sure that Darling was dead.

Eight hours later Kansi walked into a grocery store in Herndon, Virginia, where he was a regular customer, and asked the proprietor to procure him a one-way ticket to Pakistan. The owner of the store, Mohammad Yousaf, made some phone calls and obtained the requested ticket for a flight that left the following day. When Kansi returned to pick up his $740 ticket, which he bought with cash, he asked Yousaf to order him a cab to the airport. But Yousaf, who lives near the airport, offered to give him a ride. Yousaf noticed that Kansi was wearing slacks and a sweater over a shirt, and that he had no luggage at all. He recalled asking him, "You are going to Pakistan with no gifts or anything?" and that Kansi replied that he didn't need anything. "He was quiet. Nothing special," Yousaf told the *Washington Post.* "I did not have even the slightest notion of suspicion."[3] At 5 P.M. on January 26, Kansi was already on the plane when authorities began disseminating sketches of the killer based on witnesses' accounts.

In the weeks prior to the attack, Kansi had bought ammunition, two handguns, and a Colt AR-15 assault rifle, which he subsequently exchanged for an AK-47 assault rifle. He also ordered a bulletproof vest.[4]

Kansi had spent most of his time in the United States inside the Pakistani expatriate community in northern Virginia. He rented rooms from expatriates and worked for their companies. But he never really found his way. His acquaintances described him as socially awkward. He had been involved in a militant organization dedicated to creating a "greater Pakhtunistan," a new nation-state comprised of Pashtuns from both Pakistan and Afghanistan. People in the Pakistani expatriate community didn't like that, Yousaf explained.[5]

When authorities captured Kansi four years later, he explained that he shot the CIA officers to protest mistreatment of Muslims in Palestine and elsewhere in the world.[6]

In June 1999, I wrote to Kansi at Sussex One State Prison in Waverly, Virginia, where he was then on death row, requesting to speak with him. At first Mr. Kansi said he was willing to speak with me, but only on condition that I pay him for the interview. Later, he said he would accept a donation to an Islamic charity in lieu of payment. When I told him I would not be able to pay him for an interview, even with a charitable donation, he decided to meet with me free of charge. With each exchange, Kansi grew more enthusiastic about the prospect of meeting me, telling me in his last letter before we met, "I hope your book becomes one of the best sellers in US and you become a millionaire, so rich and travel in a nice new Mercedes Benz." He also told me to request a visit in the "noncontact visiting place," so that we would be able to see one another. Otherwise, he told me, they would bring a phone into the death row block and we would not be able to see each other during the conversation.

On November 7, 1999, I travel to the prison. I arrive early. The guards inspect my identification cards and instruct me to walk through a metal detector. They direct me to walk, alone, to the noncontact visiting area, where Kansi is already waiting for me, also early. Although the visiting area is close to the entrance, I get slightly disoriented and have to ask directions a second time. A guard tells me, in a slightly patronizing tone, first door to your right, it will be open.

Kansi is in a kind of glass cell in the far-left corner of a large room. The room is freshly painted a blinding white and smells of Lysol. Kansi seems oddly happy to see me, as if he hadn't seen an outsider for a long time.

Why did you attack the CIA? I ask. Were your motivations religious or political?

"I attacked the CIA for both religious and political reasons," he says. "In 1993 the U.S. government was fully supporting Israel. Israel oppresses Palestinian Muslims. Therefore it is a religious duty for all Muslims to help the Palestinians. Also the United States was attacking Iraq. After the withdrawal of Iraqi forces from Kuwait, there was no need to persist in attacking Iraq."

He tells me that American policies are "anti-Islamic" worldwide. His opposition to the United States dates from its support of Zia-ul-Haq's military regime and its involvement in the Afghan war against the Soviet Union. "I was against foreign powers in Afghanistan," he says.

Kansi's father was a Pashtun tribal leader. Kansi became passionately involved in a series of political groups, including Pashtun nationalist ones, while studying in Quetta, Pakistan. The one fixed element, according to a relative, was his anti-Americanism.[7]

"I did not want to kill ordinary Americans," Kansi says. "Only government officials. They are not normal people—they represent the government. Therefore they are legitimate targets for attack."

I wonder whether it is just U.S. government officials whom he considers "not normal" human beings. Do you know any government officials in Pakistan? I ask. Did you perceive them as abnormal?

"Yes. My own brothers and sisters and other relatives worked for the government of Baluchistan. They were different from ordinary Pakistanis."

But the victims were human, I say.

"Yes, they are beings of God. But there is a difference between Muslims and non-Muslims. Non-Muslims deny the last prophet. They don't surrender to the orders of God. They are rebellious people. Non-Muslims work against Islamic countries."

How did you know that you weren't attacking Muslims? What if a Muslim was working for the CIA? I ask

"I was one hundred percent sure—no true Muslim would be working for the CIA."

But the officials you killed have families. The ones you killed have children, mothers, fathers, sisters, and brothers they left behind, I say.

"When I think about the family members of the victims, it troubles me," he concedes. "But when I think about the damage the U.S. government has caused Muslims, it's much worse than what I did."

Were you involved with any of the jihadi groups in Pakistan or Afghanistan?

"I met members of Harkat-ul-Ansar, Hizb-ul Mujahideen, Lashkar e Taiba. I spent a long time in Afghanistan. I know lots of these organizations. But I never joined any of these big groups."

Twenty-eight hours after a Virginia jury convicted Kansi of murder, five employees of a Texas-based oil and gas company were shot in their station wagon on the street in Karachi. Senior Pakistani police officers said the most likely motive was to avenge Kansi's conviction.[8] A previously unknown group called the Aimal Secret Action Committee took credit for the attack. Militants in Baluchistan, Kansi's home province, had vowed to seek revenge after Kansi was captured and brought to America to be tried, in violation of Pakistan's extradition law.[9] The U.S. government believes that the assailants were connected with the group Harkat-ul-Ansar, which the State Department had recently put on its list of foreign terrorist organizations.

Were you fighting a jihad when you attacked the CIA? I ask.

"No. This was a religious duty. But not jihad. I am not sure whether God will reward me for what I did. This was retaliation. It was revenge. What I did was between jihad and tribal revenge. This was like a tribal revenge. We go after people of the other tribe—not just the one who carried out an attack. Everyone in the other tribe is considered a legitimate target."

Kansi comes from a wealthy family in Quetta. His father, Malik Abdullah Jan Kansi, inherited extensive land holdings in Quetta and increased the family wealth through investments in real estate, construction, and a factory in Karachi. The father is widely believed to have helped the CIA and the Pakistani intelligence service funnel weapons to the Afghan mujahideen in the war against the Soviets. At the time, Quetta was a way station for arms shipments to the mujahideen.

According to an article published in the *New Yorker* in 1995, not only Kansi's father, but also Kansi himself may have had a relationship with the CIA. The *New Yorker* quoted a Pakistani intelligence official: "Abdullah Jan, at least one of his cousins, and two of his sons, including Aimal, were an integral part of the CIA-ISI weapons pipeline to the mujahideen."[10] Former Pakistani ISI chief Hamid Gul says, "Kansi grew up in Quetta, the southern base for the CIA's war in Afghanistan, and may . . . have been recruited by the CIA at some point."[11]

The CIA denies that it had any contact with Kansi, but officials from two Pakistani governments rejected its disclaimer. Judy Becker-Darling, the wife of one of the agents who was slain, wonders whether Kansi knew her husband and intended to murder him in particular. Darling had worked in Karachi at the height of the Afghan war. A tribal chief told the *New Yorker* that Malik Abdullah Jan Kansi (Kansi's father) had worked for the CIA for many years. "It's well-known among his friends that many of his businesses were set up by the CIA, and it's generally assumed that the Agency used them from time to time as fronts. Oh, he received a lot of goodies over the years, including the pledge that his son [Aimal] would take his place when he retired." Also suspicious is that Kansi reportedly entered the United States without being interviewed by the Immigration and Naturalization Service, which, according to INS officials, could only happen if he had been sponsored by a U.S. government agency.[12]

Kansi denies any connection with the CIA. He also says his father never worked for the Agency.[13] Perhaps he is lying.

"When I was on the run I felt really good," Kansi says. "I never thought of getting arrested. I didn't realize until the next day when I was in the newspaper that people had died. I felt normal—I didn't feel terrible. Just normal."

He tells me he went to Afghanistan to hide. "It was very easy to go over the border. There is no visa required, no passport. The best place to hide in the entire world is Afghanistan.

"I had a powerful radio. I listened to VOA. I heard that they had arrested [me] on Indian radio. They were wrong. This was so funny I couldn't resist telling a few of my friends. But I was afraid because of the reward money. The U.S. government was offering a lot of money. People in Afghanistan are very poor."

Kansi had been on America's ten-most-wanted fugitives list for four years when he was caught, reportedly with the help of information provided to the U.S. government by Afghani and Pakistani nationals in exchange for a reward of $2 million offered by the State Department's Counter-Terrorism Rewards Program.[14] President Clinton had requested Prime Minister Nawaz Sharif to allow U.S. agents to capture Kansi on Pakistani soil, and to take him directly to the United States, in violation of Pakistan's extradition law. The abrogation of the law infuriated Pakistanis—from human-rights activists to pro-jihadi groups. At least three suits were subsequently brought against the Pakistani government.[15]

Kansi's arrest remains a mystery in Pakistan, where different versions and theories abound as to who leaked information about Kansi's whereabouts, and how. According to one senior Pakistani government official, who spoke to me on condition of anonymity, members of Pakistan's National Accountability Bureau[16] (NAB) inadvertently discovered that $10,000 of the State Department's $2-million reward money had been deposited into the account of a junior ISI official in Baluchistan. According to the official who spoke to me, the junior ISI official revealed that he had been instructed by members of the ISI branch in Quetta to leak the information about Kansi's whereabouts to the United States. He apparently transferred the reward to an ISI account within twenty-four hours.[17]

Are you afraid of death? I ask.

"I don't feel afraid," Kansi says.

Kansi's father had three sons and four daughters with a first wife, and one child, Mir Aimal Kansi, with his second wife. Relatives described Kansi as a brooding and introspective young man, the loner in the family. He suffered from a seizure disorder as a child, but recovered by the time he was ten years old. After Kansi's mother died in 1982, he became even more isolated, his relatives said. When Kansi's father died in 1989, he inherited around $100,000, which he spent, in part, on his trip to the United States.[18]

I ask him to tell me about his upbringing.

"When I was a child, my friends and I used to go to the refugee camps in Pakistan. We used to shoot there. Shoot targets. Shoot in the air. I bought an AK-47 for target practice. I like guns very much. This is part of our culture. We always keep guns in our home. My father, grandfather, had guns. We practiced target shooting. There are many tribal conflicts."

What is the name of the refugee camp where you went to shoot? I ask.

"The name of the camp was Piralizai Jungle. All the refugees are Pashtun. At the camps they pray regularly, and they are trained to fire guns. I had many friends in the refugee camps. But I did not get involved in any of the big groups. They were fighting a jihad. My father did not allow me to fight in the jihad. I was completely ready to go to fight in Afghanistan, but my father would not allow me."

Later, I ask a Pakistani government official to check the name of the camp, to make sure I got the spelling right. He tells me he knows all about this camp because it was "a top den for narcotics dealers. The highest consumption of heroin is in this area." The camp, which is in Pishin, is far away from Quetta, where the Kansi family lived. It is surprising, he tells me, that Kansi's father didn't send him to a local camp for shooting exercises, since there are many such camps nearby. It is possible that Kansi's father had business at Piralizai Jungle camp, he says, pointing out that the ISI has long used drug money to fund its operations, much like other intelligence agencies in the world. Kansi's mentioning of Piralizai Jungle may well have been "a slip of the tongue," he says.[19]

What about your schooling? I ask.

Kansi tells me he went to school at Saint Francis grammar school, the best school in Quetta, his hometown. Then he earned a bachelor's degree in political science from a government-run college in Quetta. "There was a lot of cheating on exams there," he says. "And a lot of politics. I didn't study much. I was a member of the Pashtun student association. Fighting for rights of Pashtuns in Pakistan." After that he earned a master's in English literature at the University of Baluchistan. What kind of literature? I ask. "Shakespeare. Poetry of Milton," he says.

What is your favorite book?

"Macbeth."

What are you reading now?

"I'm not reading. I watch the news all day."

Whom do you admire most?

"I like Osama bin Laden. He is demanding that foreign forces leave his country. He stands up for all Muslims.

"Our society has gone away from religious values a little bit. But here is very far from religious values. The prisoners here know nothing about religion. Society becomes more and more materialist. Religious people are better people. Here everyone is very materialistic—all they care about is acquiring wealth."

Soon after the interview, Kansi sends me a letter: "After talking to you I realized that you have knowledge about Islam which made me happy. I would like to request you to come to Islam and live and die as a Muslim believer."

I send him some more questions in writing in response: What was the message you were trying to send by shooting CIA officials, and whom were you trying to reach?

"The message was this—that if you keep on supporting Israel and Israel oppresses Palestinians (Muslims) that your own government officials can also get hurt and suffer the consequences of your wrong policy toward Muslims (Palestinians)."

He also wrote, "I was more interested in attacking the Israeli embassy in Washington DC. That was my target. . . . I went to Israeli embassy in Washington in my car (pick up truck) but the embassy was no good for one-man rifle attack. It was a good place for a bomb attack, to blow it up completely, but I did not know how to make a bomb. If I had the knowledge of making a bomb, I guarantee that I would have blown up the Israeli embassy. The CIA was my second target, the outside place of CIA is big roads and that place was good for one-man rifle attack because there you can easily shoot CIA officials who are in their car on the left turning lanes on stop light outside CIA. I don't like killing ordinary American people, as they don't have big role in making policy of US toward Israel or against Islamic countries. I believe the people who should be attacked should be government officials or senators or congressmen or people in CIA, Pentagon, White House, etc."

What if a respected Islamic scholar told you it would be wrong to shoot CIA employees? I asked him in writing.

"If a respected Islamic scholar would have told me not to do it then I would have asked him questions, and if he would have satisfied me completely then I would have not done it. Otherwise I would have done it," Kansi wrote.

What if your mother asked you not to proceed?

"If my mother would have been alive, she would have got me married and I would have never been in the US. I would have been living in the Pakistan with my mother and wife."

He closed by saying, "I think I have answered your questions to the best of my ability, although I am not a journalist or politician or an Islamic Scholar."

Mir Aimal Kansi is an example of a growing trend: lone actors or small groups who commit terrorist crimes, inspired by a terrorist ideology, but not belonging to established terrorist groups. Kansi was even more leaderless than members of the save-the-babies movement. He seems to have been moved, at least in part, by the anti-American fervor he was exposed to in his youth. However, terrorists often use slogans of various kinds to mask their true motives. It is, therefore, not inconceivable that Kansi's primary motivation was to exact personal revenge against an organization he believed had betrayed his father. As one Pakistani official explained, "Baluchistan, where Kansi was born, has a very strong tribal culture, and revenge is a central part of the ethos."[20] When Kansi says he was seeking revenge, was it for some perceived slight—either to his father or to himself? We may never know. Kansi was executed by lethal injection on November 14, 2002.[21]

One of the best examples of a lone-wolf avenger is James Dalton Bell, an MIT-trained chemist who got angry with the U.S. government and wanted to take revenge. He came up with a scheme to use virtual networks to rid the world of "miscreants" and "slimeballs," his terms for government officials and other political enemies. Although he is not a religious terrorist, he is both a virtuoso lone-wolf avenger and a budding inspirational leader, and worth discussing for that reason.[22]

The scheme involves the creation of an Internet-based organization that would reward people who correctly "predict" the death of a "miscreant" in digital cash. Sympathizers could contribute to the creation of a miscreant-free world by sending charitable donations in digital cash.

The organization would not actually exist except in virtual form, and every communication would be encrypted. The plan involves "the ultimate in compartmentalization of information," Bell explains. "It is very likely that none of the participants, with the (understandable) hypothetical exception of a 'predictor' who happens to know that he is also a murderer, could actually be considered 'guilty' of any violation of black-letter law . . . in the plan I describe, none of the participants *agrees* with ANYONE to commit a crime."[23]

Bell calls the virtual terrorism scheme "the solution to wars, nuclear weapons, militaries, politicians, tyrannies, dictators, holocausts, governments, taxes, and at the very least a substantial fraction of crime. The fix. The cure. The complete and total repair job. The last correction." He describes his essay as "not really a paper; it's more like a forecast. A manifesto. A warning. A promise . . . The word *inevitable* was practically *invented* for it."[24]

When federal agents executed a search warrant on Bell's Vancouver, Washington, home, they found a variety of chemicals, including diisopropyl fluorophosphate, a chemical that could be used to make a nerve gas similar to sarin.[25] Government officials believe that Bell made the nerve agent sarin, but cannot prove it with publicly available information. They cite e-mail messages retrieved from Bell's computer in which Bell claims to have produced sarin in the basement of his residence.[26] Officials claim they have information not in the public record, which is now protected by an agreement Bell struck in exchange for a guilty plea for a series of threats and actions against the Internal Revenue Service.

One of Bell's most "ambitious" projects was to develop and market a material that would destroy enemy computer systems.[27] He discovered that nickel-plated carbon fiber is electrically conductive and that airborne fibers can short-circuit electrical equipment.[28] Bell learned about this property of the fiber from a safety sheet enclosed with some fiber that he had bought for building model airplanes.[29] He and a friend had begun mapping out a strategy for testing and marketing the fiber, which they hoped to sell to "nefarious individuals" for use against "large vulnerable target[s] like the IRS."[30] In addition to the marketing-strategy discussions, which authorities found on Bell's computer, they also discovered that Bell and his friend had already bought some of the fiber.[31]

Bell had "hypothetical" discussions about contaminating city water supplies with another friend named Greg Daly, according to what Daly told investigators.[32] At that time, Daly worked for the city of Portland's Bureau of General Services, which carries out maintenance at the Bull Run water treatment facility, and claimed direct access to the plant.[33] Daly no longer worked at the plant, but he hinted in interviews that he still had keys to the facility.[34] Daly told investigators that Bell had been trying to extract botulinum toxin from green beans.[35] He also said that Bell boasted that making chemical weapons would be easy, and that he planned to order chemical precursors from a catalog.[36] He said that Bell told him that he had acquired a few milligrams of methyl phosphonyl dichloride, a direct precursor to sarin, and that he had managed to synthesize a small quantity of chemical agent.[37]

I wrote to Bell during his first incarceration in 1998, asking whether he was willing to talk with me about the scheme. He responded over a year later, telling me I could interview him by telephone. Beginning in February 2000, we had numerous telephone conversations.

"Terrorism is an overreaction to a legitimate problem, and that problem is called government," Bell tells me immediately.[38] "A lot of people think you have to have a government. I don't think we need one—even for defense."[39]

We talk about other lone-wolf terrorists. What do you think about Theodore Kaczynski? I ask.

"I haven't read his whole essay," Bell says. "I read about the first three paragraphs. It began like too many academic papers that I've read—I got bored after three paragraphs. His primary objection seems to be technology . . . I don't agree with that philosophy. I think technology is wonderful. Computers at the time were the products of government, big business. The computer was widely seen as the product of an oppressive organization—government and big business. Now the opposite is true. Now computers are on the side of individuals."

In the academic community we talk about how the Internet facilitates the development of virtual communities. What do you think about the idea of virtual community? I ask.

"People have obscure interests and desires. The Internet allows people with unusual interests to get together."

Is the Internet increasing the strength of the antigovernment movement?

"I'm as big a fan as it is possible to be of the Internet," he says, arguing that it "dramatically increased" the strength of antigovernment movements. It has dealt a decimating blow to the government's strength, he says, a blow they haven't even noticed. "Historically people couldn't talk to others around the world. To get your story out—maybe you'd write a letter to the editor. Today, anybody can get his or her word out." The Internet means that "the story can't be killed," Bell tells me.

Bell seems to understand intuitively that a good story is a critical component of inspirational leadership and of building a virtual network. The story Bell tells is that tax authorities are stealing our money: "Think how much the IRS is stealing from you. On a per dollar basis [they] victimize you far more than street crime." He wants to rid America of IRS "terrorists" who steal ordinary Americans' hard-earned cash, only to waste it on unnecessary projects like national defense.[40] Like the inspirational leaders in the save-the-babies movement, Bell has developed his own language. The villains are the "slimeballs" and "miscreants" who work in the government. The heroes are the people strong enough to take action against them.

Bell occasionally participated as a "juror" in "trials" at a common-law court in Multnomah County, Oregon. Bell and his fellow jurors found a number of IRS employees and government officials guilty of theft and conspiracy, and of violating amendments to the American Constitution. They demanded an award of $100,000.[41] In November 1996, Bell sent a letter to the IRS's Ogden Service Center to demand a large tax refund. He gave the IRS two months to expedite his refund, warning them that he was prepared to take the matter up with his local common-law court "for final disposition."[42] At a January 1997 meeting of his common-law associates, Bell distributed computer discs of his "assassination politics" essay, labeled "AP: A Solution to the Common Law Court Enforcement Problem," which he had already been publicizing on the Internet.[43]

Virtual networks enable violent individuals who are socially ill at ease to work together on a common political or religious cause without having to meet face-to-face. Experts claim that schizophrenics and sociopaths may *want* to commit acts of mass destruction, but they are probably the least likely to succeed because of their difficulty functioning in groups.[44] Such individuals are often prone to "political paranoia," tending toward extreme suspiciousness, megalomania, and grandiosity.[45] They tend also to feel victimized. They are often persuaded that the enemy camp—whether the government or a rival religious group—is not only out to get them but is monitoring their every move. Once lone-wolf avengers prove themselves to be dangerous, their conviction that the government is out to get them is likely to become true, at least to some extent. But that does not make them less paranoid.[46] As their paranoia increases, such individuals may become more violent. Until now, individuals have been unable to do a great deal of harm. But methods for producing crude weapons of mass destruction are now widely available.

The prospect that well-trained lone-wolf avengers or small networks could get involved in biological weapons attacks is especially worrisome, especially in light of the fall 2001 anthrax attacks in the United States, which infected eighteen people, five of whom died.[47] Inputs to biological weapons are inherently "dual-use." Unlike special nuclear materials, which are man-made and produced only at government-sanctioned facilities, biological agents (with the single exception of variola virus, the causative agent of smallpox) exist in the environment. John Collier, a leading expert on anthrax at Harvard Medical School, points out that virtually any microbiologist could isolate anthrax spores, which persist in soil for decades. "You are never going to be able to eradicate them from nature," he says.[48] Listed pathogens are used in thousands of clinical and diagnostic laboratories. The same equipment used to produce beer, for example, could be used to produce biological agents. The underlying research and technology base is available to a rapidly growing and increasingly international technical community.[49]

In response to a neo-Nazi's acquisition of *Yersinia pestis* from an American germ bank in 1995, the U.S. government tightened up the rules for shippers and receivers of select agents, the pathogens that the government considers especially dangerous.[50] But cultures are also available from germ banks outside the United States. In the fall of 2001, the World Federation for Culture Collections urged its 472 members to tighten access to dangerous microbes, but the organization is not empowered to demand compliance.[51] More than a thousand germ banks around the world do not belong to the federation, and few of them are adequately regulated or secured.[52] And because of the difficulty of detecting freeze-dried pathogens, the ability of U.S. Customs to stop illegal imports of small quantities of pathogens, such as seed cultures, is minimal.[53]

Within a week after his release from prison, Bell commenced a new campaign against his enemies, which he called a "Thug hunt." He was determined to locate the home addresses of his "slimeball" enemies, in particular IRS and ATF employees who had been involved in his earlier arrest.[54] Although he did not find the homes of his intended targets, he continued his Thug hunt even after the IRS carried out a search of his residence.[55] Bell was rearrested on November 17, 2000, on charges of stalking government officials "with the intent to injure or harass" causing them "reasonable fear of death or serious injury."[56] Bell is now in a federal prison in Lampac, California, serving a ten-year sentence.

So far, Bell has had only modest success as an inspirational leader. Carl Edward Johnson, a forty-nine-year-old man with whom Bell exchanged e-mails through a cypherpunk chat group, vowed to take "personal action" on Bell's behalf after his first arrest. Johnson established a Web site he named Dead Lucky, which offered specific amounts of "eCa$h" for the deaths of Jeff Gordon, the IRS inspector who had led the investigation of Bell, and two other IRS employees. He was ultimately convicted of sending anonymous e-mail threats to the judges involved in Bell's case, and also to Microsoft chairman Bill Gates.[57]

Authorities' biggest fear is that Bell would inspire others to develop and use chemical or biological agents as a means for creating a miscreant-free world. After the anthrax letter attacks of fall 2001, it is a risk that cannot be ignored. Perhaps the most frightening prospect is an organization that combines the strengths of virtual networks and lone-wolf avengers (resilience to law-enforcement penetration) and commander-and-cadre organizations (capacity to carry out complex, large-scale attacks), which I investigate next.

EIGHT

Commanders and Their Cadres

In the spring of 2000, I applied for a grant from Harvard University's Center for Public Leadership to study leadership of terrorist organizations. The Center for Public Leadership (CPL) provides a forum for students, scholars, and practitioners committed to the idea that effective public leadership is essential to the common good.[1] Although the leaders I was studying were hardly contributing to what the center's directors or I consider to be the common good, one of the directors told me he was fascinated by the project, and wanted to see where I would go with it. He awarded me a small grant.

A few months later, I returned to Pakistan, hoping to learn about how leadership is practiced by the commanders of the jihadi groups I had come to know, and how it differs from what American militants call leaderless resistance in virtual networks. Through conversations with leaders of several organizations, we will learn in this chapter about the incentives—both positive and negative—on offer to managers and cadres. Rewards for participating in the organization include regular salaries for managers, cash bonuses for successful operations and payments to the families of "martyrs," various levels of training, "glamour," the opportunity to be part of a tight-knit community and to serve the group, and, from the operatives' perspective, to serve God. Penalties for disobedience can include corporal punishment or death.

The chapter begins with a discussion with one of the leaders of Harkat-ul-Mujahideen (HUM), a member of the International Islamic Front for Jihad against the Jews and Crusaders, the umbrella organization formed by Osama bin Laden in 1998. Despite its membership in bin Laden's organization, the group was essentially unknown to the American public until after September 11. Harkat and its splinter groups are suspected in connection with a series of major attacks in India, the kidnapping and murder of a number of Westerners, the hijacking of an Indian airliner in December 1999, and the murder of *Wall Street Journal* reporter Daniel Pearl in 2002. John Walker Lindh, the "American Taliban," trained at a Harkat camp, as did many Al Qaeda members. In this chapter we learn how leaders of Harkat and other jihadi groups established links with a leading Indian organized criminal, a relationship that benefited both sides. Ansari, the Indian mafia don, is suspected of transferring $100,000 to a member of a Harkat splinter group, who, in turn, wired the funds to Mohammad Atta, the lead hijacker in the September 11 strikes.

The terrorists discussed in this chapter boast about successful fundraising efforts not only in the Gulf but also in Iran. One of the leaders reports that he has raised more money than he knows how to use, much of it from Islamist nongovernment organizations. He admits to having put sleepers in place in various countries and claims informal linkages with Hamas, Hezbollah, and other terrorist groups worldwide. Several managers concede that they joined the "jihad" for religious reasons, but that, over time, the salaries they earn have become more important in explaining their loyalty as holy warriors. Several talk about their disenchantment with militancy upon realizing that their leaders were less committed to the cause than to their own financial well-being. The chapter ends with a visit to several extremist *madrassahs,* or religious schools, where young men—usually from very poor families—are persuaded to join jihadi organizations.

* * *

I had been hoping to meet the leader of Harkat-ul-Mujahideen, Fazlur Rahman Khalil, ever since I had met an operative from the group in February 1999, on my first trip to Lahore. By the summer of 2000, my guide, Muzamal Suherwardy, said he would be able to arrange a meeting, presumably because the ISI had instructed Khalil to talk with me.

Fazlur Rahman Khalil considers Osama bin Laden a friend. They met during the Afghan War against the Soviet Union. His organization has offices all over Pakistan, including in Muzaffarabad, Karachi, and Multan.

I meet Khalil in his Islamabad office, which is close to—but not in—a wealthy part of Islamabad. A servant directs me to a large receiving room in the back of the house, instructing me to wait there. The room has no furniture and I am expected to sit on the floor. The office is extraordinarily dirty. The gray-white walls are oppressively dingy, and the gray rugs are tracked with grease. On the walls, brightly colored posters depicting shiny Kalashnikovs provide the only visual relief from the dirt and gray.

I feel drained and grimy from the heat and dirt of the city. I lean against the wall, but notice that my ankles are dangerously revealed. I pull down my *kameez* to cover them and then pull my knees closer to my chest. The combination of the atmosphere in the house and the jihadis' neglecting me and urging me to sit on the dirty floor makes me feel humiliated, like a child being disciplined. I get tired of waiting, and nobody stops me from wandering through the hallways. I meander into an inner courtyard where a jeep apparently awaits rebuilding. Some of its metal flesh is torn off completely; the roof is caved in and its doors are torn out with only shards remaining. The jeep is a grim reminder that not all of Pakistan is as peaceful as Islamabad, and that HUM considers itself to be at war. Still, why is the jeep here in Islamabad rather than in Muzaffarabad? I wonder. It seems unlikely that it was bombed in Islamabad. Is this some kind of theater?

Six or seven men walk by, all with impressive beards, all looking angry. I have the strong impression that they have summoned up this fierce attitude to intimidate me, and that it would be difficult for them to maintain such looks for long. They wear the headgear and vests of mujahideen, which contain pockets for ammunition.

Finally Khalil enters the reception room, accompanied by a burly, bearded, black-haired guard. Khalil's beard is graying, but his shoulders are broad and his body highly muscled. He looks to be in his mid to late fifties, and I am shocked to discover he is actually younger than I: thirty-seven. His vest is different from his comrades'. Theirs are made of khaki cotton, but Khalil's is made of netting with camouflage accents. I think to myself, this is *mujaheed* couture—the trappings of aging power. Carlos the Jackal reportedly got liposuction in his later years.

I offer Khalil a Harvard pen as a sort of peace offering. I mean this pen to express certain complicated feelings. I don't approve of your tactics, especially your practice of persuading young men to donate their lives to a losing battle. But I am ready to listen to you, to try to understand you, and to write about you as objectively as I can. Khalil jokes that he would like to offer me a machine gun in return. I tell him I might have trouble at customs. He starts by giving me the party line about a list of subjects he knows interest Americans. "We have no camps in Afghanistan. If Afghanistan tries to shut down its training camps, that is a good thing, if the camps exist," he says, mimicking a recent, obviously false statement by the Afghan government. By this time, the U.S. government had already targeted one of Harkat's camps in a retaliatory raid against bin Laden for his August 1998 African-embassy bombings.[2] "Nor," he claims, "do we have camps in Azad [Pakistan-held] Kashmir. The Pakistani army is not facilitating the jihadi groups' crossing into India. It is a long border covered with moth-infested forests. If the Indian army can't stop the mujahideen from crossing into India, how can the much smaller Pakistani army stop them? We have no relationship with the ISI. The ISI is in cahoots with the CIA anyway. And the ISI has no involvement in Afghanistan or in Kashmir. America should not be afraid to talk to us. We are not terrorists. If being a Muslim means I'm a terrorist, then I'm proud to be a terrorist."

A servant brings a pitcher of water and glasses. My guide asks him to bring tea for me because I cannot drink unboiled water. The servant, who appears to be untroubled by concerns about cleanliness, returns with a dirty-looking thermos filled with sweetened tea and buffalo milk. He avoids looking at me as he pours the sweet liquid into a chipped green china cup. I wonder if my hosts would poison me, but immediately banish that thought from my mind. I tell myself that they are unlikely to kill a woman, even one dressed in boy's sandals. And that there would be easier ways to kill me if they chose to. Next I turn to worrying about whether the cup was washed or the milk pasteurized. Eventually my fear of offending my host takes hold, and I take small sips, hoping, absurdly, to sip around the germs.

HUM claims to be active in Bosnia, Chechnya, India, Myanmar, the Philippines, and Tajikistan. U.S. government officials say that HUM has targeted Western military officials in Bosnia, and India accuses HUM of carrying out "dirty tricks," including murders in India on behalf of Pakistan's Interservice Intelligence Agency (ISI). (In turn, the ISI accuses India's intelligence agency of similar activities in Pakistan, usually in connection with the violence in Sindh.)

Before 1997, HUM was known as Harkat-ul-Ansar, an organization formed in 1993 with the merger of two smaller groups. After an apparent Harkat front group, calling itself al Faran, admitted its involvement in the kidnapping and killing of Western tourists in 1995, the State Department listed Harkat as a foreign terrorist organization (FTO). At that point the group took the name of one of its earlier subsidiaries, Harkat-ul-Mujahideen, which had been founded in 1985 to fight Soviet forces in Afghanistan. One of HUM's predecessor organizations, Harkat ul Jihadi-i-Islami (HUJI), is still active and considered particularly violent. In the rest of this chapter I will refer to all the Harkat splinter and merger groups as Harkat, rather than HUA, HUM, or HUJI. Because the splinter group known as Jaish-i-Muhammad, which broke off from Harkat in 2001, has become important in its own right, Jaish-i-Muhammad is the one splinter group I will identify by name.

The various Harkat groups are suspected by the State Department of carrying out a series of kidnappings and killings of Western tourists in Kashmir, as well as killing two American diplomats in the Pakistani coastal city of Karachi. Harkat was suspected of murdering four American oil company workers after the conviction of Mir Aimal Kansi. Many Harkat members are alumni of bin Laden's training camps in Afghanistan's Khost province known as Al Badr I and Al Badr II, which, when the Taliban took over Afghanistan, were transferred from Jamaat-i-Islami to the Harkat groups.[3]

In December 1999, Indian Airlines flight IC 814 was hijacked while en route from Kathmandu to Delhi. The U.S. State Department reports that members of Harkat "were associated with the hijacking."[4] The plane made a series of stops in Pakistan and Dubai, ultimately stopping in the Afghan city of Kandahar. Devi Sharan, captain of the jet, recalls the hijackers praising bin Laden and the fight in Kashmir.[5] They were armed with box cutters and knives. They killed one passenger by slitting his throat, threatening to kill the rest if their demands were not met. Their principal demand was that India release thirty-six Pakistani militants then held in Indian prisons, eventually settling on three: Maulana Masood Azhar, Harkat's chief ideologue; Ahmed Omar Saeed Sheikh, a British-born Harkat operative indicted in a U.S. court for the kidnapping of an American citizen in India in 1994; and Mushtaq Ahmed Zargar, leader of another jihadi group known as al Umar Mujahideen. The name Ahmed Omar Saeed Sheikh would later become well-known to Americans because of his involvement in the kidnapping and murder of *Wall Street Journal* reporter Daniel Pearl in early 2002.

While the plane was on the ground in Kandahar, the hijackers used cell phones to communicate with their coconspirators. Indian government officials who intercepted the phone calls say that the group spoke with contacts in Bombay, who helped to oversee the operation, and with contacts in Dubai, who forwarded communications to Pakistan. India eventually arrested the group of coconspirators in Bombay. They discovered that the group in Dubai had not only facilitated communication with Pakistan, but had also helped to fund the operation. Mullah Omar, the Taliban leader, was reportedly personally involved. The Taliban provided guns to the hijackers after the plane landed.[6]

India's foreign minister flew to Kandahar personally to deliver the three terrorist prisoners whose release the hijackers were demanding, two of whom were prominent members of Harkat. Taliban forces surrounding the plane smuggled the hijackers and the three terrorists away, rather than detaining the terrorists as the Indian government expected.[7] One of the five hijackers released by the Taliban in 1999 is suspected of being involved in the September 11, 2001, attacks, which also involved box cutters.[8] Soon after the exchange of prisoners for hostages, the three released prisoners were reportedly living in "lavish" safe houses in Pakistan, enjoying a "rollicking" lifestyle, according to an Indian government official responsible for monitoring the jihadi groups.[9] Two of the released prisoners, Maulana Masood Azhar and Ahmed Omar Saeed Sheikh, Khalil's former colleagues, are worth discussing in their own right.

Indian security forces arrested Maulana Masood Azhar for his militant activities in Indian-held Kashmir when he was twenty-six years old. He told his Indian interrogators that he was born in the Punjabi village of Bahawalpur to a family of eleven children. After completing the eighth grade, he began his religious training at the Jamia Islamia, a radical Islamist seminary in Binori Town in Karachi. Six years later, in 1986, Azhar graduated with distinction and took a position as imam of Choti Mosque in Delhi colony in Karachi. In 1989, Fazlur Rahman Khalil, chief of HUM and my host in Islamabad, visited Azhar's former school, the Jamia Islamia in Binori Town, and invited Azhar to join Harkat, which was just getting off the ground at the time. Azhar accepted the offer and soon rose through the ranks of Khalil's organization to become the group's principal agitator, recruiter, propagandist, and fund-raiser. He visited the Harkat office near Multan, Pakistan, and from there was sent to a training camp in Khost in southern Afghanistan. At the camp, he learned to use an AK-56 rifle and light machine gun. Khalil seems to have understood early on that Azhar would be more valuable as a leader and a propagandist than as a fighter—his Indian interrogators described him as "an obese person of medium build."

Azhar told his interrogators that his first assignment was to open a Harkat office in Karachi. He became editor and publisher of a magazine for Harkat known as *Sada-e-Mujahid.* During this period, Azhar was still working as a teacher in the mosque, but he also traveled all over Pakistan, recruiting youth for jihad. In May 1990 he went on the hajj to Saudi Arabia, together with Khalil, my host. From there the two traveled to Zambia, giving speeches and fund-raising. They raised roughly $45,000,[10] a significant sum in the South Asian context.

Between 1991 and 1993, Azhar traveled to various countries in Europe, Africa, and the Middle East and raised a substantial amount of funds.[11] Azhar also told Indian police that he had met with groups linked to Al Qaeda that had attacked U.S. troops in Somalia in 1993, and U.S. embassies in Africa in 1998.[12] Azhar was captured by the Indian military in February 1994 in Kashmir during a routine traffic stop and remained in prison until the hijackers demanded his release in 1999.[13]

Three months after his release, Azhar announced the formation of a splinter group called Jaish-i-Muhammad (Army of Muhammad). He gave speeches at the mosque in Karachi where he had been trained. He traveled with large posses of armed bodyguards, said to be HUM militants who had switched their allegiance from Khalil to the splinter group under Azhar's leadership. Azhar openly recruited volunteers for jihads in Chechnya and Bosnia with massive recruitment and fund-raising drives, often at the Binori Mosque. For the purpose of fund-raising, he joined forces with an anti-Shia sectarian group known as Sipah e Sahaba Pakistan, and its military arm, Lashkar e Jhangvi.[14]

Azhar's splinter group is suspected in connection with the October 1, 2001, attack on the Srinagar legislature in Indian-controlled Jammu-Kashmir that killed thirty-eight people, as well as the December 13, 2001, attack on India's parliament, which resulted in fifteen deaths.[15] Indian officials claim that Azhar maintained close links with Al Qaeda. Brigadier Abdullah,[16] head of the Kashmir department of the ISI, is believed to have played a critical role in promoting Azhar's new splinter group, at the expense of Harkat and its leader, Fazlur Rahman Khalil. But others in the ISI continued to promote Khalil, my host. As is common for Pakistani jihadi groups, Azhar's Harkat splinter group would itself splinter. After the war in Afghanistan against the Taliban and Al Qaeda, Pakistani police said that Al Qaeda was attempting to absorb a splinter group from Jaish. They also said that Lashkar e Jhangvi, the sectarian group closely aligned with Jaish, had already come under the Al Qaeda umbrella and was providing safe houses and intelligence information to fighters fleeing Afghanistan.[17]

Azhar met Ahmed Omar Saeed Sheikh at a training camp in Afghanistan in 1993. They apparently solidified their friendship while incarcerated near one another for five years in the Tihar Jail of Delhi, until they were released in the exchange for hostages in December 1999.[18]

Sheikh was secretly indicted by a grand jury in Washington, D.C., in November 2001 for his role in the 1994 kidnapping of four Westerners, including an American, in India. On at least two occasions prior to the kidnapping and murder of *Wall Street Journal* reporter Daniel Pearl, the United States formally requested Pakistani officials to arrest Sheikh.[19] In March 2002, following Pearl's killing, a grand jury in Trenton, New Jersey, indicted Sheikh on two counts: conspiracy to commit hostage-taking resulting in the death of Daniel Pearl, and hostage-taking resulting in the death of Daniel Pearl.[20]

Ahmed Omar Saeed Sheikh was born in 1973 to an upper-middle-class family of Pakistani immigrants then living in the UK. He attended elite private schools, where he excelled at math as well as chess. His teachers considered him a model student. But he was apparently unhappy and frequently got involved in fights. He told a British musician, arrested on charges of possessing marijuana, with whom he shared a cell in Tihar Jail, that the students at his school were racist and taunted him, calling him a "Paki bastard."[21] His family returned to Pakistan in the 1980s. Sheikh's fellow students in the elite high school he attended in the eastern Pakistani city of Lahore remember him as strident and argumentative. He returned to England to attend the London School of Economics in 1992, with the intention of eventually finding a job in the City—London's Wall Street—but quit, apparently bored, in his second year of college. After seeing a documentary film about Bosnia, *Destruction of a Nation,* he traveled to Bosnia with a humanitarian organization known as Convoy of Mercy. He reports that he was profoundly moved by what he saw there. He made contact with Pakistani jihadi groups operating in the region and, hearing their stories, decided to join a jihadi group himself.[22] In March 1993, he visited Pakistan and met with a number of Islamist leaders, including from Jamaat-i-Islami and Harkat. When he returned to the UK, he joined a UK-based Islamist organization known as Hizb-ut-Tehreer. But he returned to South Asia almost immediately, to receive military training. He was already a black belt in karate, and he received four months of additional training in Pakistan and Afghanistan. He told his Indian interrogators that his instructors were members of an elite military unit, the Special Services Group (SSG)[23] of the Pakistani army, who went by the names Salim and Abdul Hafiz. Within a year of his training, he traveled to Delhi on an assignment to kidnap foreign nationals to use them as leverage to demand the release of two Harkat leaders, Maulana Masood Azhar and Nasarullah Ranzoor Langaryal, who were then incarcerated in Indian jails.[24] Sheikh masterminded the kidnapping of three British and one American national during the fall of 1994 and was captured by Indian security forces soon after that.[25]

Sheikh made an important contact in the jail, according to Indian police investigators: an ambitious Indian gangster named Aftab Ansari.[26] Organized criminals like Ansari have expertise in money laundering, forgery, abduction, and killing; and the jihadi groups have access to training camps and relationships with intelligence agencies that are useful to criminal gangs. One member of Ansari's criminal gang who was also incarcerated in Tihar Jail reportedly told investigators that he received military training at a Lashkar e Taiba camp in Khost, Afghanistan. Asif Reza Khan, who is described by investigators as the "chief executive of Ansari's India operations," told interrogators that a relationship between Ansari's gang and Sheikh's jihadi contacts was forged when Sheikh was moved to Jail No. 1 in November 1998, where Ansari was incarcerated for minor offenses. Ansari jumped bail in 1999 and fled to Dubai, where, after Sheikh was released, the two renewed contact. According to Indian police files, Ansari's deputy Asif Reza Khan stated that "Aftab [Ansari] confirmed to me that leaders of different militant outfits in Pakistan were trying to use his network for the purpose of jihad whereas he [Ansari] was trying to use the militants' networks for underworld operations." The don would identify sources of funding and provide hideouts. In return, Sheikh would send trained terrorists and arms to carry out operations. According to the interrogation report, the two sides agreed to share personnel and to combine forces for the purpose of raising cash. It was agreed that kidnapping and extorting wealthy businessmen were the best way to raise funds. Khan also said that Sheikh had asked him and Ansari to "recruit Muslim youth who could be trained in Pakistan and would be sent back to India to participate in jihad. These youth/mujahideen could be used for underworld operations as well."[27]

Ansari was heard bragging about his involvement in the January 2002 attack on the American Center in Calcutta. Indian officials are persuaded that Sheikh was also involved.[28] They also maintain that Sheikh requested that Ansari donate $100,000 for a "noble cause," and that the money was ultimately wired to Mohammad Atta, lead hijacker in the September 11 attacks, who was then in Florida.[29] Ansari's deputy Asif Reza Khan told Indian investigators that he had always been in awe of Sheikh's commitment to jihad, despite his affluent background, but that Khan was reluctant to part with so much money. But Ansari persuaded him that the $100,000 was an "investment" in a valuable relationship.[30] Indian police agents intercepted e-mail exchanges among the various parties involved in Ansari's purported donation to Atta, which have since been shared with Interpol.[31] Ansari reportedly claimed that his objective was to become "bigger than Dawood [Ibrahim]," the most famous gangster in India, big enough to extort money by merely announcing his name, not by kidnapping targets, and he apparently felt that an alliance with jihadi groups could further his goals.[32] Ibrahim, likewise, has close connections with jihadi groups as well as with Pakistan's intelligence agencies.

The U.S. government demanded Sheikh's extradition after Daniel Pearl's murder, but Pakistan refused to comply with the U.S. demand for extradition, because it has no legal means for doing so. Although it is against Pakistani law, Pakistan has given in to U.S. pressures for extradition in the past, extraditing a number of terrorists, including Ramzi Yousef, the mastermind of the 1993 World Trade Center bombing, and Mir Aimal Kansi.

There are a number of puzzles about the Sheikh case. Sheikh reportedly turned himself in to a former ISI official a week before his "capture" was announced. The ISI reportedly did not inform General Musharraf that it had Sheikh in custody. One version of the story suggests that Sheikh wanted to trade Daniel Pearl for three of his jailed colleagues, but was unable to strike a deal with the ISI.[33] He admitted his guilt during his first court appearance, but the confession, which he later retracted, could not be used by the prosecution because such a confession, according to Pakistani law, is valid only if given voluntarily in front of a magistrate after completing certain procedures. Pakistani law requires that the dead body must be found before a murder trial can officially begin. General Musharraf was quoted as saying that Pearl had tripped upon "intelligence games" because he was "overinquisitive," to explain Pearl's abduction and murder.[34]

Sheikh was sentenced to death in a secret trial that was widely seen as legally flawed. There is widespread suspicion that Pakistani authorities fear that if Sheikh were tried in open court or extradited to the United States, he would reveal too much about his relationship with the ISI. According to a high-ranking Pakistani government official, it is certain that Sheikh worked for the ISI prior to the hijacking of Indian Airlines flight IC 814, although it is unclear whether he continued to work for the ISI in the aftermath of the hijacking.[35] Others suspect that Sheikh may have worked for the ISI in India and may know about purported ISI links with Al Qaeda.[36] Indian government authorities observed that he had continued to use the same phone from the time of his release into 2001, suggesting to them that he felt utterly safe in ISI hands.[37] On the last day of his trial, it was revealed that one of Sheikh's co-accused was a junior-level police officer, who was still working for the Special Branch at the time of Pearl's murder.[38]

Khalil preempts my questions about the hijacking incident by telling me that Azhar left his organization when he was incarcerated in India. "He was not a member of our organization when the hijackers demanded his release. Therefore America's claims that we were involved is obviously false."

I try to cut Khalil short, not wanting to waste time on such obvious lies and propaganda. What was the saddest moment of your childhood? I ask him. When he heard that the Soviets entered Afghanistan on December 27, 1979. He was sixteen. That was about the time he completed his secular education and transferred to a *madrassah*. In 1982 he volunteered to fight as an independent *mujaheed*.

"I was trained in Afghanistan," he tells me. "We formed HUA [now HUM] in 1993 to fight in Kashmir.

"America should reconsider its policy of interfering in other countries' business. The whole world is beginning to hate your country. America has become a negative symbol. Its name has become a curse. China, on the other hand, is also a big power. But no one hates China because it leaves other countries alone."

I ask him to tell me his favorite books, in addition to the Koran. He recommends the writings of the late Sheik Abdullah Azzam, a Palestinian lecturer at the Islamic University in Islamabad, who, together with bin Laden, created the organization that ultimately became Al Qaeda. Azzam left his university post to build a network of organizations that trained Arab volunteers for the jihad in Afghanistan and assisted Afghan mujahideen and refugees. Khalil also recommends the history of Hitler. He admires Hitler because Hitler understood that Jews and peace are incompatible. "Experience has proved that Hitler was right," he tells me. "Jews are the main cause of the problems in today's world." I decide that revealing my Jewish background at this point is unnecessary.

I ask him what he is most proud of. "I am most proud that I am a *mujaheed.* I want to be a *mujaheed* the rest of my life."

Why did he become a *mujaheed* rather than a soldier? "Soldiers fight for a salary. I wanted to fight for God. To be a *mujaheed* means you have a spiritual life. A person addicted to heroin can get off it if he really tries, but a *mujaheed* cannot leave the jihad. I am spiritually addicted to jihad."

He continues, "I am physically fit, and if my organization demands that I join the jihad again, I am prepared to fight again at any moment."

He is feeling old, I think to myself. He wants to feel addicted to the kind of jihad that requires muscle and physical courage.

Are you married? I ask, hoping to disarm him with an unexpected, seemingly innocuous question. He tells me yes, his second wife is living here in town. May I meet your wife? I ask. Much to my surprise, he agrees. There is a subtle shift in mood. My femaleness has entered the room.

There is a project at Harvard to learn about the nature of leadership, I tell him, and it involves interviewing leaders around the world. I'd like to include you, I tell him, but if I do, you'll have to tell the truth. If you continue to lie to me as you did today, we can't include you in the project. He doesn't bother to deny the charge. He smiles and agrees. He is prepared to tell more of the truth next time, under the right circumstances, he says. My guide tells me later that Khalil told him that Pakistan had denied possessing nuclear weapons for years, and the United States was willing to accept this obvious lie. He is following the same policy in lying to me.

Khalil gets up to leave. He needs time to prepare his wife for her unexpected visitor.

The bodyguard, agitated, remains in the receiving room with me. Ten minutes later he rises, gesturing me to follow him. We walk in silence. The streets are jammed with trucks, donkey carts, buses, and bicycles. Crossing the intersections is alarming. Five minutes later we are in front of Khalil's house. He lives in a mansion. The house, I notice, has been freshly painted a pristine, bright white. A young servant woman leads me into a sitting room. Everything is astonishingly clean. The sofas are covered with new woven cloths. In the corner is a desk with a new-looking computer.

Khalil's wife enters the room, wearing a crisp blue-and-white *shalwar kameez*. It is the blue of a tropical sea. Her skin is the color of café au lait. She looks twenty years younger than her husband. She is utterly lovely, and I find myself immensely relieved to be in her presence. She tells me, in perfect English, that she was raised in South Africa, but her family had moved to Saudi Arabia at the time she and Khalil were introduced. Saudi Arabia is a major source of funding for the jihad groups, and I wonder whether Khalil was on a fund-raising mission when he met her parents.

A servant brings a tray with cold Coca-Cola, clean glasses, and sliced mango. Mango is in season and it is perfectly ripe. She wants to know about my project. What is your understanding of the difference between jihad and terrorism? she asks me. I tell her that I am not a religious scholar, but I have read that the jihad doctrine prohibits targeting children, women, and old men, making random violence and terrorism clearly off-limits. Terrorists, in contrast, deliberately target innocents, often with the aim of instilling fear. She is satisfied with what I say and, to my surprise, offers to exchange e-mail addresses. She tells me that her husband's militants are defending innocent Muslims, who are being murdered and raped by Indian government forces. They never target innocents, she tells me. I see no point in arguing with her, as she is unlikely to believe me.

A year later, I have several reasons to travel to Pakistan, this time on projects not obviously related to terrorist leaders. I had two missions. Finance Minister Shaukat Aziz had requested that the Kennedy School design an executive-training program for senior civil servants, similar to a training program designed for the World Bank. This was my second trip to meet with Minister Aziz in connection with this project in the space of four months, and I was optimistic that we might make some headway.

Although the World Bank was prepared to provide a grant that would cover a large percentage of the cost, Musharraf was (rightly) concerned about the expense, and part of my job was to learn how we could make the program more cost-effective, in particular, by training Pakistani professors, who would then be prepared to offer a similar curriculum to public servants at home.

The second purpose of my trip was to investigate whether Pakistan would be willing to accept assistance in securing its nuclear weapons and materials. Ted Turner had recently established a new foundation called the Nuclear Threat Initiative, and his foundation was funding this effort, which included India in addition to Pakistan. My colleague Professor Scott Sagan of Stanford University was responsible for the Indian part of the project. Our goal was to return to the United States with a sense of the two countries' interest in receiving technical assistance and, if possible, a list of the kinds of equipment or training they would most like to receive. Prior to my arrival, I send a note outlining the foundation's interest to a friend, President Musharraf's chief of staff, Lieutenant General Ghulam Ahmed Khan. I also send a note to Lieutenant General Mahmud, the chief of Pakistan's intelligence agency, informing him that I will be visiting Pakistan in July in connection with these two projects, hoping for his support.

Lieutenant General Mahmud's office responds immediately. I had mentioned to Lieutenant General Mahmud, when I'd met him four months earlier, that I had grown interested in the topic of intelligence ethics and had prepared a lecture on it for a course I was coteaching at the Kennedy School with the historian Ernest May. Lieutenant General Mahmud had mentioned at the time that he would like me to give that lecture to his staff, but I hadn't followed up, in part because it seemed such a bizarre idea. Why would the ISI listen to me? Why would my thoughts about ethics be of interest to them? But Mahmud was clearly determined that the lecture be scheduled. The general's office called me repeatedly, as did the embassy in Washington and the consulate in New York. Before I left for Islamabad, the Pakistani embassy informed me that ISI personnel would retrieve me from my hotel at 9:30 A.M. the day after I was scheduled to arrive, and deliver me back to the Finance Ministry that afternoon.

What does a woman wear when she has been invited to lecture the ISI staff on intelligence ethics? I have a small collection of *shalwar kameez.* The woman's version is usually worn together with a long scarf called a *dupatta.* I discover, in talking with friendly hotel personnel, that my outfits are too informal. I reject a black pantsuit and a loose, long dress in my suitcase, which I brought to wear in Bangkok and Indonesia. Although neither would be considered revealing in a Western context, they nonetheless feel inappropriate. I decide instead to invest in a new Pakistani outfit in the overpriced store in our hotel. This time I get lots of assistance. High heels, my advisers instruct me, not the beat-up walking shoes I always wear while traveling. Silk, they tell me. I select an outfit of orange, gold, and red.

The scarf on this dress nearly reaches the floor. It is worn looped around the neck, with two long tails falling on the back. I constantly have to adjust it to prevent one tail from falling lower than the other, getting caught under a heel or snagging the furniture. I am not skilled at walking in it.

A young officer retrieves me at my hotel at the appointed hour. He is in the army, on temporary assignment to the public-affairs office at the ISI. He takes me to a jeep designed for persons significantly taller than myself, even in high heels. My scarf nearly trips me as I hoist myself into the seat beside the driver. It is hot and humid, and I try to avoid crushing the silk of my long dress, which the heat and humidity will wrinkle instantly, even though the jeep is air-conditioned.

Ten minutes later we arrive. A fence opens, apparently automatically, as our car approaches the ISI's offices in Islamabad. We pass a guard with a gun who salutes as we drive past the gate. Another guard in ceremonial dress salutes us upon our arrival. I am whisked into Lieutenant General Mahmud's office, where Major General Javed Alam, the head of the analysis wing, sits with me while we wait. A servant in crisp white linens and a bright red headdress brings sweet tea with warm milk and a variety of dainty pastries, the sort you might expect over tea at the Ritz. Mahmud soon arrives. He outlines our schedule. Twenty of his topmost deputies will be attending my lecture, he says. The process will be orderly. There will be no questions until I am done, and each deputy will have a single chance to ask a question or to comment.

When I lecture my students about intelligence ethics, I try to provoke them by asking difficult questions. When you join government service, I tell them, you may find yourself with dirty hands. Intelligence and military personnel are in the business of stealing secrets, stealing weapons-related information or even the weapons themselves, lying, and killing. You may be forced to take actions to defend your country, I tell them, that violate ordinary moral rules. An intelligence officer who acted upon the maxim that it is wrong to steal secrets would harm his country's security interests. Secretary of State Henry Stimson closed the State Department's office of cryptography in 1930, asserting that "gentlemen do not read each other's mail," leaving the U.S. government unprepared to break the Japanese codes that, if understood, could have prevented the surprise attack on Pearl Harbor.[39]

We expect our leaders to make decisions to defend the nation's interest and to take actions that would be considered immoral in ordinary life. We expect them to defend us by stealing other countries' secrets, to discover not only their capabilities but their plans and intentions. We expect them, on occasion, to enter into arrangements with unsavory characters to make such actions possible. But there is extreme danger on this path for the government officials themselves. What happens to them when they approve such actions? They must cleanly divide their professional and personal moral personae. Stealing other people's secrets and reading their mail is not acceptable in ordinary life. And there is a risk that these morally problematic activities will backfire. When operatives and spies engage in clearly immoral acts, we are responsible. The moral imperative to defend one's nation can lead to surprisingly troubling moral outcomes.

I was acutely aware that Lieutenant General Mahmud was highly unlikely to actually be interested in my views on intelligence ethics. I guessed that the ISI was irritated by an article I had written criticizing the Pakistani government's covert support of the jihadi groups as dangerous not only for the international community, but also for Pakistan itself. Some of these groups, I wrote, are funded or affiliated with bin Laden. They train at his camps, sign his fatwas, and promise to wage holy war with him against the crusaders and the Jews. I suspected that the real reason for the ISI's invitation was to try, if at all possible, to make me more sympathetic to their point of view. I was also aware of the possible dangers of this meeting. Almost anything I might say could one day be twisted to justify actions that violate moral or legal norms. And the ISI's ability to state publicly that they had invited a Harvard professor to share her views on intelligence ethics was a kind of public relations coup for them. Still, Milt Bearden, station chief in Pakistan in the late 1980s, strongly encouraged me to comply with Mahmud's request, arguing that the ISI, as far as he knew, never met with Western academics and that it was possible that such meetings, if they were to continue, "could actually make a difference in the way things play in Pakistan."

I decided to focus on the problem of hiring morally and politically unreliable surrogates, as Pakistan has been doing for some time in its proxy war with India in Kashmir, and as the CIA has also routinely done, with equally problematic results. While I knew that discussing such a delicate matter would be unlikely to win me friends in the ISI, at least it would not come back to haunt me.

After our tea and pastries, Mahmud leads me to a nearby meeting room, where twenty men are already sitting in a circle. The general invites me to sit to his left, and to deliver my remarks while seated. I deliver a short lecture, hoping to stimulate thought.

Afterward, twenty intelligence officers made twenty comments. Each officer was allocated one question or remark, and each spoke in order, going around a circle in an order that had obviously been prearranged. The comments were clearly designed to prove to their chief that they had read my work on terrorism and were prepared to challenge me, even if exceedingly politely, in his presence. One officer, the most senior one, refused to follow instructions. He insisted on asking a second question after the twenty were complete. He was humorous and seemed quite irritated with his boss. That was Major General Javed Alam, with whom I had just shared tea and pastries.

After the officers finish, Mahmud formally presents me with a gift. He tells us that it is Alastair Lamb's history of the partition of Kashmir, which Pakistanis see as sympathetic to their point of view. This book, he says to his deputies, reveals the true story of Kashmir. I leave the book in the box and put it in my suitcase. When I open the box later, it contains a paisley scarf of the finest, softest wool. My office in Cambridge is cold and I am wearing the scarf now as I type.

The next day I meet with my friend, Lieutenant General Ghulam Ahmed Khan, Musharraf's chief of staff, known as GA. He asks me earnestly and surprisingly fervently, why is Islam so misunderstood in the West? Why do so many Americans equate Islam with terrorism? He also wants to know, urgently, what, in my view, could be done to rectify this situation? He tells me that his son has suddenly grown enamored of bin Laden. Bin Laden is increasingly appealing to Pakistani youth, he says, even among the most privileged children. Pakistanis feel that America has left them behind, he tells me. We need to have good relations with Afghanistan—America can't blame us for that. Our borders with Afghanistan are porous. We can't afford to be enemies. America pursues its national interest—we must pursue ours. GA admits that his son's obsession is deeply troubling to him. He sees his own child's fascination with bin Laden as emblematic of the problem between Islam and the West, he tells me. America's double standards in its application of humanitarian law and human rights norms are to blame, he said, especially in regard to Israel.

In the wake of September 11 the urgency in GA's words seems prescient. But the lieutenant general did not live to see September 11, which would have been extremely painful for him to witness. He died in a car crash. Some Pakistanis believe that he was murdered.

GA's driver drops me off at the hotel, where my translator is waiting for me. Even though I had not planned to meet with any jihadi groups during this trip, Ameer-ul-Azeem, public spokesperson for Jamaat-i-Islami (JI), Pakistan's largest religious party, had heard of my arrival and has arranged for several meetings on his own initiative. He wants me to meet with the head of the United Jihad Council, the famous Syed Salahuddin. He also wants me to meet with the head of another jihadi organization, on condition that I not reveal the organization's name. Ameer-ul-Azeem would become known in the West after an Al Qaeda leader, Khalid Shaikh Mohammed, was discovered to be hiding in the Rawalpindi home of a member of Azeem's organization, Jamaat-i-Islami.

Azeem told a *New York Times* reporter that Mohammed, whom the U.S. government described as a senior Al Qaeda leader, should be considered a "hero of Islam." He said that his group opposed the killing of innocents by terrorists. "But we have no reason to believe that Al Qaeda even exists," he said. "Osama bin Laden is a hero, too, because we have no evidence that he was involved in the World Trade Center attacks."[40]

In July 2000, the largest Kashmir-based militant group, Hizb-ul Mujahideen (HM), called for a cease-fire in its dispute with India. Syed Salahuddin, supreme commander of the group, was immediately sacked from his position as chairman of the United Jihad Council, a conglomerate of over a dozen jihadi groups. The cease-fire, the first ever called by a Kashmiri jihadi group, was intended to last three months. But two weeks later, Syed Salahuddin held a press conference in Islamabad, announcing the cease-fire's end. An unprecedentedly violent series of attacks followed, for which both HM and LET took credit. Salahuddin was reinstated as head of the United Jihad Council.

I had long been curious to meet Salahuddin, who is a fabled commander and considered one of the architects of the Kashmiri jihad. So when my guide informed me that JI wanted me to meet with the commander, I accepted.[41]

We drive to a large house near Islamabad where several years before I had met with Abdul Majid Dar, then chief commander of HM. According to Indian government sources, Majid Dar opposed the involvement of the ISI in the Kashmir dispute, and he and Salahuddin had a falling out.[42] Majid Dar was assassinated in March 2003.

Salahuddin comes from a prosperous farming family in the Indian part of Kashmir. He has a master's degree in political science from Kashmir University. In 1987 he decided to run for a position in the state assembly. The elections that year were widely believed to have been rigged by the party in power, and Salahuddin lost. When he protested, he was arrested and beaten. That was the beginning of Salahuddin's career as a militant, as well as the beginning of the violent phase of the Kashmiri struggle.

Salahuddin is tall and stout. He wears a green beretlike cap, set at a rakish angle. A woolly black beard covers much of his face. He is cheerful and obviously intelligent. He wants you to see the world the way he does, and it is hard to resist doing so in his presence. He takes my hand to shake it, inviting me to sit on a sofa. There are obviously limits to his fundamentalism, although he has long been a member of the fundamentalist party JI, which sponsors the group he leads. I have never shaken the hand of a Muslim militant, and normally I am invited to sit on the floor—most militant leaders don't have sofas in their offices.

Salahuddin is fifty years old. His hands are powerful and beefy, reminding me, oddly, of President Clinton's. The young men who serve us Coca-Cola, tea, and mangoes seem to view him with awe. He urges me to eat more mangoes, which are good for you in this heat, he says.

Salahuddin begins by recounting for me his version of the history of the Kashmiri struggle, which closely follows the Pakistani version. It is a story I have heard many times. I am surprised to see the extent to which Salahuddin is conforming with the Pakistani position, at least in my presence and, perhaps, the presence of ISI minders. I wonder whether he feels the need to prove his loyalty to the ISI after his apparent support for a cease-fire, which rumor has it was actually engineered by Indian agents who had penetrated the group.

I allow him to finish in the hope he will then feel more at ease. How do you raise money? I ask.

"Through international charities," he says. "Our organization is involved in relief and rehabilitation. We are helping widows, providing orphanages. We raise a lot of money from our own people—there are five and a half million Kashmiris living abroad. People contribute, often with a specification—for schools, for feeding the hungry."

I will later learn from Indian government sources that Salahuddin does a lot of fund-raising for the group. They say that he traveled to Malaysia in September 1997 to raise funds and that he maintains a resident visa in Dubai, where he is reportedly associated with a company called Kashmir Master Computers. He reportedly visits Dubai frequently.[43]

Where are you getting weapons and ammunition? I ask.

"Many people are in business in Kashmir. Indian officers have become very cynical—they are in this just for the money. The same is true in Sri Lanka and was true in East Pakistan. The officers loot the people. They come to people's homes, accuse them of helping the terrorists, and then demand that they pay a fine. We buy our weapons from the officers and from organized criminals. NGOs also help us with weapons and funding."

What do you mean by NGOs? I ask. Can you give me some names? He refuses.

The question of NGOs that supply funding and weapons comes up again in my next interview. As we drive across town, my hosts tell me that the ISI chief, Lieutenant General Mahmud, meets with Syed Salahuddin and the United Jihad Council at least twice a month, always at midnight. The ISI is much closer to the jihadi groups under Mahmud, they tell me, than it has been for a long time. As usual, I wonder who is running the purported jihad in Kashmir—the ISI or these commanders. I have the sense that each feels it is running the other.

We meet next with the commander of another jihadi group who asked to be kept anonymous. We sit on the floor. A young man brings Coca-Cola, tea with milk, and two kinds of cake. The commander sees that I like the cake and urges me to take more, which I do. It is delicious.

He tells me he recently escaped from three and a half years in prison, without immediately explaining how. "They pressure us," he says. "They arrest a militant, for example, and then arrest his sister. They make her stand naked. Then they rape her. The houses are set to fire. Many people break under the mental pressure. My son was martyred at age eighteen. I could have lost my nerve, but I didn't."

How do you raise money? I ask.

"Some comes from Kashmiris living in Pakistan. We also get a lot of money from NGOs that were set up by the Americans to fund the Afghan war—they still exist. We are getting so much money. You cannot even count it because it's so much.

"Last week I was in Iran raising money. I raised money and recruited operatives in Iran. I also raise money and recruit in UAE and Bahrain. I am talking about money from people, not governments. Governments have their limitations."

It occurs to me that he seems to agree with the proposition that the rise of nongovernmental organizations (including terrorist groups) is weakening the relative power of states.[44]

How often do you go on fund-raising missions? I ask.

"Maybe four times a year on average. We have *battel mal,* a fund-raising department—they do most of the fund-raising. This year I went for hajj, and while I was in Saudi Arabia, I also gave speeches."

Do you have to pay fighters?

"We pay very good fighters. For average fighters, we provide for their basic needs and that is all."

How do you recruit operatives?

"We have our own monthly publication. Once a young man subscribes to our journal, we know he is mentally prepared. We prefer to recruit children at the age of eleven or twelve. We start preparing them mentally and physically. They are usually not ready to fight until age eighteen or twenty, although some children develop muscles early. Lots of Saudis have joined the jihad. Osama bin Laden is a great force. He goes from organization to organization, persuading people to donate money or donate their lives to jihad. Many foreigners have offered to help us—not just from Saudi Arabia. Some come from Indonesia, other countries. We select people on the basis of character."

I know that some of these jihadi publications advertise the kind of training recruits will receive as a way to lure them in. The Pakistani group Lashkar e Taiba, "The Army of the Pure," discussed in chapter 5, describes the weapons its recruits will learn to use at its training camps. The group offers to "introduce" trainees to weapons from "the Kalashnikov up to the missile" in its three-week "Suffah Tour," which it describes in a press release on the Internet. In the three-week training session called the Special Tour, trainees are taught to blast mines, fire rockets, and other guerrilla tactics. After this, "some of the boys are selected for specialization in making remote control bombs and missiles." There are no age or other restrictions on who can be trained. Eight-year-old boys are taking part in the jihad. Lashkar prides itself on attracting better-educated youth than rival organizations. It is beginning to specialize in computer technology, advertising its high-tech prowess to attract youth to join the cause, much as in U.S. army recruitment advertisements. "Mujahideen have got access to the Indian army Web site where they worked against the Indian forces," says its literature. "Lashkar e Taiba also made a remote control airplane that was caught in Occupied Kashmir. We are developing the modern technology. We can make modern devices."[45]

What about muscles? I ask.

"Muscles are also important, but muscles can be built through physical training."

Do you cooperate with other groups?

"We have person-to-person contacts with other groups, but no formal relations."

With which groups? Do you cooperate with Laskar Jihad in Indonesia?

"We have only person-to-person contacts. Sometimes fighters from Hamas and Hezbollah help us."

Where do you meet them?

"A good place to meet is in Iran. We don't involve other organizations. Just individuals. Militancy has different dimensions—different skills are required for a variety of guerrilla operations. Some militants are particularly good at making explosives, for example. So if a fighter has no work in his organization, he might come to work for us, he comes to help. A militant cannot sit idle," he explains, repeating a theme I have heard many times before.

Do you have to pay militants that are on loan? Who pays their airfare?

"The underground international market pays—the NGOs set up during the Afghan war. These are people who specialize in transferring money—they also know how to transfer people."

In other words, you are involved in illegal transfers of money and people, I say, encouraged by this extraordinary openness.

"We don't feel this is illegal. A money changer who works for us does it because he believes in our cause. He is not doing it to earn profits." The commander seems to assume that by *illegal* I mean counter to Islamic law. "It works like this. We might give a loan to a silent worker [a sympathizer who is not actively involved in militancy] for his business, and we fix a percentage. The percentage goes to support our work."

So you also function as a bank?

"We operate like an Islamic bank."

Where do you get weapons?

"We buy them from corrupt Indian officers. Sometimes we raid Indian military camps. We also buy arms from the Indian underworld. If you want to see the power of the underworld, look at what happened to the hijackers of IC 814. How do you think they escaped? The Indian foreign minister would do anything to get those hostages released. He was begging for help from the Taliban. India didn't even recognize the Taliban. People were paid off.

"The Indian officers are completely corrupt," he continues, obviously interested in this topic. "They try to buy militants, but we buy them also. Sometimes they arrest a militant, and then they sell us back his gun. They steal our valuables. This war in Kashmir is business for them, but it's a holy mission for us. The basic difference between them and us is that they are afraid to die."

How do you collect intelligence?

"A successful operation requires intelligence in advance. Every jihadi organization has an intelligence wing. We have our own department of intelligence, called the Secret Intelligence and Analysis Wing [SRAJ]. They are local people in Kashmir—they might be drivers, or shopkeepers. We also have sleepers. When our militants are arrested, they tell us what is going on in jail.

"I am in this field for the last thirteen years. I know my counterparts on the Indian side very well. We engage in a kind of bilateral trade with RAW.[46] Every time something is about to happen, they always warn me. That is how I got out of jail."

Do they sympathize with you?

"No, but we have some plants in RAW that help us, and we also make trades with them. How do you think Lashkar e Taiba was able to reach to Red Fort in Delhi? We have sources and sleepers inside India.

"Kashmir is not our ultimate goal. We have two primary goals—remove India from Kashmir and convert Pakistan to an Islamic society. But we'll go to any other place for jihad until we conquer the whole world. We don't omit America from this list," he says, looking slightly apologetic about referring to my country in this way. "America is not our enemy, but whoever works for the system that oppresses us must be conquered."

* * *

In Lahore, I meet a senior manager for Harkat-ul-Mujahideen whom I will call Yusuf. He used to work for Sipah e Sahaba Pakistan, the virulently anti-Shia party. He left that job to join HUM. He would like to give me some recruitment posters as a souvenir. We drive through narrow streets to a small storefront, his office, in a sad-looking neighborhood. Yusuf runs in and returns to bring me some key chains decorated with pictures of Kalashnikovs, posters—also of Kalashnikovs—and calendars displaying a variety of guns. I give him a Harvard pen. Now that we've exchanged gifts, Yusuf becomes more serious. I want to leave my organization, he says. I ask him why. He tells me that he no longer believes in the cause. We reach the hotel before he has a chance to tell me very much, but he agrees to return later with my guide.

The next day, my guide informs me by telephone that he and Yusuf are waiting for me in the lobby. I take the glass elevator downstairs, walk past the frail-looking guard who protects hotel guests with a nightstick, and wend my way through the crowd of wheelers and dealers who always gather in the lobby of my hotel. The piped-in piano music, which usually sounds so cheerful, sounds tinny and false to me now. I spot my militant. I am frightened—for myself and for him. He works for an organization that has killed many people. But now he claims to have changed his mind. Is the ISI setting a trap for me? Do they want to see how I will react to this situation? Could it be that, despite my complete openness with them about my interviews, they still think I work for the CIA? Alternatively, if he is truly a disgruntled militant, does he realize that the ISI is probably watching me—and will observe his confession? I decide not to invite him for coffee, but to remain in the lobby, where I assume it is harder to hear us.

Yusuf begins telling me about his frustrations. He is making a good salary—better than he could make in the civilian sector, he says. But he sees his bosses getting rich off jihad and has come to feel disgusted. They have dirty offices and serve you bad food just to prove they have no money. But they live in mansions, he says. Jihadi organizations receive a lot of donations, and a lot of the money ends up going to the leaders. Four industrialists in Karachi donate 1 crore (10 million) rupees annually. People donate big houses, and the bosses live in them. In the previous government, under Nawaz Sharif, all the jihadi bosses had VIP status. In the Musharraf government they have an even higher status. What does that amount to? I ask. They go to Saudi Arabia or other Gulf States for fund-raising or to recruit, and when they return, the customs authorities don't check their luggage. The Pakistani embassy in Saudi Arabia will help them, facilitate their fund-raising and recruitment efforts. Foreign boys come to our camps to get training, then they go back and join militant groups in their own countries. We had a camp for training foreigners in Afghanistan, but it was given to Jaish-i-Muhammad. Now we provide training in Azad Kashmir (Pakistan-held Kashmir).

Do operatives resent the wealth of the leaders? I ask.

"The operatives think this is religious work. They have complete faith in their bosses."

Do you believe that the leaders are truly motivated by religion?

"They preach about proper behavior for Muslims, but they don't abide by the rules themselves.

"I have worked with jihadi organizations for twelve years. Most of the people who join these groups are from the poorest classes. Eighty-five percent come from below the poverty line, twelve percent are from the middle class, and around three percent from the rich. I feel very demoralized now. I want to write a book."

What happens when workers disagree with the policy of the organization?

"It depends on the nature of the disagreement. If an operative is in Indian-occupied Kashmir and rejects his orders, he could be shot. Three months ago eleven militants were killed in IOK [Indian-occupied Kashmir] for disobeying orders. This is never printed in the papers. There are probably thirty-five to fifty militants killed per year. This year a father realized that his son did not die a martyr but was killed by Lashkar e Taiba managers. The father went to the Jurga (Islamic council). He had given Lashkar e Taiba three hundred and fifty thousand rupees. He said his son had been murdered by Lashkar e Taiba.

"If somebody writes a book, that could be very serious," he says, finally turning to the key question. "I'll take preventive measures—I'll try to seek asylum somewhere."

What is the morally correct thing to do with this information? He seems to be half-hoping I will tell the embassy he might like to defect. I am reluctant to test out this idea. What if he doesn't actually want to defect, and Pakistani intelligence officials see him meeting with U.S. officials? That would be dangerous for both him and me. What if he does want to defect, but our government wouldn't want him? I do nothing, hoping that is the right thing to do, hoping he will approach the U.S. embassy on his own if that is what he truly wishes.

Eventually, Yusuf decides to leave HUM, despite his inability to find an acceptable job as a "civilian." He wants to write a book that tells his story, and he starts e-mailing me regularly. He wants me to help him get the book published. He thinks he will be able to support himself through writing if only I can find him a publisher. This is a tall order, however, since his English is bad. I try to explain to him that most authors don't make a living by writing books. Most have other jobs—as professors, as doctors, as reporters, as waiters. I expect I earned about minimum wage with my last book, I tell him, and the university press that published it considered it to be a big seller. Yusuf is undaunted. He is determined to tell the world his views of jihad, at any cost. He proposes writing an article for a liberal paper in Pakistan, one that is read by the intellectual elite. He is disappointed to discover that the paper will not provide for his safety, nor are they willing to pay him what he believes the article is worth. His friends are warning him that publishing the article or the book will mean certain death, and that he should take a job and fade into the background, at least for a few years. But he is determined to martyr himself for this new cause. It is as though, even leaving the jihad, he can't give up the life of danger and sacrifice—this time for a jihad against the jihadis.

Yusuf is not the only member of a jihadi organization who seems disillusioned with holy war. Another leader said about his bosses, "At first I thought that they are serving a religious cause, but now I feel they are running a business. They are . . . suppliers of human beings. They use poor and illiterate boys for their own private cause and call it jihad. This 'jihad' has nothing to do with religion." Asked how his organization receives its funds, he said, "The . . . real methods for raising funds is smuggling of goods through Afghanistan, Iran, and India. This includes drug trafficking, in some cases to India. Mujahideen cross the borders and carry drugs, delivering them to the Indian underworld mafia. Similarly, the mujahideen bring with them many smuggled items such as cosmetics and . . . electronic goods from Afghanistan to Pakistan to raise funds."[47]

For many of the managers interviewed for this book, jihad is a lucrative undertaking. Some of the younger operatives in Pakistan said they are not paid for their services (although their basic needs are covered). But managers and trainers are relatively well paid. Cash bonuses for "excellence in fund collection and recruitment" are provided, according to a former manager of one group.[48] Another reported cash bonuses of five thousand rupees for successful operations in the field.[49] A manager for Harkat-ul-Mujahideen reported a salary of nine thousand rupees per month. He said that the operatives "do jihad" for a "spiritual reward," while those in managerial positions are working for a "material reward."[50] A trainer who teaches mujahideen foreign languages said he received ten thousand rupees—more than twice the amount that he made as a teacher prior to joining the organization.

One long-term militant now working as a district commander in Indian-controlled Kashmir said that salaries for operatives in his group, Hizb-ul Mujahideen, depend on the operatives' needs. He reported that operatives earn 1,225 rupees per family member per month. In his family, he said, "We are ten persons, so we get 12,250 rupees per month. Bachelors, or anyone who has no one to feed, gets 515 rupees per month as pocket money in addition to food, clothing, and other basic needs." Most of the money for operations in Kashmir comes from the Pakistani government, he said.[51]

Several militants in Pakistan reported changing their views on jihad over time. One reported, "Initially I was of the view that they are doing jihad, but now I believe that it is a business and people are earning wealth through it. . . . The public posture is that we are doing jihad in Kashmir, while the real thing is that it is a business empire." Regarding the leaders of his organization: "I thought they were true Muslims, but now I believe that they are fraud, they are selling Islam as a product." And his reasons for staying in the organization also changed. "First I was there for jihad, now I am there for my financial reasons."[52]

Not only money is important: emotional satisfaction and status are also critical. Operatives describe the emotional satisfaction of their work, and the status they earn in their community. "One becomes important due to his work. Successful operations make a militant famous and glamorous among his fellow men," a trainer said.[53] The emotional satisfaction of the work and the charge it gives operatives becomes "addictive." A former Pakistani operative and trainer said, "Once a person gets involved in fighting, he cannot leave." After a while he will only "feel relaxed with a jihadi organization. Even if you make him work in a religious political party, he will not feel easy, he will feel disturbed."[54]

A Pakistani charity called the Shuhda Islam Foundation offered to take me to meet families of the "martyrs," the boys who have died fighting in the "jihad" in Kashmir. Pakistan's leading Islamist party, the Jamaat-i-Islami, established the foundation to support the boys' families. My hosts from the charity also agreed to take me to visit a few *madrassahs,* the religious schools where hopeful mujahideen from around the world receive mental training for jihad. Although *jihad* means "to strive" and the purification of the self, I wanted to visit some of those *madrassahs* that focus on what the Prophet called the "lower jihad," recruiting children for holy war.[55]

Atif Abbassi, a twenty-seven-year-old former *mujaheed* working for the foundation, has arranged to introduce me to martyrs' families who are living in a poor rural region about an hour outside Lahore. Abbassi picks me up from my hotel in Lahore in a chauffeur-driven, air-conditioned car. His black beard is neatly trimmed and his pristine white *shalwar kameez* is perfectly ironed. He is utterly, perfectly polite. Abbassi is obviously proud of his work and pleased to introduce me to the families he helps support. The Shuhda Islam Foundation provides financial support to nine hundred families, he tells me. Sometimes the foundation pays off their loans. Sometimes it sets them up in businesses. And sometimes it helps them with housing. It also provides emotional and spiritual support, by constantly reminding the families that they did the right thing by donating their children to assist their Muslim brothers in Kashmir.

We drive for an hour, past rice paddies, orchards, farms, and brick factories. Pakistan's wealthy feudal lords own most of the farms and factories, my host tells me. The workers live in the mud huts by the road. I see men, women, and children working in the fields, their skin browned from the sun. The children are dirty but beautiful, the bright softness of their near black eyes visible even from a distance. I ask why the children aren't in school. Their parents need them to help in the fields, my host tells me. Besides, the workers can't afford the fees for schoolbooks or the bus fare to school. Some of the children attend *madrassahs* at night when they finish their work. There is no plumbing, and we see men and boys relieving themselves in the fields. I pull my scarf tighter over my hair and pretend not to notice.

Two Kashmiri militants have joined us on the trip. They follow in a second car. They will be living in the village for some time, Abbassi tells me, inspiring local children to join the jihad.

A small group, including local Jamaat-i-Islami officials, gather to welcome our party to the village. I am encouraged to speak first to the two Kashmiri mujahideen. Ibrahim is a slight but strong twenty-two-year-old boy. His father owns a fruit shop on the Indian side of Kashmir. Ibrahim completed junior high school and then transferred to a *madrassah*. He joined Hizb-ul Mujahideen (the Kashmir-based militant wing of Jamaat-i-Islami) when he was fourteen. During his six years of fighting, Ibrahim was arrested twice by Indian security forces. What happened when you were arrested? I ask him. "I was made to take off my clothing. My father was forced to stand nude in front of me. Then they tortured me with electric shocks. And then they dragged me behind a car." The first time he was arrested, he was incarcerated for two months. The second time he escaped after fifteen days, a nearly impossible feat, the others tell me. He had to reach a window ten feet above the floor. "The guards with guns didn't see me. The ones who saw me didn't have guns. They chased me, but I escaped. It was a miracle." He is anxious to get back to fighting, he tells me. "A *mujaheed* can't sit idle, he can't stop. Just like you can't stop writing."

He crossed the mountains to the Pakistani side of Kashmir to receive commando training. There is no commando training available on the Indian side. Sixty boys participated in the phase of training he has already completed. They learned how to live in the forest, how to survive without food, and how to detonate explosives. (One of my hosts asks my translator not to mention the explosives, but he told me about it later.) Ibrahim will cross back to the Indian side. But first he is taking a tour of Pakistan to visit his Muslim brothers, the Pakistani mujahideen. The saddest moment of his life was when a friend was martyred. The Indian army took his friend's entire family to the jail and tortured all of them, even the women. How does your mother feel about your fighting the Indian army? I ask him. "She is with me," he claims. "She tells me, 'Don't come home until you are a martyr.' " Ibrahim's thirteen-year-old younger brother also wants to be a *mujaheed.* "I have already taught him to use a gun," Ibrahim tells me.

Junaid is also from the Indian side. He joined Hizb-ul Mujahideen when he was eighteen. From the age of nine he wanted to be a *mujaheed,* "but didn't know how." He has been through one round of a ninety-day course training. Sixty mujahideen received the training with him, and twenty-five of those were strong enough to be "launched" back into Indian-held Kashmir. Guides lead the trained commandos over the mountains to the Indian side. They return to the Pakistani side with a new crop of trainees. Some boys take the ninety-day training course a number of times to build up their courage and strength. How hard was it to cross the Line of Control, I ask Junaid. "It takes eight days to cross. I was very hungry," he tells me. "I was eating grass." What are you doing here in this village? I ask. "I am receiving mental training," he says. "I want to meet the families of the martyrs. I want to be a martyr too." The boys are taken to Pakistan, one of my hosts tells me in surprising moment of candor, to persuade them to support Pakistan's preferred solution: that Kashmir be part of Pakistan and not independent.

My hosts take me next to a poorer area. The street is lined with open sewers. A family I will call the Mirs moved here from a farming town when their son Zafar Iqbal became a mechanic. The house, made of unpainted concrete, is a step up from their earlier home, which was a mud hut. After Zafar Iqbal died in Kashmir, the foundation helped pay off the family's substantial debts. And the foundation set up Habeeb, the martyr's father, in business. He now owns two shops in the village. I notice the new rugs on the floor and the new cloth bedspreads. When Zafar died, eight thousand attended his funeral in Kashmir, his mother tells me. God is helping us out a lot, she says, pointing to her home and smiling. Habeeb and Hanifa had six sons, one now dead. "Two look after me now," Habeeb explains. One was trained as a *mujaheed* but is now home. We are donating our youngest to God, Habeeb tells me, pointing to his ten-year-old son. After completing fifth grade he will study in a *madrassah* full-time, to prepare himself mentally and physically for jihad. I ask the boy what he wants to do when he grows up. Be a *mujaheed,* he tells me.

We drive over rutted dirt roads to a third house. Nargis and Nisar Ahmed had three sons, now two. Nisar is a furniture repairman who was barely able to make ends meet, my host tells me. Nargis, the martyr's mother, tells me how proud she is to have donated her son, but she starts sobbing as she says this. I am uncertain how to respond. All the other mothers seem to have hypnotized away their pain, and I am unready to face this anguish. I put my arm around her and feel a birdlike frame. The bones feel light, exposed, and vulnerable. Like most of the martyrs' homes, the walls are decorated with glamour shots of AK-47s. Their second son, now fifteen, wants to be a martyr too. This family, even poorer than the others, needed a lot of financial help, my host tells me, and the foundation is helping them out a lot.

Syed Qurban Hussain is a hakim (a traditional doctor of herbal medicine) and the father of seven sons. All seven were trained to be mujahideen in Afghanistan. "They wanted to go," Hussain tells me, "but I also encouraged them." His fifth son, a district commander in Kashmir, spent two years in what the Pakistanis call Indian-occupied Kashmir (IOK). This son had been educated through the eighth grade and had learned the Holy Koran by heart in a *madrassah*. He was killed at age twenty-six by Indian security forces. Hussain's seventh son has just returned from IOK. He slid five hundred feet down a mountainside and his backbone is fractured. But, his father tells me proudly, he is planning to go back as soon as he is healed. Like his martyred brother, Hussain's youngest son also knows the Koran by heart, which gives him the "right mental approach" for jihad, his father tells me.

Hussain tells me he is happy to have donated a son to the cause of jihad. Whoever gives his life in the way of Allah lives forever and earns a place in heaven for seventy members of his family, to be selected by the martyr. "Everyone treats me with more respect now that I have a martyred son. And when there is a martyr in the village, it encourages more children to join the jihad. It raises the spirit of the entire village," Hussain tells me. His wife reiterates that when a son dies in the jihad, the family's standing in the community goes way up. I sent my son to fight in the jihad myself, she tells me proudly. And I would send all seven sons. I would be happy if all seven sons should be martyred. They will help me in the next life, which is the real life, she says. It is their Islamic duty to be martyred. We do this to create justice in the world, she tells me sternly. This mother is better educated than the others are. She attended fifth grade. She is proud. She showers me with gifts. Gold-painted bangles, prayer beads. I offer her the only thing I have with me that might interest her, a Harvard pen, in return.

Near this mother's home is a neighborhood flooded with water—the houses encircling what has become a small pond. The smell of sewage is overpowering. The pool of sewage and garbage is a breeding ground for mosquitoes, one of my hosts points out, and the government is remiss for not having drained it. There is a *madrassah* here that my hosts want me to see. On a bulletin board outside the principal's office is a child's portrait of a nuclear weapon. It was drawn with extreme care, colored a bright green—a picture of "the Islamic bomb." We pass a classroom. Twenty small boys are reciting the Koran, trying to memorize it. They look cute and vulnerable, like all ten-year-old boys. I wonder how many of these boys will end up dead in Kashmir, and how soon. As we return to Lahore, my host insists on playing a tape that he says is a recording of "the song of jihad," which is a sad song indeed.

In Islamabad I meet with a number of Pakistani officials to ask them about the *madrassahs*. The State Department recently asked Pakistan to close certain *madrassahs* "that actually serve as conduits for terrorism." Pakistan is responding, its officials claim, because some of these *madrassahs* are bad for Pakistan itself. In addition to the international terrorists the State Department worries about, *madrassahs* produce sectarian terrorists like the Lashkar e Jhangvi killers that tried to assassinate Nawaz Sharif, Pakistan's former leader.

Pakistan's then interior minister, Mr. Moinhuddin Haider, outlined his plan of attack.[56] First, all *madrassahs* will be required to register with the government. This is not the first time that the federal government has made such a demand, but the job is still far from complete. Estimates of the number of *madrassahs* located in Pakistan range from forty thousand to eighty thousand, and fewer than five thousand had registered so far. Each of the conservative religious parties has its own set of *madrassahs*. They are gaining influence among the poor by educating their children for free, a service not provided by the Pakistani government.

Second, Mr. Haider said, the *madrassahs* must expand their curriculum to include not only the study of the Koran and Islamic law but also math, science, computer skills, and other subjects. The goal should be to produce "balanced persons" able to find productive employment in Pakistani society—not just as mujahideen or village preachers. Third, they will have to list their sources of financial support. Fourth, when a foreign student applies to study at a *madrassah,* the principal will be required to give the Pakistani government three months to secure permission from that student's government before the student will be allowed to matriculate. Arab governments have been complaining that Pakistan's *madrassahs* are providing mental training in jihad, and that the graduates cause problems for the governments of their native countries when they return. Fifth, the director of the seminary and the child's parents must sign a form in which both parties agree that the school will not encourage the student to enter a militant training camp. These forms already exist, the minister said, but many *madrassahs* are not honoring them. He recounted a story about a young man whose parents had signed such a form, only to discover that their child had been sent to a training camp in Kabul without their permission. "The brand of Islam they are teaching is not good for Pakistan. Some, in the garb of religious training, are busy fanning sectarian violence, poisoning people's minds," Haider complained. This was neither the first, nor the last time the Pakistani government would make such pronouncements. As I write these words in 2003, little progress has been made.

In the morning we drive to Akora Khattak, located near Peshawar in the North-West Frontier Province. This *madrassah,* called Darul Uloom

Haqqania, is most famous for having played a critical role in the creation of the Taliban. The word *taleb,* in fact, means "student" in Arabic. Maulana Sami ul-Haq is chancellor of the school and the son of its founder. He is a former senator and an important figure in Pakistan's right-wing religious opposition. He would later come to play a significant role as the leader of one of the six religious factions that, under a coalition named Mutahida Majlis e Amal (MMA), achieved an unprecedented victory in Pakistan's parliamentary elections in October 2002. The MMA won 53 out of 342 seats and became the third-largest party of Pakistan's National Assembly.[57] Prior to the 2002 elections, the religious parties' best previous showing was nine seats in the 1993 elections. In 2002, the MMA gained a majority in the North-West Frontier Province, where Sami-ul-Haq's *madrassah* is located and where some senior Al Qaeda members are believed to hide; they also did well in Balochistan, former home of Mir Aimal Kansi, discussed in the previous chapter.[58] The school is in a poor rural area, and most of the students come from extremely poor families who cannot afford to feed their children, he tells me. When they matriculate at the school, they receive not only free education (something not provided by the state in many parts of Pakistan), but also free food and lodging. The families are striking an implicit deal with the school: the school relieves the family of the expense of housing and feeding a child. The child, in turn, is "donated to God." "All over the world, it is common for poor families to donate one child to the priesthood," Maulana Sami-ul-Haq explained. "Poor people love religion. The rich love their wealth instead of God." What is less common is that these poor people often donate their children not as priests but as cannon fodder in Pakistan's proxy war with India over Kashmir. Or to fight in other purported jihads, in Afghanistan and elsewhere.

Sami-ul-Haq invites me to lunch. Some twenty religious elders are visiting Sami-ul-Haq today, as they often are, he explains. All are dressed in white and all have black beards, but they wear a variety of caps. None attempts to speak with me and they quickly look away from me if they notice me looking at them. We sit on the floor. Pillows line the walls. A large cloth laid out on the floor is covered with bowls of spicy food: okra in a peppery cardamom sauce, lamb, beef, chicken, and rice. I am given silverware but the others eat their food with a flat bread made of coarsely ground wheat. Unlike his guests, Sami-ul-Haq is jolly here, sitting at the head of the cloth spread on the floor. His beard is graying, and the henna used to dye it gives it a reddish, cheerful hue. The Taliban, Sami-ul-Haq's creation, are known for their extreme cruelty to women. But Sami-ul-Haq's daughter, who is retarded, sits in the place of honor at Sami-ul-Haq's side. When she demands more food, he indulges her with surprising sweetness. For the rest of my visit, she follows Sami-ul-Haq around, occasionally climbing onto his lap to be stroked.

Haqqania is one of the largest *madrassahs* in Pakistan, currently training over twenty-eight hundred students. Approximately four hundred five-year-olds come to the school each year, but most students begin their training at age twelve or thirteen. Many learn the Koran by heart. They cannot understand the Arabic words but they learn to chant them. (Many of the Pakistani students come from dirt-poor tribal areas and do not speak Urdu, Pakistan's official language, but Pashto, the language spoken by the Taliban.) The school trains future mujahideen from Pakistan, Afghanistan, Uzbekistan, Tajikistan, southern Russia, and Turkey. Students from China and Sudan have studied here in past years, according to Sami-ul-Haq. Approximately half the student body was Afghani at the time I visited.

Students at Haqqania rarely see women. Although I was allowed to visit classrooms and some living quarters, some areas were off-limits because the older students would become quite "angry" if they saw a woman, even one wearing traditional Pakistani dress with a sheet over her head for added protection.

The school is Deobandi in orientation. Deobandism arose in British India as an anticolonialist, reformist, intellectual branch of Sunni Islam. Its aim was to harmonize classical texts with the demands of secular life in prepartition India. Sami-ul-Haq considers Maulana Mawdudi (one of the founding fathers of Islamic fundamentalism, as Islamism is referred to in Pakistan) to be "a bit modern." The movement has its own political party in Pakistan, the Jamiat-ul-Ulema Islam (JUI), one of whose branches Sami-ul-Haq runs. The party promotes the enforcement of Shari'a under the guidance of the "righteous ulama." If JUI had its way, the ulama would have the right to determine whether laws passed by the parliament are consistent with Islamic law. By and large, the platform is anti-Shia. It promotes Hanafi (Sunni) law, arguing that the majority of Pakistanis are Sunnis. Anti-Shia fatwas (religious decrees) and texts are promoted by Deobandi *madrassahs,* and students coming out of these schools are often virulently sectarian. Personality clashes have split JUI into several camps, and the rival camps now compete on the basis of which is more anti-Shia as well as anti-American.

The U.S. cruise missile attacks against militant training camps in August 1998 "damaged the image of the United States," Sami-ul-Haq explained, and turned Osama bin Laden, an ordinary man, into a hero. America's opposition to *madrassahs* is damaging the image still further, instilling "sentiments of violence" in *madrassah* students, he tells me.

No military training takes place at this *madrassah* or, according to the Pakistani government, at any other *madrassah* in Pakistan. But many of Sami-ul-Haq's students become mujahideen after they leave. "We don't force them to join the jihad," Sami-ul-Haq told me. "They become mujahideen voluntarily."

I asked Maulana Sami-ul-Haq how he felt about Haider's plans to crack down on the *madrassahs*. The government may persuade some *madrassahs* to register, but those will be the ones found in large cities, not in Pakistan's rural areas, Sami-ul-Haq said. He has no intention of broadening the curriculum to suit the Pakistani government. "Our goal is to teach students Islamic fundamentalism, which is not taught in regular schools and must be safeguarded. We teach fourteen different subjects of Islamic law. We cannot teach all subjects—these fourteen subjects are a full curriculum in and of themselves." The government's goal, he said, is to "destroy the spirit of the *madrassahs* under the cover of broadening their curriculum." The government is failing to provide an education system worthy of the name, Sami-ul-Haq complained. "My advice to Mr. Haider is to fix Pakistan's state-school system first," before giving advice on how to run the seminaries. Sami-ul-Haq categorically rejected Haider's demand that *madrassahs* inform the Pakistani government about foreign students wishing to matriculate. With regard to the forms allegedly distributed to *madrassahs* requiring them to commit to the students' parents not to send the children to jihad, Sami-ul-Haq exclaimed that he had never heard of such a form, "other than from your mouth." He said, "Those forms do not exist. You may tell Mr. Haider that this is the biggest lie I have ever heard." Regarding Haider's desire that *madrassahs* produce balanced persons able to find employment in a variety of fields, Sami-ul-Haq advised the government to begin by ensuring that graduates of public schools are able to find employment. "Our students have much better lives than those of public schools do. Students with doctoral degrees from the government school system end up washing dishes or driving taxis in the United States. Our students donate their lives to a religious cause and are far happier than their students." Besides, "God is giving the *madrassahs* so much money" that they are able to keep expanding. This is not true for the secular schools. Regarding the requirement to report sources of funding, Sami-ul-Haq scoffed, "When the government starts giving us money, we will be prepared to report our sources of funding to them." The entire exercise of attempting to regulate the *madrassahs* is "a game of diplomacy with the West. Haider is flirting with America," Haq pronounced, clearly enjoying this opportunity to denounce a high-level government official. In his heart, however, Moinhuddin Haider "loves the *madrassahs*." Two weeks before, in a meeting with the leaders of *madrassahs*, Haider "swore on the Holy Koran he is not against the *madrassahs.*" Sami-ul-Haq lamented, "Haider was a good governor. God knows what happened to him. America has hypnotized him." And, more importantly, "America has assessed Pakistan's army wrongly. The army is now Islamic. It is committed to the *madrassahs*. . . . This is the first time I am revealing the truth to a foreigner," he tells me giddily.

Maulana Sami-ul-Haq's eldest son has been designated his father's successor as the head of the school and currently teaches here. I asked him about his own schooling. Although most of his training was in Islamic law and practice, he knows a little bit of "old style" geometry developed "six hundred years ago in Egypt."

Sami-ul-Haq's two sons took me for a tour of the school, which is immense and expanding. They showed me a lecture hall big enough to seat five thousand students if they sit on the floor, asking me whether Harvard had such beautiful classrooms. The room is airy and cheerful, with a stark white ceiling decorated with carved tiles. I had to admit I have not seen a classroom that rivals Haqqania's new lecture hall.

They introduced me to a number of students, all of whom professed the desire to become mujahideen when they leave the school. They took me to the Russia House, where Central Asian and Russian students live. Although my hosts were visibly agitated when they discovered that I speak Russian, I was allowed to question the students. Here I had control over the questions addressed to the students. Are you happy here? I asked. *"Normalno,"* answered one, meaning "everything is cool." Like their Afghan and Pakistani counterparts, these students wanted to become mujahideen. When asked why he wanted to be a *mujaheed,* a twenty-year-old Chechen student told me his goal is to fight Russians.

The school is planning to construct a new dormitory to house foreign students and to expand the number of students it accepts from Russia, Tajikistan, Uzbekistan, Dagestan, Chechnya, and China from one hundred to five hundred.

There are three factions to Sami-ul-Haq's political party, known as the Jamiat-ul-Ulema Islam or JUI. JUI is a powerful player in the religious opposition. Sami-ul-Haq's faction is known as JUI-S. Fazlur Rahman leads a second faction known as JUI-F. And Ajmal Qadri leads the newest faction, known as JUI-Q. Each party has its own set of *madrassahs,* and each has its own militant wing.

Qadri is a laughing Buddha-like figure. He wears the pristine, starched, white dress of a Muslim cleric. He has great expanses of soft brown flesh and a large, imposing belly. He has a twinkling eye, and his pointy red slippers are decorated with gold and mirrors, giving him the look of a well-fed elf. I asked him how he felt about Haider's policies. "Haider is on the wrong track," Qadri explained. "If he persists with his current policies, the Ministry of Interior will cease to exist. If he goes further, he is in danger of losing his life." The *madrassahs* have had their own traditions for twelve hundred years, Qadri explained, and will not stand for the changes the current Pakistani regime is attempting to impose. The goal of the *madrassahs* is to propagate Islam throughout the world, not to satisfy the West. "Why should we expand our curriculum?" Qadri asked. "It was designed twelve hundred years ago in Iraq" and is far superior to the curriculum taught in secular schools. Haider is demanding that *madrassah* students learn to use computers. "We already have computers *at our madrassahs,"* Qadri explained. The students use the computers to study Koranic law. There is a downside, however, which is that students sometimes look at pornography, and they sometimes chat with girls. In one of Qadri's *madrassahs* in Peshawar, a student started chatting with a young woman in Germany. The girl, who was Christian, was so taken by Qadri's student that she traveled all the way to Peshawar. This caused terrible problems for me, Qadri explained. But the girl is a computer specialist who told Qadri she could earn $70,000 per year in the West. With Qadri's approval, she converted to Islam and married his student. She observes Deobandi practices, never leaving her house without donning a burka (a mode of dress adopted by the Taliban in which a woman's head is entirely covered, with tiny slits cut out of the fabric in front of their eyes).

"I believe that a clash of civilizations is inevitable," Qadri explained. "And in this clash, the fittest will survive. We are much more cultured than America and the West. The West is bereft of the strength that comes from families. Plus, the West is run by Jews. Americans and the Jews have begun a new crusade, which is known as globalization." Because Islamic thought is far more modern and scientific than the system adopted by the West, Islam is bound to win this unavoidable jihad against the Jews.

Asked whether he had ever signed a form promising not to send his charges to fight in a jihad, Qadri said that students who don't want to go don't go, nobody is forced. I asked what percentage of his students decide not to join a jihad. "Zero percent," he said. But only 10 to 15 percent are selected to be trained. "Young people are demanding to join a jihad because of their resentment against the West." Those who are not selected become hakim (doctors of traditional herbal medicine), *qazi* (Islamic judges), or clerics. In addition to the twelve-hundred-year-old curriculum that includes the study of the Koran and Islamic law, students are trained in *dutka,* a form of wrestling developed three thousand years ago in India. Only the most physically fit of Qadri's students are selected to be trained at his camp, which is called Forward Kahuta. The problem, he told me, is that he just doesn't have enough room in the camp to take more recruits. And only 5 percent of those trained, Qadri said, get "launched" to the Indian side.

Of 700 students at Qadri's *madrassah,* 127 are foreigners, including students from Burma, Nepal, Chechnya (7), Bangladesh, Afghanistan, Yemen, Mongolia, and Kuwait.

Over dinner, a senior member of Jamaat-i-Islami (JI) explains that Musharraf and his government are making statements about *madrassahs* to please the West. They are trying to market themselves as fighting the fundamentalists, he tells me. They learned this marketing plan from Benazir Bhutto. Not a single *madrassah* has been raided. Nor has action been taken against a single *mujaheed.* The plan is a posture, not a policy, he tells me.

The next day I meet with Mujeeb-ur-Rehman Inqalabi, a senior member of Sipah e Sahaba Pakistan (SSP), an overtly sectarian, anti-Shia political party. SSP has a profound influence on all Deobandi *madrassahs,* Inqalabi tells me. The administrators of Deobandi *madrassahs* invite SSP leaders to their annual ceremonies. SSP members are fighting in Kashmir, Afghanistan, Chechnya, and Bosnia, he tells me. We are close to Harkat-ul-Mujahideen and Jaish-i-Muhammad, he tells me, and close to the Taliban. He doesn't tell me this, but his group is also close to Al Qaeda. We supply militants to these groups. Whenever one of our youngsters wants to do jihad, he joins one of these groups, he tells me.

This is where Pakistan is in trouble, I think to myself. The government is deeply worried about sectarian killings inside Pakistan. But the same organizations that supply mujahideen to Pakistan's proxy war with India are producing sectarian killers. And eventually those groups, I am convinced, will turn against Pakistan itself for its insufficient orthodoxy.

Inqalabi takes me to a large *madrassah* in the middle of Lahore called Jamia Manzoor ul Islamiya. Pir Saif-ulla Khalid is the principal of the school. The pir (the word means "mystic" or "saint") is clothed all in white. He wears an *imama,* the headdress of the Taliban, said to have been worn by the Holy Prophet. He carries a large staff when he walks and insists on holding the staff when I persuade him to be photographed.

He has a long, white beard, and the charisma of a mystic. But he is a mystic with an angry edge.

Our meeting takes place in a large air-conditioned room lined with bookshelves. I note that the shelves are entirely devoid of books. Four hundred and fifty students live at the school, and another hundred are day students. Most of them, Pir Khalid tells me, come from poor families who cannot afford to feed the children. I ask about his own training. How did he come to be the principal of a school? He studied in *madrassahs,* he tells me. He has no interest in math or science. We are not interested in creating "balanced persons" here, he tells me. We teach them for God, not to help them get a job or find a wife. When a student masters our subjects, girls will gather at his feet.

I am allowed to wander around and speak with the students. The students sit on the floor, their Korans and their elbows resting on long, narrow tables. Twenty students at each table sway as they recite the words of the Holy Koran. Their bedding is on shelves along the wall. At night they sleep in the classroom on the floor. The first student we talk to is twelve years old. He studied in school through the fifth grade. That was the end of his secular schooling because his family cannot afford any more. His elder brother also stopped going to school at the end of fifth grade. Now he is learning the Koran by heart. I ask what he wants to do when he grows up. He tells me he wants to be a *mujaheed,* to kill the non-Muslims who are oppressing Muslims. How does he feel about Shia? They are kafirs (infidels), he says, they are not Muslims. How does he feel about America? Down with America, he says. I ask why he feels that way. Everyone says that, he says. But do you know why? I ask him. No, he says.

Asked about the biggest threat to their groups' survival, a militant says that "free secular education for all" leading to an "increase in the literacy rate" is the gravest threat to the survival of the jihadi groups in Pakistan.[59] Another answers, "Wealth. It is a hard fact that poor people are more religious and more likely to join militant groups."[60] A former militant who now runs a religious school that supplies youth to various Islamic extremist groups in Pakistan worried that "the spread of secular educational institutions" is the biggest threat to the groups. "But we in *madrassahs* are providing free education and free board, while the secular educational institutions are costly and few in number." He described a kind of mutual dependency between the *madrassahs* and the jihadi groups. "To survive in this field [running a *madrassah*], you need a strong group to back you," he said. But the groups, by the same token, "cannot survive without *madrassahs.*" Asked whether there might be psychological reasons for his interest in religious militancy, he answered that there might be, he wasn't sure. He describes his childhood as miserable. "People like us cannot dream of a happy childhood. Happiness these days is deeply associated to wealth. I was from a desperately poor background. There was always a shortage of food, which was a major cause of domestic dissatisfaction." The happiest moment of his childhood was when he was admitted to a *madrassah,* he said, because "there I got food, clothing, and other necessities."[61]

Asked about the most important audience for their recruitment activities, a former public-relations officer for a sectarian jihadi group said, "There are two groups we try to reach. One is the poor. We are in contact with them through the *madrassahs.* We need their children." But he added, "We work hard to establish contacts with the upper class so that we can persuade them to donate money to the group."[62]

Later I met with a Kashmiri who decided to quit militancy. He had been a high-ranking member of a Kashmir-based group called Hizb-ul Mujahideen. He lives on the outskirts of Islamabad, in a poor part of town. My friends arrange to drive me there. Unlike the other militants I have met so far, he is Kashmiri rather than Pakistani.

Our car pulls up to an old apartment building apparently made of cement. Pieces of the building have fallen into the open sewer and no one has bothered to repair it. You have to cross the sewer to enter the building. Visitors leave their shoes in the outer hallway, as is common in observant Muslim homes.

Ameer ul-Azeem, the public affairs officer for Jamaat-i-Islami, tells me that the ISI is angry with this militant, and that he in turn feels hounded by his former supporters in the Pakistani government. They are pressuring him to help plan operations. He is reluctant to assist them but feels compelled to submit to their demands. I don't ask what they threaten him with if he doesn't comply. He is obviously nervous and reluctant to speak openly with anyone he doesn't know well. Ameer reassures him that I am an American professor, neither from the ISI nor from India's intelligence agency, known as RAW. Most of the time operatives give me names—whether true or false, sometimes requesting that I change their names for publication. But this militant doesn't bother giving me a false name, as though he has no energy left for such pretensions. It is as though he has no false names left. Later, however, he will reveal his name to me through Muzamal urging me to use his name in this book. It is Maqbool Pandit. I notice that his arms are muscled and appear strong, but that his face has a look of nearly unbearable pain. I soon learn that his daughter is still in Kashmir, and that the Indian authorities have not allowed her to join her father in Islamabad.

Pandit invites us to sit on the floor of a small, inner room with no windows. He serves us green tea with ginger and spicy chicken, in the Kashmiri style. I notice him eating hungrily.

Why did you quit militancy? I ask him.

"Most operatives are bachelors. I have a responsibility toward my children," he tells me. I don't believe that this is the main reason, but I can see he's not going to tell me any more. I ask about his current occupation. He is working at the stock exchange.

Compared with the Pakistani militant leaders I have met, Maqbool Pandit is poor. I wonder whether this is because he is Kashmiri or because he has decided to leave militancy. He earned a master's degree in history prior to becoming involved in militancy and had thought about getting a doctorate in economics. But he joined the struggle early on, right after the 1987 elections—the beginning of Kashmiri militancy. At that point he had been running a business, exporting timber and fruit. But he gave up his business to help launch the armed struggle. He is exactly my age, which was forty-three at the time.

"At first the casualty rate was high on our side," Pandit says, "but since 1992 our efficiency has greatly improved. We are far more professional than the Indian army is now. Our men are better trained and our technology is superior. For example, we use remote-controlled explosives." Indian scholars and officials deny this claim.

How did you raise money? I ask.

"We procured funds from ethnic Kashmiris in Pakistan and around the world."

Most of the money actually was from Pakistan and Saudi Arabia, other militants told me. Did Pandit receive instructions about what he was allowed to say, I wonder?

Where did you train operatives?

"In Azad [Pakistan-held] Kashmir. Trainers and organizers often end up moving to Pakistan, as do their families."

How did you feel when Pakistani youth volunteered to join your organization?

"We are happy when people come to help us; they are sacrificing their lives for us. It's a blessing for us."

Not all Kashmiri militants feel this way about the influx of Pakistani youth into Kashmir to join in their struggle. One told me that after the Soviet withdrawal from Afghanistan, the jihadis found themselves out of work. "Hundreds of trained cadres from the Afghan war were looking for new hunting grounds, creating a law and order problem for Pakistan. Sending these trained cadres to Kashmir was an efficient means to take care of the problem, while at the same time run a low-intensity war against India." Pakistan eventually gained complete control over the militancy in the valley, he said, "by encouraging religious rhetoric and frenzy. Personnel were recruited from three groups: ignorant religious extremists driven by hate, criminals, and outcasts form the lowest rungs of society." These three groups developed a vested interest in the business of jihad, and "that served Pakistan's designs well."[63] Once Pakistan took charge of the militancy, Kashmiri militants became second-class citizens, he said.

At this point I ask to use the bathroom. The militant directs me through the hallway past the kitchen. I see women behind a kind of screen. They ignore me. The heat from the kitchen stove is searing as I walk by. Outside the bathroom door are rubber slippers to protect one's bare feet as one squats on the toilet. There is soap in the sink, and the towels are perfectly clean, perhaps in my honor.

I return to find the men speaking animatedly. Pandit seems more comfortable.

Where did you find guns and ammunition? I ask.

"Most come from the Indian army," he says. "From individual officers who are corrupt. It's not a policy decision on the part of the Indian government." I asked Indian government officials and scholars to comment on Pandit's claim about the militants purchasing Indian military weapons. According to them, weapons have serial numbers, and if a significant fraction of the weapons seized from the militants were actually acquired from the Indian military, such a thing would eventually become known. "You would have a mutiny in the forces if such a thing was found out—men who are sent out to die by their commanders or leaders do not appreciate the fact that the enemy may have been armed by the same commanders or leaders," one scholar told me.[64]

This is shocking to a foreigner, I say.

"It surprised us too at first," he admits. "We are living in the poorest region of the globe. Poverty gives birth to things you can't understand in America. The Indian army is now completely corrupt. Corruption has risen a hundred times because of Kashmir. We procure arms and ammunition from them for money. We also procure information and safe passage. They arrest innocent people and demand payment to release them. They steal timber, take fruits from the garden. They steal whatever they can. The Indian army is now ruined."

Why do you think they are so corrupt? I asked Pandit.

"It's human nature. They are poorly paid. No one wants to join the army anymore. Kashmir is a bad posting. The soldiers feel they are fighting their own people. The soldiers are poor. They have no skills. Common soldiers become very unhappy. But it's mostly the officers who are corrupt.

"RAW has a strong network in Kashmir. RAW tries to create infighting among the militant groups. But we have our own people inside RAW." Indian officials deny this claim.

"The officers want to minimize fighting with us. There are also border security forces. There are paramilitaries and police forces. The various units work against each other. Each is answerable for its own performance. Sometimes they allow mujahideen to pass through their region unharmed. They may even let us live in certain areas, provided we promise there will be no activity there. We sit down with the officers over a meal, work out an arrangement. If a *mujaheed* is arrested, and only a few officers know of his arrest, sometimes they will release him for money. But our resources are limited. There is money outside [Indian-held] Kashmir, but inside we are often living hand-to-mouth."

Some of this recitation may be true, but some of it—including the claim that jihadis are living "hand-to-mouth"—strikes me as propaganda.

By this time I have some ideas about the grievances that give rise to holy wars—not just the slogans, but the deeper pain. My Kashmiri interlocutor was clearly an intellectual. As an experiment, I decide to ask him a more broad-reaching question, one that I have never put to a militant before: Perhaps if we spoke about more general issues he would be less afraid to share his true views. What do you think is the deep cause of militancy?

He is reluctant to respond. If he has been briefed before my arrival about what to say to me, perhaps the ISI did not prepare him to answer such a question. Perhaps he is modest. "I am not a political scientist." He asks me to tell him my view.

Normally I don't discuss my views with militants, but this time I feel the urge to have a real conversation.

My impression, I tell him, is that no one really cares about the Kashmiris. Neither the Indians nor the Pakistanis, nor any government anywhere. Both sides are determined to retain the entire disputed territory. It is not about religion, despite the Pakistani jihadi groups' rhetoric. Nor is it about self-determination. Perhaps it was when the conflict first began, but today these are slogans to manipulate youth. It has nothing to do with the Kashmiri people's plight. This fight is about real estate, national identity, political power, and profits—both personal and organizational. The fight is kept alive because organizations depend on it and because, on both sides, people are making a living. Smuggling goods. Selling arms. Lending money. Running camps. Running "charities." Training vulnerable young men to believe that the way to feel important and useful is by killing and getting killed in a purported holy war. The jihadi leaders live in mansions, while their operatives risk their lives. Agencies on both sides profit—professionally and financially. Why would they want this "jihad" to end? I ask.

What really counts, I say, are perceived humiliation, relative deprivation, and fear—whether personal, cultural, or both. The rest is sloganeering and marketing. I see this all over the world, I tell him, including in America. But holy wars take off only when there is a large supply of young men who feel humiliated and deprived; when leaders emerge who know how to capitalize on those feelings; and when a segment of society—for whatever reason—is willing to fund them.

Sometimes, the segment willing to fund the terrorist group is a government agency. Governments use militants for special operations against internal enemies or as informal armies against outside powers; but governments can lose control of their surrogates, especially if diasporas and "charities" are also willing to provide funding. The leaders need to make young men feel that their lives are worth more as holy warriors purifying the world than as ordinary citizens, a task that is presumably easier when ordinary jobs are hard to find and ordinary life doesn't make the young men feel valuable and important. And holy wars persist only when organizations and individuals profit from them—psychologically or financially.

After a silence that stretches almost to the point of discomfort, Pandit says, "This is exactly right. Sometimes the deprivation is imagined, as in America. In Kashmir, it's real. But it doesn't really matter whether it's real or imagined." And it doesn't seem to matter whether it's contemporary or historical. A skilled terrorist leader can strengthen and harness feelings of betrayal and the desire for revenge.

After visiting the leaders and their cadres, I had a pretty good sense of how a cadre organization works. In the next chapter we study an organization that combines the advantages of both cadre organizations and leaderless resistance.

NINE

The Ultimate Organization: Networks, Franchises, and Freelancers

One of the surprises of September 11 was that some of the suicide bombers had been living and studying in the West for years. We like to think that our way of life and the freedoms we enjoy are so attractive that anyone who lives among us will inevitably become pro-Western. The globalization of Al Qaeda—its recruitment of locals to participate in attacks, and its careful grooming of operatives, were discussed by the terrorists themselves in a New York City courtroom, where four of the 1998 African-embassy bombers were tried a year and a half before September 11. It is too bad that the terrorists' revelations, including about the organization's vast business holdings, its detailed planning of operations, its emplacement of sleepers, and its attempts to acquire weapons of mass destruction, didn't receive more attention. If they had, perhaps we would not have been so astonished by Al Qaeda's ability to operate inside America.

This chapter begins with a discussion of a terrorist who participated in the bombing of the U.S. embassy in Dar es Salaam, Tanzania, in August 1998. His story is important for two reasons. First, he was a sleeper. A "talent scout" noticed that he attended a radical mosque regularly, and that he was increasingly agitated about the plight of Muslims around the world. Told that he would have to be trained at a camp to earn the trust of his new Islamist friends, he spent his own money to travel to Afghanistan. The real purpose of his training was to assess his potential. He was found to be barely educated, with few skills. But he had something else critically important to Al Qaeda at the time: language skills and Tanzanian citizenship. This is exactly the kind of operative that Americans are beginning to fear—a confused young man who thinks he is helping Muslims by serving as a sleeper for a terrorist group, whose principal value to the terrorists is his country of residence. Now we fear that the terrorist sleepers may be our next-door neighbors.

The second reason this operative's story is important is that he comes from Africa, an area of the world that may well become an enclave of Islamist extremism and anti-American sentiment in the future. Americans tend to fixate on enemies that can be fought with military might. We have a much harder time seeing failing states, where terrorists thrive, as a source of danger. We need to assess why bin Laden's and other extremists' ideas spread. And we need to look for clues globally, not just in the Middle East.

America has had the luxury of ignoring countries at far geographic remove throughout most of its history. This is no longer possible. Nor is it sufficient to concentrate exclusively on one or two villains in a given decade. We have to be alert to the possibility that the villain may be a seductive, hateful idea about Us versus Them, rather than an individual; and that the hateful idea may be taking hold—in seemingly obscure or remote locations. The growing availability of powerful weapons, porous borders, and the communications revolution make it possible for smaller and smaller groups to wreak havoc almost anywhere on the globe.

In the spring of 2000 two American defense attorneys contacted me to ask whether I would be willing to serve as an expert witness in the trial of Khalfan Khamis Mohamed, an Al Qaeda operative who was involved in the bombing of the U.S. embassy in Dar es Salaam, Tanzania, in August 1998. That attack, and the simultaneous bombing of the American embassy in Nairobi, Kenya, killed 224 people, most of them Africans, and injured thousands.

Mohamed had already admitted his guilt at the time his lawyers called me. He had told the FBI that he had rented the house where the bomb was built, bought the truck used to transport components, bought a grinder for grinding the explosive, and ground some of the TNT himself. After the bombing, he fled to South Africa with a new identity, a new passport, and $1,000 in cash, this last procured for him by Al Qaeda.

After a worldwide manhunt lasting longer than a year, South African authorities found Mohamed in Cape Town, working at an Indian fast-food restaurant called Burger World. The South African government extradited him to the United States. The U.S. government wanted him executed for his crimes. Mohamed's lawyers wanted my help in arguing that his punishment should be to spend the rest of his life behind bars in a maximum-security federal prison, but that he should not be put to death.

Khalfan Khamis Mohamed was born in 1973 on the island of Pemba and grew up in the village of Kidimni on Zanzibar Island. His twin sister, Fatuma, was born in the evening, but he didn't arrive until morning, giving his mother a lot of trouble, she recalls. But from that point on, she says, "He was just an ordinary child who went to school. . . . After school he performed the normal domestic chores and liked playing football, like all youth. He didn't indulge in any antisocial behavior."[1]

The family was poor. They lived in a mud hut with a thatched roof. His father died when Mohamed was six or seven years old. People on Zanzibar don't pay close attention to dates, and Mohamed's mother doesn't recall exactly when her husband died. After the death of his father, Mohamed helped his mother support the family by working on the farm, harvesting fruits that grow wild in the forest, and taking care of a neighboring farmer's cows.

Mohamed comes from a very different sort of place than many of the terrorists discussed so far in this book—a place that, ironically, benefited from globalization long before the term become popular. Zanzibar consists of two islands: Zanzibar (known locally as Unguja) and Pemba. The islands are in the Indian Ocean, twenty-five miles off the coast of Tanzania, six degrees south of the equator. Clove, jackfruit, mango, and breadfruit grow in the valleys of Pemba Island. Coconut trees, brought by Indian traders centuries ago, now grow wild. Monkeys, civets, bushpigs, and mongooses thrive in the forests. Some one hundred species of birds live in Tanzania, and thirteen species of bats have been identified on Pemba. The islands are also famous for their butterflies and the great variety of game fish found in the waters between them. Fishing and agriculture are Zanzibar's main industries.

Today, Pemba and Zanzibar are largely isolated from the rest of the world. Foreign visitors tend to be adventurers attracted by the lush, undisturbed reefs or the profusion of game fish found in Pemba Channel. Visitors describe an extraordinarily friendly people who seem utterly mesmerized by their foreign looks and ways. They write of the remarkable melee of cultures—African, Arab, Persian, and Indian—magnificent Arabic architecture, abundant fruits and fishes, but also poverty and squalor, the scent of spices rising above the stench of sewage and rotting fish.

Although it is relatively isolated today, Zanzibar was once the trading center for all of Africa, with trade links to Arabia, China, India, Persia, and Southeast Asia. The nineteenth-century English explorer Richard Burton described Pemba as an "emerald isle" in a "sea of purest sapphire." The scent of cloves, he said, was enticing even from the sea. The people were a mixed race who had retained, despite their conversion to Islam, the skills of divination and other "curious practices palpably derived from their wild ancestry."[2] The traditional dhow, a single-masted ship with a lateen sail, used by Arab merchants for two millennia to sail on the monsoon winds, is still in use today and is still built in the same way—with a hull of mangrove or teak, and ribs of acacia—with no nails.

A succession of invading powers left remnants of their cultures and languages. Shirazi Persians, who settled on the coast of East Africa in the tenth century, intermarried with the locals, giving rise to an Afro-Persian race.[3] Omani Arabs, who settled on Zanzibar some six centuries later, have had the largest influence on the culture and language. The name Zanzibar is the Arabic expression for "land of blacks." Kiswahili, Tanzania's official language, contains a substantial fraction of Arabic, Farsi, and Hindi words, as well as some Portuguese and English ones.

Tanzania was formed as a sovereign state in 1964 through the union of Tanganyika, on the African mainland, and Zanzibar. Zanzibar and Pemba Islands have a separate government administration from the rest of Tanzania. Zanzibaris are seeking greater autonomy for their archipelago. They would like to reap more of the profits of the export of cloves, which the central government taxes heavily, and to control more of the tourist trade.

Tanzania's ruling party, and Tanganyika itself, are predominantly Christian. The ruling party refers to any threat to its rule as motivated by Islamism, which, ironically, may incite precisely the kind of extremism the ruling party fears. During the last decade, elections have been declared fraudulent by multiple international observers, and protests have been met with violence perpetrated by the police, who are predominantly Christian, against Zanzibaris, who are predominantly Muslim. To the extent that Islamism is indigenous in the region, it is found more on the mainland than on the islands, as well as in neighboring Kenya, although this could change. Zanzibaris are deeply disappointed that the United States did not protest Tanzania's tampering with the election results of 1995 and 2000 or the violence that ensued, although the government's crimes were published widely.[4] Although the region is remarkably tolerant historically, stimulated by its longtime exposure to multiple cultures, anti-Western Islamist sentiment could easily take root here if democracy fails and state repression continues.[5]

Muslims represent 97 percent of the population of Zanzibar, most of them Sunni. Shia represent 12 percent of the population. As in Indonesia, Islam coexists with Zanzibar's traditional religions, including animism. Zanzibar is famous for its sorcerers, seers, and witch doctors. Spells often involve Arabic texts, and witches often dress in traditional Arab garb. Evelyn Waugh wrote that novices came to Pemba from as far away as Haiti to study magic and voodoo. A cult of witches "still flourishes below the surface," he wrote, expressing his frustration that "everything is kept hidden from the Europeans."[6] Zanzibar is the home of a secret sect known as the Wachawi, who practice their arts even today. They are said to be able to take on the shapes of animals and birds. Haitian voodooists learned to animate corpses for labor in the fields by studying with the Wachawi, who reportedly developed the technique to escape their masters' notice when they fled bondage. The Wachawi are said to be able to bring the recently deceased back to life, with personality and memory intact. Locals describe their neighbors returning from midnight meetings in the bush, pale and speechless, having seen their recently deceased loved ones restored to life.[7] Early-twentieth-century visitors said that natives told them of powerful witch guilds, which required prospective members to offer up a near relation—a spouse or a child—to be eaten by other initiates.[8]

As a child, Mohamed attended a *madrassah* in the afternoons. The family described him as serious and quiet—more observant than his siblings, but also a better student. When he was in the middle of tenth grade, his older brother, Mohamed Khalfan Mohamed, asked Mohamed to come to live with him and his family in Dar es Salaam on the mainland to help out in the family dry-goods store. Mohamed intended to complete his schooling in Tanzania, but his time was taken up with his work at the shop and attending mosque. He had always been somewhat of a loner, his siblings recounted, but he became even more isolated after dropping out of high school, spending time only with his family and people he met at the mosque.[9]

The mosques in Dar es Salaam were more political than the one Mohamed attended in Zanzibar. There was a great deal of discussion about the plight of Muslims in Chechnya and especially in Bosnia. Worshipers were told that it was their duty to help fellow Muslims around the world in any way they could.[10] One of Mohamed's new friends was a man named Sulieman. Sulieman was from Zanzibar, but he worked on a fishing boat based in Mombasa, Kenya, owned by a man whom Mohamed knew only as "Mohamed the fisherman." Mohamed the Fisherman turned out to be Mohamed Sadiq Odeh, a Saudi of Palestinian origin who was a member of Al Qaeda. Odeh would play an important role in the embassy-bombing conspiracy.[11]

Sulieman introduced Mohamed to Fahid, who would also participate in the bombing, who visited Dar es Salaam only occasionally. Mohamed started spending much of his free time with Fahid and Fahid's friends, who were very religious. Sometimes they met in Dar es Salaam, and sometimes in Mombasa, Kenya. Mohamed says that they mainly talked about how to help Muslims around the world. Often, he said, they would meet in cars.[12]

By 1994, Mohamed began to despair at his own life, family members said. He spent more and more time at the mosque. He was radicalized in that mosque, his sister-in-law recalled.

Mohamed told Fahid he wanted to go to Bosnia to fight against the Serbs. Fahid told him that you cannot become a soldier for Islam without training. Fahid also told Mohamed that he did not trust him, and that he could earn Fahid's trust only if he went to Afghanistan to be trained. Mohamed saved his earnings from the dry-goods shop and in 1994 traveled with Sulieman to Pakistan. Fahid had given them a contact in Karachi, who arranged for their trip to the camp. Fahid had been at the camp for around a month when Mohamed and Sulieman arrived. Mohamed told FBI investigators that the camp was called Markaz Fath, and that it was run by a Pakistani jihadi group called Harkat-ul-Ansar (the group we discussed in the previous chapter). He said his teacher was a Pakistani named Abu Omar. Mohamed said that he met a lot of people at the camp, one of whom was an American known as Sulieman America. The people he met were interested in helping Muslims around the world, Mohamed said, and in waging a jihad against America and against conservative Muslim states. He said he had never heard the name Al Qaeda.[13]

During the first two months at the camp, the group was trained to use light weapons (handguns and rifles), launchers, and surface-to-air missiles. Mohamed and his friends Sulieman and Fahid were selected for advanced training, which included learning how to manufacture explosives and how to join detonators and wires. Mohamed was not trained in the use of chemical weapons, although he said that other members of his group were. Afternoons were taken up with Islamic studies—including films of atrocities perpetrated against Muslims in Chechnya and Bosnia—and sports. Mohamed stayed at the camp for nine or ten months, he says.[14] At the end of his training, Mohamed wanted to go to Bosnia, but he was not selected. He was told to leave a number in case he was needed at a later date. Mohamed went back to Dar es Salaam, bitterly disappointed that he had not been allowed to join the fight against the Serbs.[15]

Mohamed continued to spend time with the "brothers" he had met in the mosque or had gotten to know at the camp. He went to Somalia twice in 1997—once to teach Somali fighters what he had learned in Afghanistan, and once for a meeting with the men who would ultimately bomb the American embassy.[16] Just before his first trip to Somalia, Fahid introduced him to a man named Hussein, who would later lead the group that bombed the U.S. embassy in Dar es Salaam. Fahid told Mohamed that Hussein is our brother, that he is a good man who had been trained to be a *mujaheed.* Odeh, explaining how Mohamed fell under Hussein's influence, described Hussein as "persuasive, authoritarian," and "a very strong leader, a man of compelling personality." Mohamed was impressed by Hussein's knowledge of Islam. Sometime after this meeting, Hussein moved to Dar es Salaam with his family. They stayed with Mohamed in a small flat.[17]

Three years after he returned from Afghanistan, Hussein approached Mohamed to invite him to participate, in a "jihad job." Mohamed said that he would like to participate, although he was not informed about what the "jihad job" would entail. Eventually Hussein asked Mohamed to take certain actions. He instructed him to buy a truck, which Mohamed did in his own name. He paid for the truck, a white Suzuki, with cash that Hussein gave him. Fahid accompanied him and drove the truck because Mohamed did not know how to drive. The group used the truck to transport equipment needed for the bomb, including cylinder tanks, detonators, fertilizer, and TNT. Hussein also asked him to rent a house, large and private enough to conceal the group's activities. Mohamed remembered Hussein telling him that he wanted the house to be hidden from the street, but that it should also be "nice." Mohamed found a house with a high wall, which he rented in his own name. The owner insisted that Mohamed pay a year's rent in advance, which he did, with money Hussein gave him.[18]

Mohamed, Hussein, and Hussein's family moved into the house in the Ilala district of Dar es Salaam. Other team members came to the house, but no one ever discussed his role in the plot. Hussein instructed Mohamed to remain in the house most of the time, so that if any neighbors came by, there would be someone who could speak to them in Swahili. Other team members arrived soon before the bombing: an engineer named Abdul Rahman, whom Mohamed described as working with "all confidence"; and "Ahmed the driver," whom Mohamed thought was Egyptian. Ahmed was the suicide bomber who would drive the truck into the embassy. Some five days before the attack, Hussein told Mohamed that the target of the bombing would be the American embassy. Mohamed helped load the tanks, boxes of TNT, and sandbags into the back of the truck. When the truck got stuck in the sand behind the house, Mohamed helped the driver dig it out.[19]

Hussein and the rest of the team left several days before the bombing. Most of them said they were going to Mombasa, without specifying their final destination. In fact, they had been instructed to return to Afghanistan before the bombing took place. Hussein asked Mohamed to remain in Dar es Salaam, to help the driver with any last-minute details, and to remove incriminating evidence from the house. Mohamed did as he was told, with one exception. He did not like the idea of throwing away the food grinder he had used to grind the TNT, since it was still usable. So he gave it to his sister Zuhura, asking her to clean it well and to pass it on to his mother.[20]

When he was captured by the FBI in October 1999, Mohamed told investigators he was not sorry that Tanzanians were killed, which he said was part of the business. He said he had bombed the embassy because it was his responsibility, according to his study of Islam. He said he thought the operation was successful because the bomb worked, it sent a message to America, and because it kept American officials busy investigating it. He also said that if he had not been caught, he would continue participating in the jihad against America or possibly against Egypt, and that if the U.S. government were to release him from custody, he would bomb Americans again. He told his investigators that he thought about jihad all the time. He told them he wants Americans to understand that he and his fellow warriors are not crazy, gun-wielding people, but are fighting for a cause.[21]

I travel to New York to watch Mohamed's trial. Security is tight. The taxi drops me several blocks from the entrance to the courthouse because the street is blocked to traffic. You must pass through several layers of security before you get to the room where the trial is being held. There are metal detectors and guards on the first floor, and you have to show identification and sign in outside the courtroom. A guard is suspicious about why I am here. I explain that I am a defense-team visitor, and an agent instructs me to sit in the third wooden bench on the right. I can see from the back of the room that the bench is already full. When she sees that I mean to sit there, a woman pulls a child onto her lap and slides closer in toward her neighbor on the hard wooden bench. This is Mohamed's family, I realize. The women wear bright Zanzibar cottons. The boys and men wear prayer caps. The little boy immediately to my right is wearing pressed white cotton. He stares at me with velvety eyes, not at all shy, seemingly delighted with the opportunity to examine such a strange foreign creature, whom good fortune has brought conveniently near at hand. His mother is too distracted to notice his staring and he is free to inspect every inch of me, which he does with obvious pleasure. It is a hot day. I notice the smell of anxiety in my benchmates' sweat, but also the pleasant scent of spices. I see Mohamed's mother at the far end of the bench. She sits tall, with dignity, but she looks modest and kind. She appears surprisingly calm, at least for now. There are brothers, sisters, children, and spouses also sharing the bench, as well as the family with whom Mohamed lived when he fled to South Africa.

A social worker has been called up to the witness stand to provide Mohamed's social history. She has traveled to Zanzibar twice and shows the court pictures of Mohamed's school, the neighborhood where he grew up, and the take-out restaurant where Mohamed worked as a chef in Cape Town. When she is done, various members of Mohamed's family are called up to the stand. Each is asked what they remember about Mohamed. An older brother remembers him as good in school and good at soccer. Mohamed was kind and peaceable, he said, and would always try to break up fights. A younger sister recalls him helping her with her schoolwork. Another says that Mohamed played games with her children, his nieces and nephews. The mother of the family for whom he worked in

Cape Town recalled how patient and kind Mohamed had been with her children and her elderly parents. He even taught her elderly mother to read the Koran. She said that she would gladly have given up her daughter in marriage to Mohamed. All but one of Mohamed's family members said it was their first time traveling by airplane or traveling abroad.

The last witness was Mohamed's mother, whose name is Hidaya Rubeya Juma. There was a hush in the room as a large lady dressed in bright cottons and a turban took the stand. I saw Mohamed looking down as his mother took her seat. It seemed to me that Mohamed had a harder time facing his mother than he did facing his victims or accusers. There was jolt of pain in the room, as though the air had been ionized with terror—his and ours. Not a fear of death, but the recognition of evil. The recognition that this person who had killed so many has a mother who loves him, despite his crimes, and that he is afraid to look her in the eye. That despite his evil actions, he is human, just like us. It is one thing to understand this intellectually. It is another to see a mother face her killer son, with his many victims looking on, seeing her fear, her agony, and her loss. The loss of her son—first to evil, and maybe to death.

Mohamed's attorney, Mr. David Ruhnke, asked Mohamed's mother, "After you leave and return to Africa next week, do you know whether you will ever see your son again?"

"I don't even know," she answered quietly.[22]

"Do you know what this is about, and that the people here have to decide whether your son is to be executed or put in prison for life? And I want to ask you a very difficult question, which is, if your son were executed, what would that do to you?"

"It will hurt me. He is my son."[23]

Soon after this, the court was adjourned. Hidaya Rubeya Juma was the last witness to appear in the penalty phase of Mohamed's trial. Closing arguments began at the next session.

In his closing arguments, the prosecutor, Mr. Fitzgerald, emphasized what he referred to as Mohamed's two-sided personality. "I submit to sit before you and tell you that Khalfan Mohamed's personal characteristics as an individual human being include the following: one, Khalfan

Mohamed has exhibited responsible conduct in other areas of his life; two, Khalfan Mohamed has shown himself to be a person capable of kindness, friendship, and generosity; and three, Khalfan Mohamed lost his father at an early age and worked to help his family, which struggled financially after the death of the major breadwinner." Mohamed can be very kind, Fitzgerald adds. "You want him to marry your daughter. You wouldn't think he would hurt an ant. The next day he is in custody, saying, 'Yeah, I bombed people and I'll do it again.' That's what he is. He's got two faces. . . . He fooled his family. . . . He is capable of savagery."[24]

Jury members concluded that, if executed, Mohamed would be seen as a martyr and that his death could be "exploited by others to justify future terrorist acts." He received a life sentence without parole.

When authorities interrogated Mohamed Sadiq Odeh in Pakistan, where he had flown on the day of the bombing, he admitted that he was a member of Al Qaeda and gave his interrogators the names of some of the Al Qaeda members involved in the plots. He also referred to "two or three locals," whose names he appeared not to know, who had been left behind in Dar es Salaam and Nairobi to finish the job. One of those expendable locals was Mohamed.

According to several Al Qaeda members who testified at the trial, Al Qaeda is highly "tiered," and for the most part, Africans were not admitted to the upper ranks. Mohamed was recruited as a sleeper because he had a passport, language skills, and would not stand out as a foreigner in Dar es Salaam. Odeh explained to the FBI that there are several types of Al Qaeda operatives: sophisticated operatives who are involved in intelligence collection, choosing targets, surveillance, and making the bombs. But another category of operatives includes "good Muslims" who "are not experts in anything that would have a long-term benefit to the rest of the group."[25] The main thing they have to offer is their knowledge of the local languages and customs.

These dispensable young men, recruited to act only in the implementation phase of an attack, are unlikely to join Al Qaeda in a formal sense. They are often identified in the mosque, Odeh said. Atrocities against Muslims—anywhere in the world—help to create a climate that is ripe for recruiting young men to become soldiers for Allah. It is not even necessary to mention the name Al Qaeda to recruit them, Odeh told Jerry Post, a psychiatrist who interviewed him.[26] It is possible that many of the American, British, and Southeast Asian sleepers that law-enforcement authorities continue to discover all over the world were recruited to play a similar role. Like Mohamed, the group of Yemeni Americans taken into custody in September 2002 apparently went to Afghanistan for a relatively short course of training. In the camp, potential recruits' skills and commitment can be closely observed so that trainers can funnel them into the appropriate tier of the organization. Because of Al Qaeda's strict policy of sharing information only on a need-to-know basis, sleepers—who serve as a kind of reserve army in the targeted country—are unlikely to know precisely for what they have been recruited until immediately before an attack.

Some of the most important revelations of the trial were contained in an Al Qaeda instruction manual called the "Declaration of Jihad against the Country's Tyrants," which was entered into evidence. The manual makes clear that intelligence and counterintelligence (avoiding detection by the enemy intelligence agencies) is a priority for Al Qaeda. It instructs sleepers in the art of disappearing in enemy territory by shaving their beards, avoiding typical Muslim dress or expressions, not chatting too much (especially with taxi drivers, who may work for the enemy government), and wearing cologne. Sleepers are urged to find residences in new apartment buildings, where neighbors are less likely to know one another. Found by the Manchester (England) Metropolitan Police during a search of an Al Qaeda member's home, the manual was located in a computer file described as "the military series" and was subsequently translated into English.[27] In the "first lesson," the manual describes the "main mission for which the Military Organization is responsible" as "the overthrow of the godless regimes and their replacement with an Islamic regime."[28] The second lesson spells out the "necessary qualifications and characteristics" of the organization's members, which include a commitment to Islam and to the organization's ideology, maturity, sacrifice, listening and obedience, keeping secrets, health, patience, "tranquillity and unflappability," intelligence and insights, caution and prudence, truthfulness and counsel, ability to observe and analyze, and the "ability to act."[29] Subsequent "lessons" teach the trainee how to forge documents, establish safe houses and hiding places, establish safe communications, procure weapons, and gather intelligence. A large number of training manuals have been discovered in Afghanistan and elsewhere.[30]

Witnesses at the trial explained the structure of the organization in some detail. Bin Laden was known as the "emir," or leader. Directly under him was the Shura Council, which consisted of a dozen or so members.[31] The Shura oversaw the committees. The military committee was responsible for training camps and for procurement of weapons. The Islamic Study Committee issued fatwas and other religious rulings. The Media Committee published the newspapers. The Travel Committee was responsible for the procurement of both tickets and false-identity papers and came under the purview of the Finance Committee. The Finance Committee oversaw bin Laden's businesses.[32] Al Qaeda had extensive dealings with charitable organizations. First, it used them to provide cover and for money laundering. Second, money donated to charitable organizations to provide humanitarian relief often ended up in Al Qaeda's coffers. Finally, and perhaps most importantly, Al Qaeda provided an important social-welfare function. It was simultaneously a recipient of "charitable funds" and a provider of humanitarian relief, a kind of terrorist United Way.

In this sense, Al Qaeda is similar to the Pakistani and Indonesian jihadi groups we have examined in earlier chapters. Al Qaeda has a clear hierarchy. There are commanders, managers, and cadres; and cadres consist of both skilled and unskilled labor. Foot soldiers are likely to be found in schools or mosques, and only the best and brightest make it to the top. Some midlevel operatives are paid enough inside the organization that they may find it difficult to leave, while for others—generally those who come from wealthier families—the spiritual and psychological attractions of jihad are sufficient. Information is shared on a need-to-know basis, as in an intelligence agency.

Several Al Qaeda functions are worth discussing in somewhat more detail: planning operations, relations with states, recruitment, training, developing the mission, and weapons acquisition.

PLANNING OPERATIONS

Some Al Qaeda operations take years to plan and implement, and sometimes the group reattempts attacks that failed the first time around. The idea to attack the World Trade Center appears to have originated well before the 1993 attack. Ramzi Yousef, who spent three years in a safe house provided by bin Laden prior to his arrest,[33] made clear to the FBI that he intended to knock the two buildings down, but that lack of funds had prevented him from achieving his ambitious goals. He had also plotted, together with his right-hand man, Abdul Hakim Murad, as well as Khalid Sheikh Mohammed, his uncle, to destroy eleven American airplanes midair, a plot that was successfully tested on a Philippine airliner in December 1994, killing one passenger and injuring at least six others.[34] The plot became known as the Bojinka Plot, which is Serbo-Croat for "the explosion."[35] Numerous reports have emerged that Al Qaeda had considered using airplanes as weapons before, including the widely reported plot to attack the CIA headquarters. Bin Laden admitted on videotape that he had not expected the Trade Center buildings to collapse, but that he had rejoiced in the surprising effectiveness of the attack.

For some operations, leaders are involved in detailed planning. Ali Muhammad, an Egyptian-born naturalized U.S. citizen who admitted conducting photographic surveillance of the U.S. embassy in Nairobi, told American investigators that bin Laden himself had looked at surveillance photographs and selected the spot where the suicide truck should explode in the 1998 attack.[36] But not all plots receive this level of oversight. Members of Al Qaeda in Jordan, for instance, who were arrested while preparing for attacks to be carried out during the millennium, were providing for themselves, rather than receiving lavish sums. Ahmed Ressam testified that he had been given what amounted to seed money for his planned attack in Los Angeles during the millennium. During the trial of Mokhtar Haouari, a coconspirator in the "millennium plot," Ressam testified that he had had to raise most of the funds on his own, which he did by making use of his long-standing expertise in credit-card, immigration, and welfare fraud; as well as other criminal activities such as theft and robbery.[37]

The attack on the USS *Cole* was originally planned on another U.S. destroyer, *The Sullivans*. The suggested target date for the attack on *The Sullivans* had been January 3, 2000, at the height of Ramadan. This first attempt to sink an U.S. warship failed when the explosives-laden boat sank.[38]

Al Qaeda is patient. A senior counterterrorism official of the FBI observes, "They plan their operations well in advance and have the patience to wait to conduct the attack at the right time. Prior to carrying out the operation, Al Qaeda conducts surveillance of the target, sometimes on multiple occasions, often using nationals of the target they are surveying to enter the location without suspicion. The results of the surveillance are forwarded to Al Qaeda HQ as elaborate 'ops plans' or 'targeting packages' prepared using photographs, CADCAM (computer-aided design/computer-aided mapping) software, and the operative's notes."[39] This sophistication, coupled with a wealth of financial and material resources, allows bin Laden's terrorist network to stage spectacular attacks.

RELATIONS WITH STATES

The jihadi groups we have discussed in previous chapters built up strong relationships with individual politicians, intelligence agencies, or various factions of divided governments. The Pakistani jihadis were long sustained by Pakistan's ISI and are still assisted by former ISI agents, who serve as trainers at terrorist-training camps. It is likely that some current ISI agents still support the jihadi groups, even after President Musharraf's post–September 11 promise to force pro-jihadi elements out.[40] As we discussed in chapter 3, active-duty military personnel helped to train Laskar Jihad mujahideen in Indonesia and have had a long-standing relationship with the leader of Jamaah Islamiyah, now closely associated with Al Qaeda.[41] As we discussed in chapter 2, Saddam Hussein offered cash payments to the families of Palestinian suicide bombers, and Saudi charities, purportedly unconnected to the government, do the same. Iran provides funding to a variety of jihadi groups around the world, including Sunni ones, as well as safe haven. Ali Mohamed, a witness for the U.S. government in the African-embassies bombing trial held in 2001, testified that Al Qaeda maintained close ties to Iranian security forces. The security forces provided Al Qaeda with bombs "disguised to look like rocks," he said, and arranged for the group to receive training in explosives at Hezbollah-run camps in Lebanon.[42]

But bin Laden went beyond cooperating with states and state agents. He made himself so indispensable to leaders willing to provide him sanctuary that the assets of the state became his to use. He built a major highway in Sudan. Bin Laden's businesses became major employers of Sudanese citizens. For example, Al-Damazine Farms, which manufactured sesame oil and grew peanuts and corn, employed some four thousand people.[43]

Bin Laden established a close personal relationship with Hassan al-Turabi, leader of the National Islamic Front in Sudan and a leading Islamist intellectual who was educated in the West. Al-Turabi was trying to establish an Islamic state in Sudan based on a strict interpretation of Islamic law. Bin Laden also worked closely with Sudan's intelligence agency and military. As a result of these relationships—and Sudan's financial dependence on bin Laden—he was able to build training camps, establish safe houses, and plan terrorist operations from Sudanese territory. The National Islamic Front supplied bin Laden with communications equipment, radios, rifles, and fake passports for his personnel.

Bin Laden made important foreign contacts while living in Sudan. During an Islamic People's Congress in Sudan in 1995, he met leaders of other radical Islamist groups, including Hamas and PIJ (Palestinian Islamic Jihad), as well as extremist organizations from Algeria, Pakistan, and Tunisia. Al Qaeda further extended its worldwide network of contacts through training, arms smuggling, or providing financial support to groups based in the Philippines, Jordan, Eritrea, Egypt, Yemen, and elsewhere.

After the U.S. government pressured Sudan to expel bin Laden in mid-May 1996, he moved his operation to Jalalabad, Afghanistan. He reportedly lost $300 million in investments that he was forced to leave behind. Despite these losses, soon after his arrival in Afghanistan, bin Laden began buying the services of the Taliban. He offered up members of his elite unit, the 055 Brigade, to assist the Taliban in its efforts to destroy the Northern Alliance.[44] Over five years, he gave the Taliban regime some $100 million, according to U.S. officials.[45] In return, he received the Taliban's hospitality and loyalty. According to Mohammed Khaksar, who served as the Taliban's chief of intelligence, then as deputy minister of the interior prior to his defection to the Northern Alliance in 2001, "Al Qaeda was very important for the Taliban because they had so much money. . . . They gave a lot of money. And the Taliban trusted them."[46]

Does Al Qaeda need the services of a state to continue to function as it did prior to September 11? I think the answer is that it probably does. But there is no reason to think that Al Qaeda and the International Islamic Front (IIF)[47] can't change their way of functioning so that the services of a state are no longer as critical. The IIF is a learning organization. The movement is beginning take on some of the attributes of groups we've studied in previous chapters, encouraging leaderless resisters, virtual networks, and lone-wolf avengers. The IIF is also increasingly relying on what I will call franchises—groups that have their own regional agendas, but are willing to contribute (including financially) to Al Qaeda's global, anti-American project when invited; and groups or individuals who may not be formal members but were trained at Al Qaeda's camps and are willing to work as freelancers.

WEAPONS ACQUISITION

Conventional

The Al Qaeda body responsible for the procurement of weapons is the Military Committee—one of four committees that are subordinate to the *shura majlis,* the consultative council of the network. Apart from being responsible for the development and acquisition of both conventional and unconventional weapons, the Military Committee is also in charge of recruitment and training, as well as the planning and execution phases of Al Qaeda's military operations.[48]

Al Qaeda acquires weapons and explosives from a variety of sources, depending on the type of operation and its location. The 055 Brigade, for instance—Al Qaeda's guerrilla organization that fought alongside the Taliban against the Northern Alliance—used weapons left behind by the Red Army. It also received weapons from the Taliban and the Pakistani intelligence service, the ISI.

During the 1990s, many of Al Qaeda's procurement officers obtained weapons in Western countries. During bin Laden's stay in Sudan, from 1991 to 1996, the establishment of businesses in the East African country provided much of the cover for the network's procurement of weapons.[49] Al Qaeda's global reach has enabled it to establish a worldwide network of procurement officers. One of them, according to terrorism expert Rohan Gunaratna, was bin Laden's personal pilot, Essam al-Ridi, a U.S. citizen who obtained communication equipment from Japan; scuba gear and range finders from Britain; satellite phones from Germany; night-vision goggles, .50-caliber sniping rifles, and a T-389 plane from America.[50] Al Qaeda has also procured weapons from Russian and Ukrainian organized criminal rings. Al Qaeda's and the IIF's links with organized criminal groups are likely to grow stronger in the aftermath of September 11, as many Western states are stepping up the pressure against Al Qaeda cells operating in some of these countries.

Unconventional Weapons

Bin Laden has repeatedly made clear his desire to acquire unconventional weapons. In January 1999 he told a reporter, "Acquiring weapons for the defense of Muslims is a religious duty. If I have indeed acquired these weapons, then I thank God for enabling me to do so. And if I seek to acquire these weapons, I am carrying out a duty. It would be a sin for Muslims not to try to possess the weapons that would prevent the infidels from inflicting harm on Muslims."[51] After September 11, he pronounced that he already possessed chemical and nuclear weapons.[52] Bin Laden's deputy Ayman Zawahiri wrote in his memoirs that "the targets and the type of weapons must be selected carefully to cause damage to the enemy's structure and deter it enough to make it stop its brutality," probably in reference to unconventional weapons.[53]

Chemical and Biological Weapons Iraqi chemical-weapons experts shifted some of their operations to Sudan after the Gulf War, according to CIA assessments released to the press. Bin Laden moved to Sudan at about the same time. Beginning in 1995, the CIA began receiving reports that Sudanese leaders had approved bin Laden's request to begin production of chemical weapons to use against U.S. troops stationed in Saudi Arabia.[54] Khidhir Hamza, the director of the Iraqi nuclear weapons program from 1987 to 1990, claimed that bin Laden's agents had contacted Iraqi agents with the aim of purchasing weapons components from Iraq. Saddam Hussein reportedly sent Ansar al-Islam, the terrorist group that attempted to assassinate the prime minister of the Kurdistan Regional Government, Barham Salih, to train in Al Qaeda camps.

Ahmed Ressam, one of the Al Qaeda operatives apprehended in the millennium plots, described crude chemical-weapons training at camps in Afghanistan, including experiments on animals.[55] In December 2000, special units of the Italian and German police arrested several Al Qaeda agents based in Milan, Italy, and Frankfurt, Germany, who had plotted to bomb the European Parliament building in Strasbourg, France, using sarin, a nerve agent.[56] Other evidence of the group's interest in chemical and biological weapons includes a manual that provides instructions for using chemical weapons;[57] a manual that provides recipes for producing chemical and biological agents from readily available ingredients;[58] and intercepted phone conversations between Al Qaeda operatives who were discussing unconventional agents.[59]

In August 2002, CNN bought a cache of Al Qaeda videotapes in Afghanistan that showed Al Qaeda's gruesome chemical-weapons experiments, substantiating earlier reports about experiments on animals. On one of these videotapes, several men are seen rushing from an enclosed room, shouting at each other to hurry; they leave behind a dog. After the men leave, a white liquid on the floor forms a noxious gas. The dog is seen convulsing and eventually dies.

A large cache of documents and other materials was found during the raid that led to the capture of Al Qaeda's operational planner, Khalid Shaikh Mohammed, in March 2003. The seized documents revealed that Al Qaeda had acquired the necessary materials for producing botulinum and salmonella toxin and the chemical agent cyanide—and was close to developing a workable plan for producing anthrax, a far more lethal agent.

Mohammed had been staying at the home of Abdul Quoddoos Khan, a member of Jamaat-i-Islami. Khan is reportedly a bacteriologist with access to production materials and facilities.[60]

The greatest worry, however, is that the International Islamic Front, possibly working together with Hezbollah or other terrorist groups, will acquire assistance from persons who have access to a sophisticated biological-weapons program, possibly, but not necessarily, one that is state run.

Nuclear Weapons The U.S. government has been concerned about Al Qaeda's interest in acquiring nuclear weapons since the mid-1990s. In early February 2001, Jamal Ahmad al-Fadl admitted that one of bin Laden's top lieutenants ordered him to try to buy uranium from a former Sudanese military officer named Salah Abdel Mobruk. The uranium was offered for $1.5 million. Documents described the material as originating in South Africa. Al-Fadl received a $10,000 bonus for arranging the deal. He testified that he does not know the outcome.[61]

U.S. government officials reportedly believe that Al Qaeda successfully purchased uranium from South Africa.[62] Mamdouh Mahmud Salim, a senior deputy to bin Laden, was extradited from Germany to the United States in 1998. The U.S. government accuses Salim of attempting to obtain material that could be used to develop nuclear weapons.[63]

Numerous reports have emerged that bin Laden has forged links with organized criminal groups based in the former Soviet Union, Central Asia, and the Caucasus in his attempts to acquire nuclear weapons.[64] Russian authorities suspect the August 2002 murder of a nuclear chemist may have been linked to a clandestine effort to steal the country's nuclear technology.[65] They also report that they had observed terrorists staking out a secret nuclear-weapons storage facility on two occasions, and that they had thwarted an organized criminal group's attempt to steal 18.5 kilograms of highly enriched uranium.[66] This last claim is unusual and alarming, in part because of the quantity—enough to make several nuclear weapons—and in part because the material was actually weapons-usable. Most press reporting about nuclear thefts turn out, after investigation, to refer to caches of low-enriched uranium or radioactive but not nuclear-weapons-usable materials.

American officials are suspicious about the activities of two Pakistani nuclear scientists, Sultan Bashiruddin Mahmood and Abdul Majid, who reportedly met with bin Laden, Ayman Zawahiri, and two other Al Qaeda officials several times during August 2001. Pakistani officials insist that despite Mahmood's experience in uranium enrichment and plutonium production, the two scientists had "neither the knowledge nor the experience to assist in the construction of any type of nuclear weapon."[67] The two scientists, who were eventually released, reported that during one meeting, Osama bin Laden declared he possessed "some type of radiological material" and was interested in learning how he could use it in a weapon.[68]

If Al Qaeda builds a nuclear weapon or already has one, it is probably a relatively crude device. An extensive study conducted by the Institute for Science and International Security in Washington found "no credible evidence that either bin Laden or Al Qaeda possesses nuclear weapons or sufficient fissile material to make them," but that if Al Qaeda obtained sufficient nuclear-weapons-usable material, it would be capable of building a crude nuclear explosive.[69]

RECRUITMENT

In the years following the Soviet invasion of Afghanistan, Al Qaeda's recruitment was conducted by the Maktab al-Khidamat (MAK—Services Office). Osama bin Laden and his spiritual mentor, the Palestinian head of the Muslim Brotherhood, Abdullah Azzam, established the MAK in 1984. The MAK recruited young Muslims to come to Afghanistan to fight the Soviet infidels. With branches in over thirty countries, including Europe and the United States, and a sizable budget, the MAK was responsible for propaganda, fund-raising, and coordinating recruitment. While bin Laden covered the costs for transporting the new recruits, the Afghan government provided the land, and training camps were soon established.[70]

Most Al Qaeda operatives appear to have been recruited by Islamist organizations in their home countries. A Spanish investigation in November 2001, for example, concluded that a group known in Spain as Soldiers of Allah gradually assumed control over the Abu Bakr mosque in 1994. It had financial ties with Al Qaeda and regularly sent volunteers for training in Bosnia, Pakistan, and the Philippines.[71] Surveillance of a key recruitment officer based in Italy, Abu Hamza, revealed a tightly linked network of Al Qaeda recruitment officers in Europe, which included Abu Hamza and Sami Ben Khemais in Italy, Tarek Maaroufi in Belgium, and Abu Dahdah in Spain.[72] In Germany, in addition to recruitment through mainstream Islamic associations and charitable agencies, Al Qaeda recruiting officers used amateur videos of fighting in Chechnya to attract recruits.[73] One two-hour-long recruiting video that was probably produced in the summer of 2001 showed a mock assassination of former president Clinton, along with footage of training bases in Afghanistan. Methodically, the film moves from picture frames of Palestinian children killed or wounded by Israeli soldiers and Muslim women being beaten, to pictures of "great Muslim victories" in Chechnya, Somalia, and against the USS *Cole*. The video concludes with a call for Muslims to embark on the hegira, or migration, to Afghanistan.[74]

In Pakistan, Indonesia, and Malaysia, seminaries are often fertile ground for recruitment. Many of them promote the excitement of joining the jihad as much as they do the horror stories of atrocities against Muslims. In Malaysia, a school associated with Al Qaeda issued brochures exhorting young radicals to forgo Palestine for Afghanistan, where they were promised three thousand kilometers of open borders and the friendship of many like-minded colleagues, who had made Afghanistan the international center of Islamic militancy. Abu Bakar Ba'asyir, the spiritual leader of Jamaah Islamiyah, a Southeast Asian terrorist group closely affiliated with Al Qaeda, championed bin Laden and exhorted students in Indonesia and Malaysia to carry on a "personal jihad" following bin Laden's lead.[75]

The way Khalfan Khamis Mohamed was recruited is typical for foot soldiers. Recruiters locate raw talent in a seminary or a mosque. The raw talent is then sent to a camp, where it is assessed on various dimensions: commitment to Islam, psychological reliability, intelligence, and physical prowess. Identifying reliable recruits is considered the most difficult job. Among Al Qaeda's most well-known and successful recruiters of elite operatives are Muhammad Atef, who was reportedly killed by U.S. bombs in November 2002, and Abu Zubaydah, a Palestinian born in Saudi Arabia, now in U.S. custody.

TRAINING

Osama bin Laden provided training camps and guesthouses in Afghanistan for the use of Al Qaeda and its affiliated groups beginning in 1989. Western intelligence agencies estimate that by September 11, 2001, between 70,000 and 110,000 radical Muslims had graduated from Al Qaeda training camps such as Khalden, Derunta, Khost, Siddiq, or Jihad Wal.[76] Of those, only a few thousand graduates—who distinguished themselves spiritually, physically, or psychologically—were invited to join Al Qaeda. The difficulty of making the cut as a full-fledged recruit meant that Islamists from all over the world regarded joining Al Qaeda as the highest possible honor, Gunaratna explains.[77]

The exact number of training camps in Afghanistan that are associated with Osama bin Laden is unknown, and estimates range from a dozen to over fifty such camps.[78] In the mid-1990s, Al Qaeda shifted its headquarters to Khartoum and established or assisted in the establishment of an estimated twenty training camps in Sudan. Other training camps have been identified in lawless corners of Somalia, Yemen, Indonesia, Chechnya, and other countries. The camps serve a variety of purposes in addition to training members and reserves. They create social ties, so that operatives feel committed to the cause on both ideological and solidarity grounds. Specialists then funnel recruits into the right level of the organization and into the right job: public-relations officer, regional manager, trainer, sleeper, or other.

John Walker Lindh told investigators that the camp he attended near Kandahar offered both basic and advanced training. After the basic training course, trainees can select different tracks to follow, one involving battlefield training and the other "civilian warfare training." The battlefield course includes "advanced topography, ambushes, tactics, battlefield formations, trench warfare . . . practicing assassinations with pistols and rifles, and shooting from motorcycles and cars." The civilian warfare course includes "terrorism, forgery of passports and documents, poisons, mine explosions, and an intelligence course which teaches trainees how to avoid detection by police." Most of the trainees were Saudi, he said. He also said that the leader of the camp approached all foreign trainees to recruit them for "foreign operations." The foreign recruits were instructed not to discuss the conversation about foreign operations with their fellow trainees, and they were not given any details about what the foreign terrorist operations might entail.[79] Trainees were also asked whether they were willing to work in their own country. Lindh said that the leader of the camp, Al Musri, interviewed him personally.

Tapes reportedly captured by the U.S. army in Afghanistan show Al Qaeda members training to carry out operations in the West. The tapes show a level of professionalism that suggests that Al Qaeda had received significant assistance from a professional military, according to an analyst who read the army's assessment and viewed the tapes himself. On one tape, operatives are trained to carry out an ambush near a six-lane highway similar to those that are found in the United States and Europe. Hostage scenarios include raids of large buildings with many occupants. Trainees playing the role of terrorists dictate commands to the hostages in English, and the trainees playing the hostages respond in English. Operatives are trained to determine whether soldiers or other armed personnel are among the hostages so that those with weapons can be segregated from the rest. The armed hostages are then executed in front of television cameras. Another scenario prepares operatives for assassinating dignitaries—possibly national leaders—on a golf course. It is clear from the tapes that Al Qaeda is training its operatives to maximize media coverage, according to the army's assessment.[80]

The most important aspect of training, however, is mental training and religious indoctrination. Religious indoctrination includes Islamic law and history and how to wage a holy war. The story that recruits must learn is about identity—it is about who *we* are as distinct from *them,* to whom Zawahiri, bin Laden's deputy, refers to as the "new Crusaders."[81]

Most importantly, camps are used to inculcate "the story" into young men's heads. The story is about an evil enemy who, in the words of Zawahiri, is waging a "new Crusade" against the lands of Islam. This enemy must be fought militarily, Zawahiri explains, because that is the only language the West understands. The enemy is easily frightened by small groups of fighters, and trainees learn how to function in small cells.[82]

THE MISSION OF TERRORIST ORGANIZATIONS: THE TERRORIST "PRODUCT"

A professional terrorist chooses his mission carefully. He is able to read popular opinion and is likely to change his mission over time. Astute leaders may find new missions—or emphasize new aspects of the mission—when they realize they can no longer "sell" the old one to sponsors and potential recruits, either because the original mission was achieved or, more commonly, because the impossibility of achieving the mission has become obvious.

Terrorism grows out of seductive solutions to grievances. When revolutions succeed, which happens occasionally, the imperative to address the problems of the aggrieved group comes to be accepted by a wider population. But the techniques of terror—the deliberate murder of innocent civilians—are counter to every mainstream religious tradition. This is why the mission—the articulation of the grievance—is so important. It must be so compellingly described that recruits are willing to violate normal moral rules in its name.

The people on whose behalf the terrorists aim to fight must be portrayed as worthy of heroic acts of martyrdom. In his memoir, Zawahiri says that an alliance of jihadi groups and "liberated states" is anxious to seek retribution for the blood of the martyrs, the grief of the mothers, the deprivation of the orphans, the suffering of the detainees, and the sores of the tortured people throughout the land of Islam. He says that this age is witnessing a new phenomenon of *mujaheed* youths who have abandoned their families, countries, wealth, studies, and jobs in search of jihad arenas for the sake of God.[83]

The enemy must be portrayed as a monstrous threat. Zawahiri warns his followers that the new Crusaders respect no moral boundaries and understand only the language of violence. The enemy is characterized by "brutality, arrogance, and disregard for all taboos and customs." He urges jihadis to choose weapons and tactics capable of inflicting maximum casualties on the enemy at minimal cost to the mujahideen. He warns followers that the enemy makes use of a variety of tools and proxies, including the United Nations, friendly rulers of the Muslim peoples, multinational corporations, international communications and data exchange systems, international news agencies and satellite media channels. The enemy also uses international relief agencies as a cover for espionage, proselytizing, coup planning, and the transfer of weapons.[84] John Walker Lindh told interrogators that he had decided to "join the fight of the Pakistani people in Kashmir" when he was in a *madrassah* in Pakistan, where he heard reports of "torture, rape, and massacre of the Pakistani people by India." He said that he was overwhelmed by the "guilt of sitting idle while these atrocities were committed," and he volunteered for training, first in Pakistan, then in Afghanistan, ultimately ending up fighting with the Taliban.[85] A trainer for HUM who was interviewed for this book said that he decided to join the jihad when he was in eleventh grade, after hearing about two Muslim women who were raped by Indian forces.[86] Ironically, the enemy's existence—and even his atrocities—help terrorist groups prove the importance of their mission. The Lashkar e Taiba public-affairs director told me he felt "happy" about the growth of the Hindu extremist group Bajrang Dal, the arch-nemesis of the Pakistani militant groups. It provides a raison d'être for Islamic fundamentalism in Pakistan, he said. "What is the logic for stopping the jihadi groups' activities if the Indian government supports groups like Bajrang Dal?" he asked.[87]

Peter Verkhovensky, a character in Dostoyevsky's 1871 novel *The Demons,* claims to be a socialist but is ultimately exposed as a cheat and a fraud. But the real villains in the novel are the bad ideas that seduced young men to join revolutionary movements. Leaders, who may have been true believers in their youth, cynically take advantage of their zealous recruits, manipulating them with an enticing mission, ultimately using these true believers as their weapons. Joseph Conrad described terrorists as "fools victimized by ideas they cannot possibly believe. . . . While they mouth slogans or even practice anarchist beliefs, their motives are the result of self-display, power plays, class confusion, acting out roles."[88]

Both Dostoyevsky and Conrad understood that the prospect of playing a seemingly heroic role can persuade young men to become ruthless killers in the service of bad ideas, but the bad ideas must be seductively packaged. Terrorist groups have to raise money by "selling" their mission to supporters—including donors, personnel (both managers and followers), and the broader public. Selecting and advertising a mission that will attract donations—of time, talent, money, and for suicide operations, lives—is thus critically important to the group's survival.

Zawahiri observes that the New World Order is a source of humiliation for Muslims. It is better for the youth of Islam to carry arms and defend their religion with pride and dignity than to submit to this humiliation, he says.

Violence, in other words, restores the dignity of humiliated youth. This idea is similar to Franz Fanon's notion that violence is a "cleansing force," which frees the oppressed youth from his "inferiority complex," "despair," and "inaction," making him fearless and restoring his self-respect.[89] Fanon also warned of the dangers of globalization for the underdeveloped world, where youth, who are especially susceptible to the seductive pastimes offered by the West, comprise a large proportion of the population.[90]

Part of the mission of jihad is thus to restore Muslims' pride in the face of a humiliating New World Order. The purpose of violence, according to this way of thinking, is to restore dignity and to help ward off dangerous temptations. Its target audience is not necessarily the victims and their sympathizers, but the perpetrators and their sympathizers. Violence is a way to strengthen support for the organization and the movement it represents. It is a marketing device and a method for rousing the troops.

In this regard, Zawahiri is conforming also with the views of Sayyid Qutb, whom Zawahiri describes as "the most prominent theoretician of the fundamentalist movements" and Islam's most influential contemporary "martyr." Qutb's outlook on the West changed dramatically after his first visit to America, where he was repulsed by Americans' materialism, racism, promiscuity, and feminism. Americans behave like animals, he said. They justify their vulgarity under the banner of emancipation of women and "free mixing of the sexes." They love freedom, but eschew responsibility for their families.[91] He saw the West as the historical enemy of Islam, citing the Crusades, European colonialism, and the Cold War as evidence. Qutb emphasized the need to cleanse Islam from impurities resulting from its exposure to Western and capitalist influence.

Western values have infiltrated the Muslim elites, who rule according to corrupt Western principles. The enemy's weapons are political, economic, and religio-cultural. They must be fought at every level, Qutb warned.[92] The twin purposes of jihad are to cleanse Islam of the impurifying influence of the West, and to fight the West using political, economic, and religio-cultural weapons—the same weapons the West allegedly uses against Islam.

ADVERTISING THE MISSION

Like more traditional humanitarian relief organizations, terrorists have to advertise their mission to potential donors and volunteers, and they tend to use similar techniques. As we have seen, they hold auctions, fund-raising dinners, and press conferences. They put up posters and put out newspapers. They cultivate journalists hoping for favorable press coverage. They openly solicit donations in houses of worship, at least where the state allows it. They send leaders on fund-raising missions abroad and arrange for private meetings between leaders and major donors. They make heavy use of the mail, the telephone, and the Internet, often providing their bank account numbers and the bank's address. They demonstrate their effectiveness with sophisticated Web sites, often including photographs or streaming-video recordings of successful operations and of the atrocities perpetrated against the group they aim to help. All of these techniques are practiced by humanitarian organizations. As we discussed in the previous chapter, terrorist groups also advertise the kind of weapons that recruits will learn to use, in some cases including cyberwar. Person-to-person contacts, however, remain a critical component of fund-raising and recruitment drives.[93]

CHANGING THE MISSION

Astute terrorist leaders often realize that to attract additional funding, they may need to give up their original mission. The original mission of

Egyptian Islamic Jihad, for example, was to turn Egypt into an Islamic state. By the late 1990s, the group had fallen on hard times. Sheik Omar Abdel Rahman was imprisoned in the United States for his involvement in a plot to bomb New York City landmarks in 1993. Other leaders had been killed or forced to move abroad. Zawahiri reportedly considered moving the group to Chechnya, but when he traveled there to check out the situation, he was arrested and imprisoned for traveling without an entry permit.[94] After his release in May 1997, Zawahiri decided that it would be practical to shift his sights away from the "near enemy," the secular rulers of Egypt, toward the "far enemy," the West and the United States. Switching goals in this way would mean a large inflow of cash from bin Laden, which the group desperately needed. Islamists see Egyptian president Hosni Mubarak, who is supported by the United States, as a traitor to Islam on numerous grounds. He has continued his (assassinated) predecessor's controversial policy of appeasing Israel at the expense of the Palestinians. His administration is widely viewed as corrupt and repressive. He has expelled or imprisoned most members of the Islamic resistance to his rule. Egyptian human rights organizations estimate that some sixteen thousand people with suspected links to Islamic organizations remain jailed in Egypt.[95]

The alliance between Zawahiri and bin Laden was a "marriage of convenience," according to Lawrence Wright. One of Zawahiri's chief assistants testified in Cairo that Zawahiri had confided in him that "joining with bin Laden [was] the only solution to keeping the jihad organization alive."[96] "These men were not mercenaries, they were highly motivated idealists, many of whom had turned their backs on middle-class careers. . . . They faced a difficult choice: whether to maintain their allegiance to a bootstrap organization that was always struggling financially or to join forces with a wealthy Saudi who had long-standing ties to the oil billionaires in the Persian Gulf," Wright explains.

After Zawahiri shifted his focus away from Egypt, some of his followers left in protest, forming a splinter faction named Vanguards of Conquest (Talaa' al-Fateh), which was weakened as a result of the Egyptian government's clampdown on Islamists. In return for bin Laden's financial assistance, Zawahiri provided him some two hundred loyal, disciplined, and well-trained followers, who became the core of Al Qaeda's leadership. Zawahiri describes the new mission as a "global battle" against the "disbelievers," who have "united against the mujahideen." He adds, "The battle today cannot be fought on a regional level without taking into account the global hostility towards us."

Another example of a group that changed its mission over time to secure a more reliable source of funding is the Islamic Movement of Uzbekistan. Its original mission was to fight the post-Soviet ruler of Uzbekistan, Islam Karimov, whose authoritarian rule is characterized by corruption and repression.[97] When Juma Namangani, leader of the Islamic Movement of Uzbekistan, was forced underground, together with his followers, they eventually made their way to Afghanistan, where they made contacts with Al Qaeda. Abdujabar Abduvakhitov, an Uzbek scholar who has studied the group since its inception, explains that the group found that by adopting Islamist slogans it could "make more money and get weapons."[98] The IMU shifted its mission from fighting injustice in Uzbekistan to inciting Islamic extremism and global jihad, thereby gaining access to financial supporters in Turkey, Saudi Arabia, Pakistan, and Iran, Abduvakhitov explains. The group's new literature promoted the Taliban's agenda, reviling America and the West, but also music, cigarettes, sex, and drink. Its new slogans made the movement repulsive to its original supporters in Uzbekistan, however.[99]

When the IMU terrorists returned to Uzbekistan in 2000, they had medical kits, tactical radios, and night-vision goggles. "All of this speaks to better funding, it speaks to better contacts," an unnamed intelligence officer told the *New York Times.* "They made an impression on bin Laden."[100]

In the spring of 2001 the group entered into an agreement with Mullah Omar, the leader of the Taliban, to delay its Central Asian campaign and to fight the Northern Alliance. Namangani became commander of the 055 Brigade, bin Laden's group of foreign fighters. After September 11, Namangani found himself at war with America. He had alienated his original supporters in his country, and the financial backers he attracted with his turn toward Islamism were no longer able to fund him because they were dispersed and largely broke. He was killed during the war in Afghanistan in November 2001.[101]

Changing the mission can cause a variety of problems. Volunteers may be wedded to the original mission and may resent the need to kowtow to donors, rather than focusing on the needs of the beneficiaries, as happened with the part of Egyptian Islamic Jihad that refused to join forces with bin Laden. Managers are vulnerable to the charge of mission creep. From the viewpoint of the original stakeholders in the organization, there is a principal-agent problem if the group's mission shifts. An important example of this is when a state (or agencies within in a divided state) fund insurgent groups in the belief that they will have total control over the groups' activities. But if a group diversifies its revenue stream, the state may find itself losing control. This is the case with regard to the militant and sectarian groups in Pakistan, which were largely created by the ISI. Now that a significant fraction of these groups' income comes from other entities, the groups are increasingly engaging in activities that are counter to the state's interests. Similarly, Indonesian jihadi groups that raise money from sources in the Gulf are slipping out of the control of their original backers in the Indonesian military. (In both these cases, it is important to point out again that the state is not a monolithic entity and that individual agents, or even agencies, may be acting in violation of state policy.)

Osama bin Laden himself has changed his mission over time. He inherited an organization devoted to fighting Soviet forces and turned that organization into a flexible group of ruthless warriors ready to fight on behalf of multiple causes. His first call to holy war, issued in 1992, urged believers to kill American soldiers in Saudi Arabia, the Horn of Africa, and Somalia. There was virtually no mention of Palestine. His second, in 1996, was a forty-page document listing atrocities and injustices committed against Muslims, mainly by Western powers. His third, in February 1998, for the first time urged followers deliberately to target American civilians, rather than soldiers. Although that fatwa mentioned the Palestinian struggle, it was only one of a litany of Muslim grievances. America's "crimes" against Saudi Arabia (by stationing troops near Islam's holiest sites), Iraq, and the other Islamic states of the region constituted "a clear declaration of war by the Americans against God, his Prophet, and the Muslims . . . By God's leave, we call on every Muslim who believes in God and hopes for reward to obey God's command to kill the Americans and plunder their possessions wherever he finds them and wherever he can," bin Laden wrote.[102] On October 7, 2001, in a message released on Al Jazeera television immediately after U.S. forces began bombing in Afghanistan, bin Laden issued his fourth call for jihad. This time he emphasized Israel's occupation of Palestinian lands and the suffering of Iraqi children under UN sanctions, concerns broadly shared in the Islamic world. While most Muslims reject bin Laden's interpretation of their religion, bin Laden felt the moment was ripe to win many over to his anti-Western cause. Bin Laden was competing for the hearts and minds of ordinary Muslims. He said that the September 11 "events" had split the world into two "camps," the Islamic world and "infidels"—and that the time had come for "every Muslim to defend his religion" (echoing President Bush's argument that from now on "either you are with us, or you are with the terrorists"[103]).

Bin Laden's aim was to turn America's response to the September 11 attack into a war between Islam and the West. With this new fatwa, bin Laden was striking at the "very core of the grievances that the common Arab man in the street has toward his respective government, especially in Saudi Arabia," Nawaf Obaid, a Saudi analyst, explained.[104] John Walker Lindh told U.S. investigators that Al Qaeda had come to believe that it was more effective to "attack the head of the snake" than to attack secular rulers in the Islamic world.

EXPANDING THE NETWORK

Al Qaeda and the IIF are not only changing their mission over time in response to new situations and new needs, but also their organizational style. With its corporate headquarters in shatters, Al Qaeda and the alliance are now relying on an ever shifting network of sympathetic groups and individuals, including the Southwest Asian jihadi groups that signed bin Laden's February 1998 fatwa; franchise outfits in Southeast Asia; sleeper cells trained in Afghanistan and dispersed abroad; and freelancers such as

Richard Reid, the convicted "shoe bomber," who attempted to blow up a plane. Lone wolves are also beginning to take action on their own, without having been formally recruited or trained by Al Qaeda.

The Al Qaeda organization is learning that to evade law-enforcement detection in the West, it will need to adopt some of the qualities of the virtual network style we discussed in previous chapters. Coordination of major attacks in the post–September 11 world, in which law-enforcement and intelligence agencies have formed their own networks in response, will be difficult. Al Qaeda is adapting by communicating over the Internet and by issuing messages intended to frighten Americans and boost the morale of followers. The leadership of Al Qaeda appears to be functioning less as a group of commanders and more as inspirational leaders. A Web site that appeared after September 11 (but is no longer available) offered a special on-line training course that teaches the reader how to make time bombs and detonate enemy command centers. The site invited visitors to read a chapter on the production of explosives, saying, "We want deeds, not words. What counts is implementation." Other sites made reference to the Encyclopedia of Jihad, which provides instructions for creating a "clandestine activity cell," including intelligence, supply, planning and preparation, and implementation.[105] In an article on the "culture of jihad," a Saudi Islamist urges bin Laden's sympathizers to take action on their own. "I do not need to meet the Sheikh and ask his permission to carry out some operation, the same as I do not need permission to pray, or to think about killing the Jews and the Crusaders that gather on our lands." He accuses the enemies of Islam of attempting to alter the Saudi education system to describe jihad as a way of thinking rather than as mode of action. Nor does it make any difference whether bin Laden is alive or dead. "If Osama bin Laden is alive or God forbid he is killed, there are thousand Bin Ladens in this nation. We should not abandon our way, which the Sheikh has paved for you, regardless of the existence of the Sheikh or his absence."[106]

An anonymous article in another Islamist forum, "the lovers of jihad," argues, "The Islamist view of the confrontation with the United States is settled. Furthermore, it is going to be the new ideology of the second generation of the Jihadi movements around the world. They do not need the existence of bin laden, after he fulfilled his role in the call and agitation for this project."[107]

As with any network, the challenge for the Al Qaeda network of groups is to balance the needs for resilience and for capacity. Resilience refers to the ability of a network to withstand the loss of a node or nodes. To maximize resilience, the network has to maximize redundancy. Functions are not centralized. (This decreases the efficiency of the organization, but terrorist networks are unlikely to optimize efficiency as they do not have to answer to shareholders and they tend to view the "muscle" as expendable.) Capacity—the ability to optimize the scale of the attack—requires coordination, which makes the group less resilient because communication is required. Effectiveness is a function of both capacity and resilience.

Network theorists suggest that a network of networks is a resilient organization. Within each cluster, every node is connected to every other node in what is known as an "all channel" network. But only certain members of the cluster communicate with other clusters, and the ties between clusters are weak, to minimize the risk of penetration.

The strength of ties is not static, however; it varies over time. Training together in camps establishes trust, the glue that holds a network together. (Recall Fahid's claim that he would not be able to trust Mohamed unless he trained in Afghanistan.) But *task ties,* the term network theorists use for relationships needed to accomplish particular tasks, are likely to be weak or even nonexistent until a leader brings a group together to carry out an operation.

In a law-enforcement-rich environment, the most effective terrorist organization probably consists of many clusters of varying size and complexity held together by trust and a shared mission rather than a hierarchical superstructure. Individual clusters may find their own funding through licit or illicit businesses, donations from wealthy industrialists, wealthy diasporas, or the relationships they develop with states or state agents. Individual groups may even compete for funds in what is known as a chaordic network.[108] They may recruit and arm their groups separately. Innovation—such as attempts to acquire or use unconventional weapons—is promoted at all levels. Some of the clusters will remain dormant until a concrete operation is being planned. Those that are active in failing states where the state either supports them or cannot fight them will be able to remain active full-time. The only thing the sub-networks must have in common is a shared mission and goals.

In this network of networks, leadership style will vary. Complex tasks require hierarchies—the commander cadre–type organization we discussed in chapter 8. For very small operations, of the kind that are carried out by the Army of God that we discussed in chapter 6, little coordination or leadership is required: small cells or lone wolves inspired by the movement can act on their own. Individual operatives can have a powerful effect, as the sniper in suburban Washington in the fall of 2002 made clear. As more powerful weapons become available to smaller groups, virtual networks will become more dangerous.

The use of sleepers can make an organization significantly more resilient. Sleepers are informed of their tasks immediately before the operation. They are likely to be told only what they need to know: information is strictly compartmentalized.[109]

Technology has greatly increased the capacity of networks. Networks can now be decentralized but also highly focused. Members can travel nearly anywhere and communicate with one another anywhere. Money is also easily shipped.[110] This is especially true for organizations like Al Qaeda, which utilize informal financial transactions and convert their cash into gems or gold.

Since September 11 and the war in Afghanistan, Al Qaeda and the IIF have been forming the kind of network of networks connected by weak ties that network theorists argue is the most effective style of organization, and making use of sleepers and freelancers, which increases the resilience of the alliance.

SOURCES OF FUNDS

As is the case for many terrorist groups, Al Qaeda raises money in four ways: criminal activities, businesses, financial or in-kind assistance from states or state agents, and charitable donations.

Businesses

Al-Fadl testified that bin Laden set up a large number of companies in Sudan, including Wadi-al-Aqiq, a corporate shell that he referred to as the "mother" of all the other companies: Al Hijra Construction, a company that built roads and bridges; Taba Investment, Ltd., a currency trading group; Themar al-Mubaraka, an agriculture company; Quadarat, a transport company; Laden International, an import-export business. Al-Fadl said the group controlled the Islamic bank al-Shamal and held accounts at Barclays Bank in London as well as unnamed banks in Sudan, Malaysia, Hong Kong, Cyprus, the United States, and Dubai.[111] According to the U.S. indictment, "These companies were operated to provide income and to support Al Qaeda, and to provide cover for the procurement of explosives, weapons, and chemicals, and for the travel of Al Qaeda operatives."[112]

Like many terrorist groups, Al Qaeda is involved in both licit and illicit enterprises. Bin Laden attempted to develop a more potent strain of heroin to export to the United States and Western Europe, in retaliation for the 1998 air strikes in Sudan and Afghanistan. He provided protection to processing plants and transport for the Taliban's drug businesses, which financed training camps and supported extremists in neighboring countries, according to the United Nations.[113] Al Qaeda used informal financial transactions known as *hawala,* which are based largely on trust and extensive use of family or regional connections,[114] and a network of honey shops, to transfer funds around the world.[115] It is now converting cash into diamonds and gold.

Charitable Donations

Charities, purportedly unaffiliated with the terrorist groups, seek funding for humanitarian relief operations, some of which is used for that purpose, and some of which is used to fund terrorist operations. Many jihadi groups use charities for fund-raising abroad or as a front for terrorist activities. Al Qaeda members testified that they received ID cards issued by a humanitarian relief organization based in Nairobi called Mercy International Relief Agency. The organization was involved in humanitarian relief efforts, as its name suggests, but it also served as a front organization for operatives during the period they were planning the Africa embassy bombings.[116]

By soliciting charitable donations abroad, groups draw attention to the cause among diaspora populations. The Gulf States, North America, the United Kingdom, and European countries are important sources of funding for terrorist groups. The U.S. government looked the other way when the IRA engaged in fund-raising dinners in the United States, but began to see the downside to such a policy when the groups being funded began killing American citizens.

But perhaps even more importantly, by soliciting money from the people, a terrorist organization (or terrorist-affiliated organization) can establish its bona fides as a group devoted to the interests of "the people." While much of the group's money may actually come from criminal activities, business operations, or government assistance, charitable donations are important as a "defining source of revenue," a point made in regard to more traditional NGOs by Mark Moore, a specialist in non-profits at Harvard University. In my interviews, leaders tend to emphasize charitable donations as the most important source of revenue for their groups; while operatives, presumably less attuned to the public-relations implications of their words, admit that smuggling, government funding, or large-scale donations by wealthy industrialists are the main sources of funding.[117] Money flows into jihadi groups through charities; but money also flows out to the needy. Sophisticated jihadi organizations function very much like the United Way.

LEADERLESS RESISTERS, FREELANCERS, AND FRANCHISES

The New World Order and its instruments—Al Qaeda's new foes—are attractive targets to a surprising array of groups. By emphasizing the New World Order as its enemy, Al Qaeda will be able to attract a variety of groups that oppose Western hegemony and international institutions.

White supremacists and Identity Christians are applauding Al Qaeda's goals and actions and may eventually take action on the Al Qaeda network's behalf as freelancers or lone-wolf avengers. A Swiss neo-Nazi named Albert Huber, who is popular with both Aryan youth and radical Muslims, is calling for neo-Nazis and Islamists to join forces. Huber was on the board of directors of the Al-Taqwa Foundation, which the U.S. government says was a major donor to Al Qaeda.[118] The late William Pierce, who wrote *The Turner Diaries*, the book that inspired the Oklahoma City bombing, applauded the September 11 bombers. Pierce's organization, the Alliance Nahad, urged its followers to celebrate the one-year anniversary of September 11 by printing out and disseminating flyers from its Web site. One of the flyers included a photograph of bin Laden and the World Trade Center and the caption, "Let's stop being human shields for Israel."[119] Matt Hale, leader of the World Church of the Creator, a white supremacist organization one of whose members killed a number of blacks and Jews, is disseminating a book that exposes the "sinister machinations" that led to September 11, including the involvement of Jews and Israelis, in particular, the Mossad.[120]

Horst Mahler, a founder of the radical leftist German group the Red Army Faction, has moved from the extreme left to radical right. He too rejoiced at the news of the September 11 attacks, saying that they presage "the end of the American Century, the end of Global Capitalism, and thus the end of the secular Yahweh cult, of Mammonism." He accuses the "one-World strategists" of trying to create a smoke screen to prevent ordinary people from understanding the real cause of September 11, which America brought on itself through its arrogance. "This is war," he says, "with invisible fronts at present, and worldwide." September 11 was just the first blow against the Globalists, whose true aim is to exterminate national cultures, he says. "It is not a war of material powers," he says. "It is a spiritual struggle: the war of Western civilization, which is barbarism, against the cultures of the national peoples. . . . The oncoming crisis in the World Economy—independent of the air attacks of 11 September 2001—is now taking the enchantment from 'The American Way of Life.' The absolute merchandisability of human existence—long felt as a sickness—is lost, along with the loss of external objects, in which human beings seek recognition and validation—but cannot find them."[121]

The racist right is also applauding the efforts of other "antiglobalists" in addition to bin Laden. Louis Beam, author of the leaderless-resistance essay we discussed in chapter 6, is urging all antiglobalists, from all political persuasions, to join forces against the New World Order (NWO). He applauds the participants of the Battle of Seattle, who, he says, faced a "real invasion of black booted, black suited" thugs, while the racist right continued talking endlessly about the impending invasion of foreign troops in United Nations submarines.

"Mark my words," Beam says, "this is but the first confrontation, there will be many more such confrontations as intelligent, caring people begin to face off the Waco thugs of the New World Order here in the United States. The New American Patriot will be neither left nor right, just a freeman fighting for liberty. New alliances will form between those who have in the past thought of themselves as 'right-wingers,' conservatives, and patriots with many people who have thought of themselves as 'left-wingers,' progressives, or just 'liberal.' "[122]

Perhaps the most articulate proponent of forming an anti-NWO coalition is Keith Preston, a self-described veteran of numerous libertarian, anarchist, leftist, labor, and patriot organizations and an active anarchist. He argues that the war between the "U.S. and the Muslim world" is one front in a larger war, "namely, the emerging global conflict between those interests wishing to subordinate the entire world to the so-called 'New World Order' of global governance by elite financial interests in the advanced countries on one side and all those various national, regional, ethnic, cultural, religious, linguistic, and economic groups who wish to remain independent of such a global order." He believes that the rapid drive to create this NWO must be reversed or it will "likely produce a system of totalitarian oppression similar to that of the Nazi and Soviet regimes of the twentieth century only with infinitely greater amounts of economic, technological, and military resources. All forces throughout the world seeking to resist this development must join together, regardless of their other differences, and provide mutual support to one another in the common struggle. The current U.S.-led 'coalition' against so-called 'terrorism' is simply a cover for continuing the process of global consolidation of power and crushing all efforts at resistance." Islamic fundamentalists, he says, are fighting the same global interests seeking to impose "global government, international currency systems, firearms confiscation, international police forces, NAFTA, and other regressive economic policies on the American people." He proposes joining forces even with Jewish fundamentalist sects, "such as the Neturei Karta, who have condemned Israeli imperialism and expansionism." He urges the "bandits and anarchists" to join together with the "tribes, sects, warlords, and criminals" to assert themselves forcefully.[123]

While the threat these groups pose is nowhere near as significant as that of current members of the Al Qaeda alliance, some of their members may decide to support Al Qaeda's goals, as lone wolves or leaderless resisters, giving it a new source of Western recruits.

The tri-border area where Argentina, Brazil, and Paraguay meet is becoming the new Libya: The place where terrorists with widely disparate ideologies—the Marxist groups FARC and ELN, American white supremacists, Hamas, Hezbollah, and members of bin Laden's International Islamic Front—meet to swap tradecraft. Authorities worry that the more sophisticated groups could make use of the Americans as participants in their plots, possibly to bring in materials.

Perhaps the best example of a freelancer—an individual trained by Al Qaeda who takes action largely on his own—is Richard Reid. In October 2002, Richard Reid pled guilty to the charge that he tried to blow up a plane with a bomb hidden in his shoe in December 2001. He also admitted that he was trained at an Al Qaeda camp and said that he was a member of Al Qaeda, a statement that some experts suspect is not literally true. Reid gave in to his interrogators almost immediately, suggesting that he had not undergone the kind of rigorous psychological training that is typical for Al Qaeda members. Magnus Ranstorp, a terrorism expert who has studied the Islamist community in London, from which Reid was apparently recruited, argues that Reid is most likely a fringe amateur inspired by what he saw in Afghanistan and by the movement in general. Others point out that Reid was in contact with Al Qaeda members by e-mail.[124]

Jamaah Islamiyah—the Franchise

The group known as Jamaah Islamiyah grew out of Islamic opposition to Soeharto's regime. Like that of Lashkar Jihad, the group we discussed in chapter 5, its goal was to establish an Islamic community, *jamaah*

Islamiyah, throughout Southeast Asia. Its spiritual leader, Abu Bakar Ba'asyir, founded and runs a *pesantren* (seminary) called Ngruki near Solo, Java, close to the *pesantren* we discussed in chapter 3. Ba'asyir and his closest followers fled to Malaysia in 1985 to escape Soeharto's suppression of the group. Some members returned after Soeharto's resignation in 1998, and some remained in Malaysia. Although some members of Jamaah Islamiyah (JI) have clear links to Al Qaeda, JI is the violent wing of a broader movement that supports Ba'asyir. The movement, known as the Ngruki network, named after Ba'asyir's school, includes a broad range of prominent individuals, some of whom are active in the Indonesian government. Many Indonesians are deeply concerned that the war on terrorism, and the U.S. push to arrest suspects without clear evidence, could radicalize the Muslim community.[125]

THE POST–INDUSTRIAL-AGE TERRORIST ORGANIZATION

Mobilizing terrorist recruits and supporters requires an effective organization. Effectiveness requires resources, recruits, hierarchies, and logistics. It requires adopting the mission to appeal to the maximum number of recruits and financial backers.[126] As we have seen, contestants often choose to call competition for natural resources or political power a religious conflict when they believe it will make their grievances more attractive to a broader set of potential fighters or financial backers. (Governments may do the same by labeling opposition groups religious extremists to win international support for crushing them.)

Money—used to buy goods and services—is a critical component of what distinguishes groups that are effective from those that disappear or fail to have an impact. The terrorists discussed in these pages raise money in a variety of ways. They run licit and illicit businesses. They auction off "relics." They run their own informal banks, which take a "charitable donation" in lieu of interest. They solicit donations on the Internet, on the streets, and in houses of worship. They appeal to wealthy industrialists, sympathetic diasporas, and to governments or their agents. By functioning as a foundation that provides social services, the groups spread their ideas to donors as well as the recipients of their largesse. Recipients of charitable assistance may be more willing to donate their sons to the group's cause. Of the religious groups discussed in this book, only the Islamists are effective in this way.

But terrorist organizations need to balance the requirements for optimizing capacity with those of resilience. Resilience (the ability to withstand the loss of personnel) requires redundancy and minimal or impenetrable communication, making coordination difficult absent cutting-edge encryption technologies. The most resilient group discussed in this book is the save-the-babies group Army of God, a virtual network whose members meet only to discuss the mission, not concrete plans. The drawback from the terrorists' perspective to this maximally resilient style of organization is that it requires individuals or small groups to act on their own, making large-scale operations difficult.[127]

The best way to balance these competing objectives is to form a network of networks, which includes hierarchical structures (commanders and cadres); leaderless resisters who are inspired through virtual contacts; and franchises, which may donate money in return for the privilege of participating.[128] The networks are held together mainly by their common mission (although some may be pursuing multiple missions, including local agendas of little interest to the rest of the network). By expanding his mission statement, bin Laden was able to expand his network to include most of the Islamist groups discussed in this book. Groups that are not Islamist but oppose globalization may be willing to donate money or operatives to the anti–New World Order cause.

The Al Qaeda network of networks is at the cutting edge of organizations today. Law-enforcement authorities will continue to discover new cells or clusters, but they will not be able to shut down the movement until bin Laden, his successors, and his sympathizers' call to destroy the New World Order loses its appeal among populations made vulnerable by perceived humiliation and violations of human rights, perceived economic deprivation, confused identities, and poor governance.

There is a trade-off for policy makers between the need to destroy the adversary that is about to strike and the need to fight the movement over the long term. Our military action becomes the evidence our enemies need to prove the dangers of the New World Order they aim to fight. It creates a sense of urgency for the terrorists seeking to purify the world through murder.

It is part of the human condition to lack certainty about our identities; the desire to see ourselves in opposition to some Other is appealing to all of us. That is part—but only part—of what religion is all about. One of our goals must be to make the terrorists' purification project seem *less* urgent: to demonstrate the humanity that binds us, rather than allow our adversaries to emphasize and exploit our differences to provide a seemingly clear (but false) identity, at the expense of peace. In the final chapter we explore these ideas in more detail.

TEN

Conclusion/ Policy Recommendations

I started this project deeply puzzled about how people who claim to be motivated by religious principles come to kill innocent people in the service of ideas. I learned that several factors—seemingly unrelated to the grievances that motivate terrorist crimes—play an important role in turning spiritual longing into murder.

As a result of my interviews, I have come to see that apocalyptic violence intended to "cleanse" the world of "impurities" can create a transcendent state. All the terrorist groups examined in this book believe—or at least started out believing—that they are creating a more perfect world. From their perspective, they are purifying the world of injustice, cruelty, and all that is antihuman. When I began this project, I could not understand why the killers I met seemed spiritually intoxicated. Now, I think I understand. They seem that way because they are. Only a few of the terrorists discussed in these pages have had visions or felt themselves to be in direct communication with God. But all of them describe themselves as responding to a spiritual calling, and many report a kind of spiritual high or addiction related to its fulfillment.

My interviews suggest that people join religious terrorist groups partly to transform themselves and to simplify life. They start out feeling humiliated, enraged that they are viewed by some Other as second class. They take on new identities as martyrs on behalf of a purported spiritual cause. The spiritually perplexed learn to focus on action. The weak become strong. The selfish become altruists, ready to make the ultimate sacrifice of their lives in the belief that their deaths will serve the public good. Rage turns to conviction. What seems to happen is that they enter a kind of trance, where the world is divided neatly between good and evil, victim and oppressor. Uncertainty and ambivalence, always painful to experience, are banished. There is no room for the other side's point of view. Because they believe their cause is just, and because the population they hope to protect is purportedly so deprived, abused, and helpless, they persuade themselves that any action—even a heinous crime—is justified. They know they are right, not just politically, but morally. They believe that God is on their side.

But God, as is His wont, is silent.[1] Even in the face of unimaginable atrocities, He doesn't answer the militants' call. The world remains contaminated by injustice and corruption. The "enemy" continues to oppress his victims—whether unborn children, helpless Kashmiris, innocent Palestinians, or redemption-seeking Jews. And this, I believe, leads to rage and even more violence. The terrorist begins to mimic his perception of the oppressor: he turns to violence. His goal is to win at any cost. Over time, in some cases, cynicism takes hold. Terrorism becomes a career as much as a passion. What starts out as moral fervor becomes a sophisticated organization. Grievance can end up as greed—for money, political power, or attention. We need to understand this dynamic and exploit it in every possible way, including by seeking to sow discord, confusion, and rivalry among terrorists and between terrorists and their sponsors.

It is part of human nature to desire transcendence—the kind of peak experience that most of us encounter all too rarely through contemplation of beauty, love, or prayer. As odd as it sounds, a sense of transcendence is one of many attractions of religious violence for terrorists, beyond the appeal of achieving their goals. More broadly, it is not just the accomplishment of their goals that terrorists seek; it is also the act of pursuing them.

Participating in terrorist violence provides a package of "goods," some of them the spiritual benefits described above, some emotional, and some material. Unless we understand the appeal of participating in extremist groups and the seduction of finding one's identity in opposition to Other, we will not get far in our attempts to stop terrorism. Whenever we face a terrorist threat, we should ask ourselves: Who stands to gain? Who is making money? Who is receiving benefits of any kind? Who is taking advantage of whom?

The terrorism we currently face is not only a response to political grievances, as was common in the 1960s and 1970s, and which might, in principle, be remediable. It is a response to the "God-shaped hole" in modern culture about which Sartre wrote, and to values like tolerance and equal rights for women that are supremely irritating to those who feel left behind by modernity.[2] Extremists respond to the vacuity in human consciousness with anger and with ideas about who is to blame. In their view, arrogant one-worlders, humanists, and promoters of human rights have created an engine of modernity that is stealing the identity of the oppressed. The greatest rage—and the greatest danger—stems from those who feel they can't keep up, even as they claim to be superior to those who can.

The terrorism we are fighting is a seductive idea, not a military target. Terrorist leaders tell young men that the reason they feel humiliated—personally or culturally—is that international institutions like the IMF, the World Bank, and the United Nations are imposing capitalism and secular ideas on them with the aim of exterminating traditional values. But these antiglobalists, whatever their stripe, are not opposed to exploiting the fruits of modernity to fight their enemies. On the contrary, they are breaking new ground in creating resilient networks that rely on high technology and are attempting to acquire and use high-technology weapons. They are also cleverly exploiting societies and individuals that are vulnerable to their message.[3]

I have come to see terrorism as a kind of virus, which spreads as a result of risk factors at various levels: global, interstate, national, and personal. But identifying these factors precisely is difficult. The same variables (political, religious, social, or all of the above) that seem to have caused one person to become a terrorist might cause another to become a saint.[4]

On a global level, the communications revolution has greatly eased spreading the viral message, mobilizing followers, and creating worldwide networks. Terrorist groups recruit, raise funds, and attract sympathizers on the Internet. They carefully stage their "remedies" to maximize press coverage. The spread of increasingly powerful and portable weapons, including components of weapons of mass destruction and related expertise, also facilitates the virus's spread.

At the interstate level or intergroup level, bad neighborhoods and failed states export crime, refugees, and grievances.[5] Refugee camps are notorious hothouses not only of disease, but also of rage and extremism. Smugglers and criminals in failed states often rely on the services of their neighbors, who thereby become involved in crime themselves. Festering interstate conflicts can breed terrorism not only in the immediate region, but also at far geographic remove. The purported mistreatment of one side or the other (or both) can become a rallying cry for terrorism elsewhere in the world. A kind of "victimization Olympics" often takes hold, with each side demonstrating through statistics or photographs or refugee flows that *it* is the aggrieved party and in need of international assistance, including in the form of terrorist volunteers.

At the national level, a government's inability to provide basic services, protect human rights, or to maintain a monopoly on violence damages the state's ability to fight extremist groups. This is true not only because terrorists avail themselves of opportunities to act unchecked, but also because a culture of violence breeds more violence and terrorism. Other factors that appear to increase a country's susceptibility to terrorism include a "youth bulge," and especially, a high ratio of men to women. Young males comprise a growing fraction of the population across the Islamic world. Studies suggest that countries with a high ratio of males to females, and with young men comprising a large fraction of the population, are significantly more prone to violence of all kinds.[6]

Poverty's role as a risk factor is controversial, but the frequently cited fact that the September 11 bombers were mostly drawn from Saudi Arabia's elite does not prove that poverty and terrorism are uncorrelated. Several studies have shown that states most susceptible to ethno-religious conflict are those that are poorer, unstable, and have a history of violence and conflict.[7] Economist Robert Barro has found that when an economy sours, the poor are more likely to become involved in crime, riots, and other disruptive activities, and that these activities increase.[8] Economist Paul Collier has found that rebellions, similarly, are more likely during periods of economic stagnation.[9] The Indonesian jihadi groups began to thrive only after that country's economic crisis, when educated young men began to have trouble finding jobs in the "civilian" sector.[10]

Perhaps most importantly, terrorists have found ingeniously cruel ways to prey on the poor and ignorant. As we have seen, extremist movements funnel young men from extremist seminaries, some of which function as orphanages for the poor, into various jihads—and into the clutches of the Taliban and Al Qaeda. These foot soldiers often function as cannon fodder, with minimal training. Khalfan Khamis Mohamed, discussed in the last chapter, did not know the name of the group that recruited him—or to what purpose—until days before the bombing of the American embassy in Dar es Salaam. In poor countries like Pakistan, militants say that their salaries play a key role, not in persuading them to join jihadi groups, but in keeping them there. Jihadi groups' social welfare activities, especially the practice of compensating militants' families in Indonesia, Pakistan, and Palestine, seem to play a role in making the groups more appealing to the poor.

Humiliation—at the national or individual level—appears to be another important risk factor. Prominent Islamists such as Sayyid Qutb and Ayman Zawahiri, the intellectual leaders of the Muslim Brotherhood and of Al Qaeda, respectively, argue that violence is a way to cure Muslim youth of the pernicious effects of centuries of humiliation at the hands of the West.[11] Globalization—and the spread of Western power and values—is humiliating to Muslims, Zawahiri says. In his view, taking up the gun is a way to restore dignity to the Islamic world as well as to individual Muslims. The Jewish terrorists discussed in this book, similarly, see the peace process and giving up the occupied territories as humiliating to Jews. Building the Third Temple, in their view, will restore their dignity in the eyes of God.

Personal humiliation—in addition to cultural humiliation—also plays a role for some terrorists. Kerry Noble, the former member of the Christian Identity group the Covenant, the Sword, and the Arm of the Lord, talked about his embarrassment at having been forced to play on the girls' side in his elementary-school physical education classes. He said that he felt strong for the first time in his life when he joined a violent, racist cult. In general, right-wing militias and Christian Identity groups see the state as emasculating and blame feminists and their nonwhite coconspirators for their humiliation. They recruit followers with an implicit promise to restore their wounded masculinity.[12]

In addition to the spiritual intoxication that may come about from participating in attempts to purify society through violence, some terrorists experience a different kind of high: they like weapons and they like to kill, and they would do so for nearly any reason. A small percentage of people take pleasure from violence in this way, perhaps as a result of a genetic predisposition.[13] They would presumably be susceptible to the seduction of a variety of terrorist movements, and they might switch to new groups over time.

WHY THE ISLAMIC WORLD IS PARTICULARLY VULNERABLE

The notion that a new world order is responsible for all of societies' ills attracts adherents all over the globe and in every religion, as we have seen in the preceding chapters; but it has spread to large numbers only in Muslim-majority states. In Egypt, Pakistan, Palestine, the Persian Gulf, Syria, Iran, Iraq, Indonesia, and increasingly, Africa, a virulent anti-Americanism is gaining ground. It is here that the mullahs and their martyrs know exactly whom to blame for their failures. And it is not just the extremists who are hostile to America. Much of the public is as well.[14]

Part of the reason for this is clearly U.S. support for Israel. But another part is that repressive Middle Eastern regimes may be good at suppressing terrorism inside their own states, but the terrorists often shift their sights to more vulnerable targets. In the previous chapter we discussed the Egyptian government's success in shutting down Islamic opposition to its rule. In response, Egyptian Islamic Jihad, whose members were extraordinarily well trained, shifted its target from the "near enemy" in Egypt to the "far enemy" in America and the West. The results include the 1993

World Trade Center bombing, a thwarted plot to blow up New York City landmarks, and ultimately, the formal merger of Al Qaeda and Egyptian Islamic Jihad.[15] The Saudi government successfully fought Islamist opposition to its rule, but allowed its citizens to support violent extremist movements around the world, including Al Qaeda. The governments of Egypt and Uzbekistan, similarly, exported the Islamist groups that opposed their rule. Most Muslim-majority states are corrupt and fragile and are unwilling (or incapable) of providing their populations with education, health care, and other resources required to create robust economies and stable polities.

Over the last quarter century, standards of living have either fallen or remained steady for most Muslim-majority states.[16] In some, extremist groups step in to offer the social services the state is failing to provide. Poor governance and inadequate protection of civil liberties have allowed extremist groups to thrive and to spread the message that the West is responsible for their plight. Weak or authoritarian governments, extremist religious groups, poverty, rage, and alienation work in concert to create a population that is furious with America, which is viewed—often rightly—as a supporter of the status quo in the Arab and Islamic world.

Four-fifths of Muslim-majority states are ruled by nondemocratic regimes. The more democratic regimes that exist in the Islamic world tend to be fragile and as plagued by cronyism and corruption as the autocratic governments. Some scholars believe that a "natural-resource curse" prevents oil-rich countries from achieving viable democracies.[17] But democratization is not necessarily the best way to fight Islamic extremism. Most states that attempt to transition from autocracy to democracy get stuck in a kind of in-between state. And electoral democracy does not necessarily imply liberal democracy, especially in the Islamic world.[18] Algeria's Islamist party won democratically, shortly after a drop in world oil prices.[19] In Pakistan, Islamist parties—some of which openly promote a "Talibanization" of Pakistan—did well in the 2002 parliamentary elections, in part because of the government's continuing failure to provide public services, but also because of anger about Islamabad's concessions to the United States in the war on terrorism.[20] In Turkey, the Justice and Development Party (AKP), a party with Islamist origins, won 363 of the

550 seats in the Turkish parliament in November 2002. Indonesia's transition to democracy and its economic downturn both seem to have played a role in facilitating the growing appeal of Islamist groups in that country. Transition to democracy has been found to be an especially vulnerable period for states across the board. The dilemma for the United States is to find a way out of this vicious cycle while still maintaining U.S. interests.

WHAT DOES THIS MEAN FOR U.S. POLICY?

Al Qaeda and affiliated groups know how to watch us closely for our vulnerabilities—psychological, spiritual, and material. They engage in psychological warfare against us, trying to maximize our fear. They take advantage of the global communication system, and of the openness of Western societies, to hide among us, plan their operations, and recruit our frustrated youth. The British diplomat Paul Schulte warns that the "Casbah" (the slum in Algiers where resentment of French colonialism finally boiled over, igniting the eight-year-long war) is spreading worldwide. Every first-world city has a third-world one within it, he observes, where residents are "emotionally very far withdrawn from the surrounding national public space," where dangerous and proselytizing extremist groups are likely to prey especially on individuals with "various statuses of official citizenship and subjective identity, identification and loyalty." To varying national extents we have to worry, he warns, "though we may not choose publicly to admit that forceful actions against external terrorist base areas may provoke these potential internal actors into decisively changing their allegiances and moving to active violence" in opposition to the West.[21] International terrorist organizations are attempting to set up shop in America and elsewhere in the West by recruiting in our prisons and developing allegiances with domestic antigovernment groups.[22] They are also buying firearms at American gun shows, studying at American flight schools, and soliciting donations from American and other Western citizens.[23] They are seeking weapons of mass destruction and have succeeded in acquiring them, at least to some extent. They aim to damage our economy. And they attack us spiritually. How can we respond to these threats, both internal and external, without falling prey to the heartrending *engrenage* that besieged both sides in the Battle of Algiers, without turning the war on terrorism into an "involuntary and mechanistic" pursuit of vengeance for its own sake?[24]

There is no easy "to do" list to respond to these risk factors. We have yet to create a technology for fixing the "God-shaped hole" in human consciousness that is a symptom of modernity, or for curing alienation, humiliation, envy, or rage. But there are still some ways we can address—if not fix—the problems that may motivate terrorism and reduce its consequences. Michael Ignatieff asks, "If terrorism is the greater evil, what lesser evil forms of violence, deception, suspensions of liberty—can be justified in combating it?" Another, equally important question often overlooked by policy makers and analysts is: How can we fight terrorist groups without making the problem—hatred of the new world order and of America—even worse?[25]

Until now, policy remedies have focused primarily on attempting to crush terrorist groups militarily by attacking their headquarters and training camps, or killing their leaders and operatives, executing those we capture, and attacking at least some of the states that harbor or sponsor them. There are several problems with this approach.

The most obvious one is that sleepers and franchises are already in place in many countries, making military strikes against a given terrorist sanctuary of limited effectiveness. By expanding its mission, Al Qaeda has been able to establish links with terrorist groups all over the world, even those whose agendas had been predominantly regional or local.

A second problem is that whenever we respond with violence of any kind, we assist the terrorists in mobilizing recruits. When the United States retaliated for bin Laden's African embassy bombings by striking at a purported chemical weapons facility in Sudan and a few crude camps in Afghanistan, supporters of bin Laden rejoiced. Fazlur Rahman Khalil, leader of the Pakistani jihadi group Harkat-ul-Mujahideen, held a press conference shortly after the American attacks, pronouncing, "Osama's mission is our mission. It is the mission of the whole Islamic world." Respondents to a questionnaire administered for this project said that recruitment had become easier after September 11 and the war in Afghanistan. Anytime military action is contemplated, its effect on terrorist recruitment and fund-raising must be weighed. To the extent that covert action is possible, it is preferred for this reason.

We need also to realize that whatever military remedies we choose, there may be long-term side effects. After the Afghan victory over the Soviets, Pakistan and Afghanistan were left to deal with the war's aftermath: refugees, criminal enterprises, and jihadis searching for new jobs. Each side took advantage of the situation to create informal but formidable armies. Those armies, together with the idea of multinational jihad and the Stingers we left behind, are haunting us today. If humanitarian concerns weren't enough to persuade us to finish the job we started in Afghanistan, national-security concerns should have been. This is a harsh lesson as we contemplate fighting new wars around the globe.

Just as Al Qaeda and the International Islamic Front have emphasized penetrating us, we need to penetrate them. Whenever and wherever possible, we should be sowing confusion and dissent among Al Qaeda and its franchises. We need to become as savvy at psychological warfare as they are. This too requires covert action, not armies. It requires emphasizing human intelligence and signals intelligence as much—or more—than satellite imagery, and hiring intelligence agents who speak local languages and are willing to face risks.[26]

Our insistence on applying the death penalty to international terrorists is causing us multiple problems. The death penalty is banned throughout Europe (and indeed, by most liberal democracies). Our European allies, irritated by what they call a growing American tendency to give in to a "unilateralist temptation," see yet another example in our insistence on employing the death penalty against international terrorists extradited from abroad.[27] Because of its opposition to the death penalty, the European Parliament has prohibited extradition of terrorists to the United States for trial without a commitment to waive capital punishment. The United Kingdom, our closest ally on most matters related to national security, has put the United States on notice that British soldiers will not turn bin Laden over to the United States if they manage to capture him, unless the death penalty is waived. Spain is refusing to extradite eight suspected terrorists without assurance that the death penalty will not be imposed.

Because of this issue, sentencing of captured Al Qaeda members tends to be haphazard, bearing little relation to the severity of their crimes. Mamdouh Mahmud Salim, a senior deputy to bin Laden whom the government says sought nuclear weapons for Al Qaeda, will not face the death penalty because the German government refused to extradite him to the United States for trial unless the death penalty was waived.[28] As we have seen, Khalfan Khamis Mohamed, described by his lawyer as Al Qaeda's "cabana boy" in Dar es Salaam, faced the death penalty.

Captured extremists and "retired" terrorists can be extraordinarily valuable assets, but only if they are appropriately handled. In his assessment of John Walker Lindh for the U.S. government, based on eight hours of interviews with Lindh, Rohan Gunaratna concluded that Lindh "is more knowledgeable and better informed about Islam, particularly the relationship between Islam and conflict, than any CIA, FBI, DIA, or INR analyst. Mr. Lindh's knowledge could improve U.S. intelligence and law enforcement officers' understanding of the factors and conditions that spawn Islamic extremism and the resultant terrorism," especially in regard to suicidal terrorism, he writes.[29] But to get this benefit, we have to be strategic. We have to offer carrots as well as sticks. If prisoners at Guantánamo are cooperating, for example, why not allow them special privileges, such as permitting their families to live nearby?

The U.S. government opposes the use of torture during interrogation. But it is not entirely averse to allowing other governments to use torture. Since September 11, the U.S. government has been discretely transporting terrorism suspects to countries that are known to torture suspects. As Secretary Powell put it immediately after the September 11 strikes, in words he may have come to regret, "Egypt, as all of us know, is really ahead of us on this issue. They have had to deal with acts of terrorism in recent years in the course of their history. And we have much to learn from them and there is much we can do together." What "we all know" is this: according to a 1996 UN report, Egypt tortures its political prisoners systematically, employing such methods as electric shocks, suspension by wrists or ankles, and threats of rape against male prisoners. In response to his newfound popularity, Mubarak told the state-owned paper in Cairo, "There is no doubt that September 11 created a new concept of democracy that differs from the concept that Western states defended before these events, especially in regard to freedom of the individual."[30] We should oppose this policy, journalist Peter Maass argues persuasively, not only for moral reasons, but for pragmatic ones as well. "Arbitrary arrests and executions, carried out by unloved governments at the bidding of the unloved United States, can lead to those governments being replaced by ones that support the terrorists instead."[31]

We should also be attempting to purchase the expertise of terrorists whose organizations go out of business. If the peace process continues to go forward in Sri Lanka, an estimated ten thousand armed cadres will soon be out of work. The Tamil Tigers terrorists are among the best trained and most disciplined in the world. They have assassinated two heads of state, Indian prime minister Rajiv Gandhi in 1991 and Sri Lankan president Premadasa in 1993, and have carried out several hundred attacks, including some two hundred suicide bombings, leading to an estimated sixty thousand deaths in the last two decades.[32] On July 23, 2001, they shut down Sri Lanka's airport entirely.[33] Operatives' future job prospects include continuing to run the Tigers' traditional businesses, which include human smuggling and shipping of licit and illicit commodities; taking government jobs in a Tamil autonomous region; or selling their expertise to the highest bidder. Counterterrorists should be seriously thinking about outbidding Al Qaeda and its sympathizers, before it is too late. The same argument applies to the thousands of Laskar Jihad fighters that were decommissioned in October 2002.

While military, intelligence, and law-enforcement approaches are necessary, we need to realize that we are fighting worldwide movements and that many individuals and organizations benefit from the terrorism we aim to curtail. We also need to recognize that for some terrorist groups, including some affiliated with Al Qaeda, survival of the group becomes more important than the grievance it formed to address. Organizations strive to persist.[34] This applies not only to terrorists, but also to the officials that fight them. Government officials involved in fighting terrorism may benefit by taking money for protecting vulnerable businesses, as in Indonesia. In an extraordinarily poor part of the world, individuals involved on both sides of the conflict in Kashmir have found ways to enrich themselves. Some of them are now living in L.A.-style mansions on either side of the border. In economies beset by rampant unemployment, such as Algeria, a position in the security service may be so attractive that the security services have little incentive to see the conflict end.

In countries where extremist religious schools promote terrorism, the best policy for outside governments may be to help develop alternative schools, rather than publicly attempting to persuade the local government to shut the extremist schools down. In Pakistan, for example, many children end up at extremist *madrassahs* because their parents can't afford the alternatives. Children who attend public schools may have to buy books or pay for their transport, while at *madrassahs* they are likely to receive free books, housing, and board. State schools may not even exist in the vicinity of their homes; and when they do, the quality of the education provided is often subpar. Assistance in developing schools that successfully educate youth to participate in modern society is a worthwhile effort, whatever its long-term impact on countering terrorism.

On the public-diplomacy front, we need to go after the slogans and the infrastructure used to mobilize recruits. We need to take public relations and public education as seriously as the terrorists do. We need to be aware that the West is reviled and try, whenever consistent with our values, to remove the thorn, for example, by loudly denouncing Israel's de facto settlement policy, which is inconsistent with the peace process. We should make sure that America's intervention in Kosovo to assist the Muslims becomes as well-known to Muslim youth around the world as the Serbs' atrocities.

We also need to make it harder for terrorists to get access to weapons—including weapons of mass destruction—and bolster our responses in case they do. Of particular importance is the need to continue upgrading security at vulnerable nuclear sites. While significant progress has been made at many nuclear facilities in Russia and other former Soviet states, many sites are still vulnerable to theft. Russia's own concerns regarding nuclear security make clear that we still have a long way to go. And Russia is not the only source of weapons-usable material. Some fifty-eight countries have nuclear-weapons-usable material on their territory, in many cases inadequately secured.[35]

Also important is to upgrade the global system of monitoring of disease around the world, since biological attacks could be difficult to distinguish from natural outbreaks of disease. Such a policy has the virtue of being useful for improving public health in general, regardless of whether terrorists ever mount a biological attack. Detection technologies, medical countermeasures, therapeutic regimens, and knowledge of the relevant organisms all need to be improved. Governments need to ensure that dangerous pathogens are adequately secured and safely stored. This applies in particular to former Soviet states, whose biological-weapons programs were most advanced. Alternative employment must be found for scientists formerly engaged in weapons work, to minimize the risk of continuing brain drain to Iran and other countries.

Will the antimodernists win? I am too optimistic to think they will. But to fight them, we need to get our own house in order.

The answer to the question "Why do they hate us?" is not only the "axis of envy" inevitably engendered by our military and economic might, but also our policies, and, more importantly, how they are perceived by potential recruits to terrorist organizations. It is not just who they are (humiliated—at least in their view—by globalization and the New World Order) and not just who we are (an enviable hegemon), but also, at least in part, what we do. We station troops in restive regions, engendering popular resentment. We maintained ineffective sanctions against Iraq, generating widespread outrage at their effect on Iraqi citizens. We demand that other countries adhere to international law, but willfully and shortsightedly weaken instruments that we perceive as not advancing our current needs. Despite our belated recognition that weak states may threaten us more than strong ones, we continue to let failed states fester.[36] The people of Afghanistan, for example, still live in a state run largely by warlords. They remain desperately poor and essentially ungoverned, scarcely better off then they were on September 10, 2001. At the time of this writing, Iraq remains a failed state. Weak economics and poor governance are making several Latin American countries attractive havens for a variety of terrorist groups.We look the other way when Israel violates human rights and in general apply "double standards."[37] We demand that other countries open their markets to our goods, even as we maintain protections on ours, applying textile quotas, for example, against countries like Pakistan, whose citizens are increasingly vulnerable to the notion that Al Qaeda is more interested in their well-being than is the United States.

In general, we describe ourselves as idealists—as if the assertion will make it so. We attach moral content to possession of weapons of mass destruction, for example, but see our own weapons in realist terms—as a sadly necessary deterrent. We demand that other countries choose idealist security policies in keeping with our vision, but act, most of the time, to further our interests, often at the expense of the rest of the world. I am not arguing in favor of cosmopolitanism, but rather, a smarter realpolitik approach. Even if we take no interest in the well-being of other states' citizens, our long-term national security interests demand that we carefully consider how our policies impact terrorists' recruitment drives. We need to take into account the inevitable trade-offs in policy-making between domestic policy objectives such as the desire for cheap oil, and long-term counterterrorism goals. In short, we need to take into account how our policies play into the hands of our terrorist enemies.

We need to accept that whatever the virtues of global markets, their benefits so far have accrued disproportionately in some parts of the world.[38] While America and the West cannot conduct a global war on poverty, we can avoid making things worse. It does not make sense, in such an atmosphere, to close our markets to Pakistani textiles or to insist on protecting our intellectual property regarding drugs when it suits us, only to consider abandoning the norm the first time we suffer an anthrax scare. We need to realize that contagion in financial markets can shatter the economy of a country like Indonesia, and we need to do whatever we can to help.

We need to concede that some of the values Americans are known for and export worldwide include relentless consumerism, atomized societies, the interpretation of freedom as no rules and no responsibility, and glorification of vulgarity and violence in film and music. To the extent that globalization and the New World Order means the spread of such values, the antimodernists' complaints are understandable, even if their violence is not.

But other values at the core of the American system and in many other parts of the world are worth defending and reaffirming in the face of assault. The first is that every human being is inestimably valuable, whatever his race, gender, or religion. Another is our commitment to freedom of religion, but not freedom to murder for religious reasons. These, alas, are values that put us fundamentally at odds with our foes.

But even if we were willing to change our policies, for our own reasons, in ways that the terrorists would like, responding to terrorists' grievances is likely to be of only marginal effectiveness. We should not delude ourselves that changing our Middle East policy or removing U.S. troops from Saudi Arabia will make Al Qaeda or the Islamist Palestinian terrorist groups roll up and die.

The religious terrorists we face are fighting us on every level—militarily, economically, psychologically, and spiritually. Their military weapons are powerful, but spiritual dread is the most dangerous weapon in their arsenal. Perhaps the most truly evil aspect of religious terrorism is that it aims at destroying moral distinctions themselves.[39] Its goal is to confuse not only its sympathizers, but also those who aim to fight it. We need to respond—not just with guns—but by seeking to create confusion, conflict, and competition among terrorists and between terrorists and their sponsors and sympathizers. We should encourage the condemnation of extremist interpretations of religion by peace-loving practitioners. We should change policies that no longer serve our interests or are inconsistent with our values, even if those happen to be policies that the terrorists demand. In the end, however, what counts is what we fight *for,* not what we oppose. We need to avoid giving into spiritual dread, and to hold fast to the best of our principles, by emphasizing tolerance, empathy, and courage.

NOTES

Introduction

1. *C.S.A. Journal* 7, 11.

2. Kerry Noble, *Tabernable of Hate: Why They Bombed Oklahoma City* (Prescott, Ontario: Voyageur Publishing, 1998), 216.

3. Robert Jay Lifton, *The Nazi Doctors: Medical Killing and the Psychology of Genocide* (New York: Basic Books, 1986), 5.

4. Albert Bandura, "Mechanisms of Moral Disengagement," in Walter Reich, ed., *Origins of Terrorism: Psychologies, Ideologies, Theologies, States of Mind* (Cambridge: Cambridge University Press, 1990), 161–91. See also Lifton, *Nazi Doctors.*

5. Kathleen Norris, "Native Evil," *Boston College Magazine,* winter 2000, last accessed 12 January 2003, www.bc.edu/publications/bcm/winter_2002/ft_evil_native.html.

6. Definitions from *Concise Oxford Dictionary,* 10th ed. (Oxford: Oxford University Press, 1999).

7. I find the relational analysts' definition of empathy as vicarious introspection to be closest to what I try to achieve in my interviews.

8. Francine de Plessix Gray, *Simone Weil* (New York: Viking, 2001).

9. Jillian Becker, "Simone Weil: A Saint for our Time?" *The New Criterion Online,* last accessed 25 December 2002, www.newcriterion.com/archive/20/mar02/weil.htm.

10. Ayman al-Zawahiri, *Knights under the Prophet's Banner,* pt. 1. Excerpts of the book were translated by the Foreign Broadcast Information Service (FBIS). See *"Al-Sharq Al-Awsat* Publishes Extracts from Al-Jihad Leader Al-Zawahiri's New Book," *Al-Sharq al-Awsat* (London), 2 December 2001, in FBIS-NES-2002-0108, document ID GMP20020108000197.

11. Al-Zawahiri, *Knights,* pt. 1.

12. In his famous discussion of "supreme emergency," Michael Walzer writes that in a context where defeat is imminent, and where defeat might be accompanied by the annihilation of a population or its way of life, ordinary limitations on fighters are temporarily suspended. Michael Walzer, *Just and Unjust Wars* (New York: Basic Books, 2000). In the aftermath of the Arab wars of conquest that followed the death of the Prophet Muhammad, religious specialists developed a set of judgments relating to the resort to and conduct of war. They distinguished between two understandings of the military aspect of jihad: first, a "collective duty" *(fard kifaya)* of Muslims to support an armed struggle in which Muslim troops—as opposed to the whole population—led by a legitimate Muslim ruler secure the boundaries of a Muslim state; and second, an "individual duty" of fighting in the name of Islam in the case that an enemy is violating Muslim rights by invading the territory of Islam, damaging Muslim property, and hence endangering Islam itself, i.e. delimiting the Muslim community's capacity to carry out its duty of serving God. In this latter case, the scholars said that fighting is *fard 'ayn,* an individual duty. *Fard 'ayn* involves the existence of an emergency condition in which "necessity makes the forbidden things permitted." It involves the suspension of traditional lines of authority so that, for instance, women and children are allowed to participate in the struggle. Citing what he says are unequivocal passages from the Koran, John Kelsay adds that without a doubt, Islam's understanding of *fard 'ayn* is inconsistent with the killing of civilians. John Kelsay, "War, Peace, and the Imperatives of Justice in Islamic Perspective: What do the September 11, 2001, Attacks Tell Us about Islam and the Just War Tradition?" forthcoming in P. Robinson, ed., *War and Justice in World Religions* (Hampshire, UK: Ashgate, 2003).

13. This section makes use of arguments and definitions I developed for my earlier book, *The Ultimate Terrorists* (Cambridge: Harvard University Press, 1999).

14. Philosopher Virginia Held argues that not all terrorists aim to terrify. Alternative objectives include gaining concessions, obtaining publicity, or provoking repression. But since terrorists attempt to achieve these objectives with the instrument of dread, this argument is more semantic than useful. Virginia Held, "Terrorism, Rights, and Political Goals," in R. G. Frey and Christopher W. Morris, *Violence, Terrorism, and Justice* (New York: Cambridge University Press, 1991), 59–85.

15. It is the means of terrorism that are morally unique, not its end. "It would be naive to pretend that *terrorism* is an innocuous term," Martha Crenshaw admonishes, "but ends must be separated from means in politics." See Martha Crenshaw, *Terrorism and International Cooperation* (New York: Institute for East West Security Studies, 1989), 5. This is precisely what this definition attempts to do: to define terrorism as a technique that can be used in the service of good or evil.

16. The definition of combatants continues to be the subject of intense debate. The Geneva Convention and its relevant protocols provide some guidance. Legitimate targets are limited to those objects, which by their nature, location, purpose or use make *an effec-*

tive contribution to military action and whose total or partial destruction, capture or neutralization, in the circumstances ruling at the time, *offers a definite military advantage*. (Geneva Convention, Article 52, paragraph 2). For a comprehensive analysis see Frits Kalshoven, Constraints on the Waging of War (Geneva: International Committee of the Red Cross; Dordrecht, Netherlands: M. Nijhoff, 1987), 88–95. Virginia Held argues convincingly that it is often impossible to differentiate categorically noncombatants from legitimate targets—in many countries children are forced to perform military service. See Held, "Terrorism, Rights, and Political Goals," 59–85. Moreover, civilian populations are often producing armaments. Russel Ewing argued in 1927, "Whereas wars once affected merely the fighting men who make up a small percent of the total population, today almost the entire population of a belligerent nation becomes engaged in wartime industries." See Russel Ewing, "The Legality of Chemical Warfare," *American Law Review* 61 (1927): 58.

17. In addition to the 125 people who were killed inside the Pentagon, 64 people were killed aboard American Airlines flight 77.

18. The description that follows is based on David C. Rapoport, "Fear and Trembling: Terrorism in Three Religious Traditions," *American Political Science Review* 78.3 (September 1984): 658–76.

19. Ibid., 669.

20. Ibid., 658–76. See also Bernard Lewis, *The Assassins: A Radical Sect in Islam* (New York and Oxford: Oxford University Press, 1987).

21. I am using the term *holy terrorism* to mean *religious terrorism* as I have just defined it.

22. Bruce Hoffman, *Inside Terrorism* (New York: Columbia University Press, 1998), 87–130.

23. Susan Nieman, *Evil in Modern Thought: An Alternative History of Philosophy* (Princeton, N.J.: Princeton University Press, 2002), 8.

24. Leibniz's answer to why God would allow a natural order that involved so much innocent suffering was that man brought such natural evils upon himself: natural evil was collective punishment for moral evil, including, but not limited, to the Fall. A massive earthquake, which destroyed the city of Lisbon in 1755, evoked a similar reaction among Enlightenment philosophers and theologians as Auschwitz did for their twentieth-century counterparts, Susan Nieman explains. Rousseau would reject Leibniz's view, ushering in a more modern conception of evil. Innocent suffering was not punishment for sin, but a symptom of ignorance. In regard to the earthquake at Lisbon, for example, it made no sense for humans to live in large cities where they were vulnerable to earthquakes. Nieman, *Evil in Modern Thought,* 1–57. Interestingly, psychiatrists are seeing new links between suffering and sin today, as we shall see.

25. It may be that to persuade themselves that the atrocities they have committed or ordered others to commit were justified, terrorists dehumanize their enemies more and more and willingly inflict pain, in some cases for profit. Lieutenant Colonel Dave

Grossman, *On Killing* (Boston: Little Brown and Company, 1995). Ervin Staub, *The Roots of Evil: The Origins of Genocide and Other Group Violence* (Cambridge: Cambridge University Press, 1989).

26. Sue Grand, *The Reproduction of Evil: A Clinical and Cultural Perspective* (Hillsdale, N.J.: The Analytic Press, 2000). Victims of repeated abuse, or children who live in violent neighborhoods or war zones, may experience PTSD. T. G. Veenema, K. Schroeder-Bruce, "The Aftermath of Violence: Children, Disaster, and Post-Traumatic Stress Disorder," *Journal of Pediatric Health Care* 16.5 (September–October 2002): 235–44.

27. Shakespeare's Richard III, for example, attributes his determination to become "a villain" (morally evil) to his having been "cheated of feature by dissembling nature" (natural evil). This might be a good example of how victimization—including by fate—can be used to justify moral wrongs. Derek Summerfield observes that "the profile of post-traumatic stress disorder has risen spectacularly, and it has become the means by which people seek victim status and its associated moral high ground in pursuit of recognition and compensation." Derek Summerfield, "The Invention of Post-Traumatic Stress Disorder and the Social Usefulness of a Psychiatric Category," *British Medical Journal* 322 (13 January 2001): 95–98.

Judith Lewis Herman, a leading authority on trauma, reminds us that it is important to keep in mind that one should not subscribe to the belief that traumatized victims become evil, since that would involve "blaming the victim," which is morally wrong. Herman points out that there is a popular literature of the "cycle of abuse" theory. This theory involves sex offenders who have undergone past traumas themselves. The theory suggests that "the sexual offense is . . . a reenactment of the trauma or an attempt to overcome it through the mechanism of 'identification with the aggressor.'" She adds that "proponents of this theory often invoke the concept of a 'cycle of abuse,' or of 'generational transmission,' whereby the sexually victimized children of one generation become the victimizers of the next." Herman points out that this theory is not empirically valid, explaining that "its most glaring weakness is its inability to explain the virtual male monopoly on this type of behavior. Since girls are sexually victimized at least twice to three times more commonly than boys, this theory would predict a female rather than a male majority of sex offenders." Judith Lewis Herman, "Considering Sex Offenders: A Model of Addiction," *Journal of Women in Culture and Society* 13.4 (summer 1988): 703–4.

28. Some studies suggest that PTSD may be at least partly heritable. Adult children of Holocaust survivors are at greater risk for PTSD—perhaps because they are more likely to expose themselves to traumatic events through life choices, or because susceptibility to PTSD is heritable. See R. Yehuda et al., "The Cortisol and Glucocorticoid Receptor Response to Low Dose Dexamethasone Administration in Aging Combat Veterans and Holocaust Survivors with and without PTSD," *Biological Psychiatry* 52.5 (September 2002): 393–403; and M. B. Stein, K. L. Jang, and S. Taylor, "Genetic and Environmental Influences on Trauma Exposure and Post-Traumatic Stress Disorder Symptoms: A Twin Study," *American Journal of Psychiatry* 159.10 (October 2002): 1675–81.

29. Judith Herman reports that the majority of trauma victims do not become perpetrators, but that trauma appears to amplify common gender stereotypes among victims of childhood abuse. Men are more likely to take out their aggression on others, while women are more likely to injure themselves or to be victimized again. Judith Lewis Herman, *Trauma and Recovery* (New York: Basic Books, 1992), 113. For adolescent males, exposure to violence and victimization is "strongly associated with externalizing problem behaviors such as delinquency, while adolescent females exposed to violence and victimization are more likely to exhibit internalizing symptoms," according to a study of African-American youth. Z. T. McGee et al., "Urban Stress and Mental Health among African-American Youth: Assessing the Link between Exposure to Violence, Problem Behavior, and Coping Strategies," *Journal of Cultural Diversity* 8.3 (fall 2001): 94–104. A 1999 study showed that male Vietnam veterans seeking inpatient treatment for PTSD were more likely to exhibit violent behavior than a mixed diagnostic group of inpatients without PTSD. M. McFall et al., "Analysis of Violent Behavior in Vietnam Combat Veteran Psychiatric Inpatients with Post-Traumatic Stress Disorder," *Journal of Trauma Stress* 12.3 (July 1999): 501–17. Combat exposure was found to have an independent positive association with interpersonal violence, when controlling for PTSD among combat veterans. F. C. Beckham et al., "Interpersonal Violence and Its Correlates in Vietnam Veterans with Chronic Post-Traumatic Stress Disorder," *Journal of Clinical Psychology* 53.8 (December 1997): 859–69. Dr. Jerrold Post argues that PTSD, secondary to living in a Palestinian refugee camp, could play a role in the creation of a terrorist. Jerrold M. Post, "Terrorist on Trial: The Context of Political Crime," *Journal of the American Academy of Psychiatry Law* 28.4 (2000): 489.

30. Nieman, *Evil in Modern Thought,* 284–85.

31. This paragraph summarizes Nieman, *Evil in Modern Thought,* 284–85.

32. The Pakistani jihadi groups, for example, are largely the creation of the Pakistani military establishment. When they fight the Indian military, the term *mercenary* might be more appropriate than *terrorist.* They are included in this book because they also target noncombatants and because they have joined forces with an organization, Al Qaeda, whose modus operandi is murdering innocents in large numbers.

33. Note that religious terrorism and just terrorism overlap, but religious terrorism is no more likely to be just than any other form of terrorism. What counts, at least for me, is that the intention is to save innocent lives that would otherwise be lost, the probability of success, the use of a discriminate weapon, balancing of other ethical goals, and the complete lack of other alternatives. I am deeply grateful to a group of theologians, ethicists, and other scholars convened by Mark Moore to help me think through these questions: Mary Jo Bane, Brent Coffin, Ronald Thieman, and Kenneth Winston. I would also like to thank Michael Reich, who strongly encouraged me to write this section and was also a careful reader.

34. They claim also to be following in the footsteps of the White Rose activists, a group of German youths who publicly protested the Nazis' attempt to exterminate the Jews. These youths were guillotined for their insistence that Jews were human beings, just as the doctor killers insist that unborn children are human beings.

35. Abraham Lincoln, "The Seventh and Last Joint Debate, at Alton, Illinois, October 15, 1858," in *The Complete Works of Abraham Lincoln,* Vol. 5, ed. John Hay (New York: Francis D. Tandy Company, 1894). Available on the Web site of the Northern Illinois University Libraries, last accessed 29 December 2002, www.lincoln.lib.niu.edu/cgi-bin/getobject_?c.2229:2./lib35/artfl1/databases/sources/IMAGE/.

36. Ibid.

37. The case I am making here would be more difficult in an authoritarian regime, where there are no legal or political institutions worth protecting, and where the polity is morally corrupted, perhaps in part as a result of fear. Examples include Cambodia under the Pol Pot regime, the Soviet Union under Stalin, or in Nazi Germany. Would it be morally appropriate to kill the person who is flipping the switch in the crematorium, for example, to save the lives of hundreds or thousands of Jews, in part by terrorizing other crematorium workers into leaving their jobs? In this case, trying to change government policy would not be an option. And while killing the tyrant would be an easier case to make, we will assume that doing so would not be possible, and certainly not possible in time to save those lives. It can be argued—especially in nondemocratic regimes—that government officials are not noncombatants. It can also be argued that although assassinating tyrants is a crime, it is sometimes morally justified. Saint Thomas Aquinas argued that because tyrannical governments are directed toward the private good of the ruler, rather than to the common good of the governed, overthrowing such governments is to be praised, unless it produces such disorder that society's suffering is increased. Aristotle and Cicero, among others, made similar arguments. But who has the right to identify a tyrant, the terrorist, or his victim? In any case, another tyrant might immediately take his place. While, in my view, it would be morally problematic to kill the crematorium worker, it would be even more problematic not to do so. If there was reason to believe that another killer would not come to take the dead one's place, or it was possible to keep killing the killers, the right thing to do, under these narrowly specified circumstances, would be to kill the killer, but then submit to the law, in my view. Here is an example of terrorism that is probably just. If the case were a contemporary one, we might demand of our terrorist that he look beyond his own country and consider whether involving international organizations might achieve the same objective. We would demand of our just terrorist that terrorism truly be a last resort, as required for a just war under the just-war tradition. The existence of CNN and of the Internet make it harder to imagine just terrorism today; it would be necessary to exhaust these avenues before contemplating murdering innocents—even in defense of a large number of other innocents. Ajai Sahni argues that it is absurd to call terrorism "just" under any circumstances, in the same way that genocide cannot ever be just, in his view. Ajai Sahni conversation with the author, Delhi, 17 January 2003. I include this example because I consider it a useful exercise to help us understand precisely the evil of terorrism.

38. Martha Nussbaum, "Making Philosophy Matter to Politics," *New York Times,* 2 December 2002, A21.

39. Cited in Lifton, *Nazi Doctors,* xvii.

40. Ibid.

41. While many terrorists are motivated by more mundane concerns, such as financial payments for killing, especially in mature terrorist groups, the groups start out by appealing to operatives' desire to purify the world—and we need to understand this.

Part I: Grievances

1. The theory of collective action was developed in Mancur Olson, *The Logic of Collective Action* (Cambridge, MA: Harvard University Press, 1965). Many people find the application of this theory to terrorism surprising or even shocking. I do not subscribe to the notion that terrorists promote social justice, but it is critical to understand that terrorists and their sympathizers believe this. Given that, the same questions Olson raised with regard to voting or pollution abatement ought to apply when individuals are deciding whether or not to contribute to terrorist organizations.

2. Joel Mowbray, "How They Did It: An 'evil one' confesses, and boasts," *National Review*, 23 December 2002; Vol. LIV, No. 24.

3. Desmond Butler and Don Van Natta, "A Qaeda informer Helps Investigators Trace Group's Trail," *New York Times* 17 February 2003, A1.

4. Political scientist Edward Banfield asserts that riots occur mainly for "fun" or "profit," not for the purported political objectives the participants claim to seek. Edward Banfield, *The Unheavenly City* (Boston: Little Brown, 1968.) Cited in James B. Rule, *Theories of Civil Violence* (Los Angeles: University of California Press, 1988).

5. Charles Tilly, *From Mobilization to Revolution* (New York: Random House, 1978), 203.

6. Clearly, people do contribute to public goods in the absence of selective incentives. Do-gooders are likely to be rewarded for their efforts with improved self-esteem, prestige, power, and (possibly) economic benefits. For example, voluntary workers acquire leadership skills and networks of influence that could be marketable in other areas, as Melucci points out. See Alberto Melucci, *Challenging Codes*, 167. But for some individuals, some of the time, factors other than individual self-interest clearly motivate behavior. Altruists, for example, take pleasure from the pleasure of others. Volunteers may also enjoy the *process* of contributing to the public good, and don't require rewards for their efforts. In this case, contributing is a rational strategy. But Jan Elster points out that for some individuals, some of the time, conscience trumps "rationality." Individuals motivated by social or moral norms will contribute to public goods even when they receive no particular utility from knowing that others benefit or from the process of contributing. While the norms we choose to follow appear partly to be shaped by self-interest (the poor often favor the norm of equality, while the rich are more likely to favor the norm of reward for effort), "norms are not fully reducible to self-interest. . . . The unknown residual is a brute fact, at least for the time being," Elster argues. See Jan Elster, *The Cement of Society: A Study of Social Order* (Cambridge: Cambridge University Press, 1989), 150. People motivated by social norms may contribute to public goods because they may be alarmed by the prospect of social disapproval, or they may like to be seen doing the "right thing," he argues. People motivated by moral norms may feel compelled to follow the rules, even if no one is watching.

7. Religion appeals to a continuum of interests concurrently. It appeals simultaneously to "rational" (selfish), teleological, deontological, and aretaic interests all at once. It encourages adherents to do the right thing for its own sake, but also in the hope of rewards after death.

8. In organizations of professional killers, such as the Pakistani jihadi groups discussed in chapter 5, material incentives play such an important role that the organization is best understood with the tools of economic analysis, analogously to an NGO or even a firm. Astute leaders become price discriminators for labor. When a prize operative no longer finds the spiritual or emotional appeal sufficient, money and power become the most important currency for encouraging him to continue his work. The higher the opportunity cost of a particular operative's time, the more he will have to be paid.

Chapter 1: Alienation

1. They were from the FBI, the Bureau of Alcohol, Tobacco, and Firearms, and state, county, and local police agencies.

2. Kerry Noble, *Tabernable of Hate: Why They Bombed Oklahoma City* (Prescott, Ontario: Voyageur Publishing, 1998), 22.

3. Robert G. Millar, "The FBI and I," unpublished manuscript.

4. Ibid.

5. Ibid.

6. Kerry Noble, interview with the author, 2 March 1998. For other studies of CSA, see James K. Campbell, *Weapons of Mass Destruction Terrorism* (Seminole, Fla.: Interpact Press, 1997); James Coates, *Armed and Dangerous: The Rise of the Survivalist Right* (New York: Hill and Wang, 1987); and Brent Smith, *Terrorism in America; Pipe Bombs and Pipe Dreams* (Albany: State University of New York Press, 1994).

7. Noble, *Tabernable of Hate,* 73.

8. *C.S.A. Journal* 7.

9. The government relied extensively on Ellison's testimony against his coconspirators to make its case, but jury members suspected that Ellison's assessment was influenced by his desire for a reduced sentence. Lamar James, "Jurors Finish 'Hard' Task; Most Decline Comment," *Arkansas Gazette,* 8 April 1988.

10. Kerry Noble, interview with the author, 2 March 1998.

11. Noble, *Tabernable of Hate,* 39. Noble, interview.

12. Ibid.

13. Noble, *Tabernable of Hate,* 145–46.

14. Kerry Noble, e-mail message to the author, 7 April 1999.

15. "Keith" (unidentified former government agent), interview with the author, 22 December 1998. He had retired from the U.S. government some ten years earlier.

16. Noble, *Tabernable of Hate,* 51.

17. Ibid.

18. Ibid., 68.

19. Robert Jay Lifton, *The Nazi Doctors: Medical Killing and the Psychology of Genocide* (New York: Basic Books, 1986), 126.

20. Ibid. Doubling is similar—but more extreme and sustained—to Melanie Klein's notion of "splitting," in which the hated parts of the self are projected onto the Other.

21. William James, *Essays in Radical Empiricism* (Cambridge, Mass.: Harvard University Press, 1976).

22. Lifton, *Nazi Doctors,* 427.

23. Lieutenant Colonel Dave Grossman, *On Killing: The Psychological Cost of Learning to Kill in War and Society* (Boston: Little, Brown and Company, 1995), 119.

24. Web site of the Aryan Nations, last accessed 8 January 2003, www.aryan-nations.org/.

25. Albert Bandura, "Mechanisms of Moral Disengagement," in Walter Reich, ed., *Origins of Terrorism: Psychologies, Ideologies, Theologies, States of Mind* (Washington, D.C.: Woodrow Wilson Center Press, 1998), 182.

26. Lifton, *Nazi Doctors,* 427.

27. Noble, *Tabernable of Hate,* 86.

28. Ibid., 120. This concept is based primarily on I Thessalonians 4:17: "Then we which are alive and remain shall be caught up together with them in the clouds, to meet the Lord in the air: and so shall we ever be with the Lord." As interpreted by Christian fundamentalists, the rapture, which lifts the chosen few out of the mayhem of end-time destruction to meet the returning Messiah, is a reward for their steadfastness. The Darbyite movement of 1830s to the 1880s was the forerunner to modern Christian fundamentalism and supplied the theological basis for the rise of fundamentalism's emphasis on biblical literalism and inerrancy and the notion of "premillennial dispensationalism"—that Jesus will return prior to his millennial rule, and that mankind has entered the end time after receiving previous "dispensations" from God in the form of Adam's banishment from Eden, the Flood, and Christ's grace, to which man has failed to respond. Anglican John Nelson Darby's seven dispensations, although reflecting earlier thinking and sources regarding the rapture, premillenialism, and dispensationalism, were original in their focus on the rapture, which became a central feature in his prophetic system, and in his ideas pertaining to ingathering of the Jews and Israel. Charles B. Strozier, *On the Psychology of Fundamentalism in America* (Boston, Mass.: Beacon Press, 1994), 183–84. James A. Aho, *The Politics of Righteousness: Idaho Christian Patriotism* (Seattle, Wash.: University of Washington Press, 1990), 53–54.

29. Pastor Robert Millar, interview with the author, 21 April 1998.

30. Noble, *Tabernable of Hate,* 88.

31. Ibid.

32. Ibid.

33. Ibid., 93.

34. Ibid., 34.

35. Ibid., 56. Noble provides some examples of where the body composed of many members is mentioned in the scriptures, but it would be hard to support the "sonship" doctrine exclusively with those passages.

36. Ibid., 55.

37. Rosabeth Moss Kanter observed many of these commitment mechanisms in nineteenth-century utopias and communes. Rosabeth Moss Kanter, *Commitment and Community: Communes and Utopias in Sociological Perspective* (Cambridge, Mass.: Harvard University Press, 1972), 83–84.

38. Noble, *Tabernable of Hate,* 74.

39. Ibid., 67.

40. Ibid., 68.

41. Ibid., 96.

42. *C.S.A. Journal* 7.

43. Leon Festinger, Henry Riecken, Stanley Schachter, *When Prophecy Fails* (New York, Harper and Row, 1956).

44. Noble, *Tabernable of Hate,* 73.

45. Ibid., 75.

46. Coates, *Armed and Dangerous,* 137.

47. Ibid.

48. *C.S.A. Journal* 7, 34.

49. "Keith," interview.

50. Noble, *Tabernable of Hate,* 63.

51. Ibid.

52. Ibid., 35.

53. Robert S. Robins and Jerrold M. Post, *Political Paranoia: The Psychopolitics of Hatred* (New Haven, Conn.: Yale University Press, 1997), 140.

54. Response of United States of America to James D. Ellison's Motion for Reduction of Sentence, *United States of America v. James D. Ellison,* Criminal Nos. 85-20006-01 and 85-200017-01, 11 May 1988.

55. Rodney Bowers, "White Radical Activities that Led to Indictments Recounted for 1983–85," *Arkansas Gazette,* 27 April 1987, A1.

56. Noble, interview; Testimony of James Ellison and Kerry Noble, *United States of America v. Robert E. Miles et al.,* 87-200008, 15 March 1988, U.S. District Court, Western District of Arkansas, Fort Smith Division, 38. Ellison's recollection of the federal building they had considered targeting was "vague," he claimed in his testimony, although Noble is explicit in his recollection both in conversation and in his book.

57. Noble, *Tabernable of Hate,* 132–33.

58. FBI Agent Knox claimed that "substantial damage was done to the natural gas pipeline." *United States of America v. Steve Scott,* 85-20014-01, U.S. District Court, Western District of Arkansas, Fort Smith Division.

59. Noble, interview.

60. Bowers, "White Radical Activities," A1.

61. As defined in Title 18, Section 2384, of the United States Code, sedition involves two or more persons conspiring to "overthrow, put down, or destroy by force the Government of the United States, or to oppose by force the authority thereof, or by force to prevent, hinder, or delay the execution of any law of the United States, or by force seize, take, or possess any property of the United States contrary to the authority thereof." Quoted in *United States of America v. Robert E. Miles et al.,* 9 February 1988, 11.

62. Ibid., 4.

63. Terrorists have rarely been accused of sedition. Another case of a sedition trial against a terrorist group besides that of The Order occurred in Chicago in 1981, when a sedition trial was brought against ten suspected members of a Puerto Rican terrorist organization. Matthew Wald, "U.S. to Try Eight on a Rare Charge, Plotting to Overthrow the Government," *New York Times,* 22 May 1987, A14.

64. "Supremacists Had Hit List, FBI Agent Says," *New York Times,* 7 March 1998, 30.

65. *United States of America v. James D. Ellison,* 85-20006-01, Affidavit for Search Warrant, Jack Knox, 6.

66. Ibid., Direct Examination of Jack Knox.

67. Ibid., Response of the United States of America to James D. Ellison's Motion for Reduction of Sentence.

68. Victoria Loe Hicks, "About-Face from Hate," *Dallas Morning News,* 16 May 1998.

69. *United States of America v. James D. Ellison,* Response of the United States of America to James D. Ellison's Motion for Reduction of Sentence.

70. Howard Pankratz, "Blast Blamed on Revenge Attack Linked to Militant's Execution," *Denver Post,* 12 May 1996.

71. April 19 is an important date for antigovernment groups. It is Patriots' Day, the day the American Revolution began in Lexington, Massachusetts, in 1775. Some Identity Christians associate the date with 1943 when Hitler began deporting the Jews of the Warsaw ghetto to concentration camps. April 19 was also the date in 1993 when followers of David Koresh died in a fire accidentally initiated in an FBI raid. Koresh had barricaded himself and his followers in a heavily armed compound in Waco, Texas. He believed himself to be "the lamb" whose death would open the seven seals, signifying the end of the world as foretold in Revelation. For many right-wing patriots, the "Second American Revolution" will begin on that date.

72. *United States of America v. James D. Ellison,* Response of the United States of America to James D. Ellison's Motion for Reduction of Sentence.

73. "Keith," interview.

Chapter 2: Humiliation

1. Hamas regards all of present-day Israel to be Palestinian territory. (Charter of Hamas, Chapter Two, Article 9). For a translation see Charter of the Islamic Resistance Movement (Hamas) of Palestine, Muhammed Maqdsi, *Journal of Palestine Studies,* Vol. 22, No. 4 (Summer, 1993): 122–134.

2. See, for example, "The Palestinian Debate over Martyrdom Operations, Part II: A Palestinian Communiqué against the Attacks Inquiry and Analysis Series," Middle East Media Research Institute (MEMRI), *Inquiry and Analysis* No. 101, 5 July 2002, available on the MEMRI Web site, last accessed 14 January 2003, www.memri.org/.

3. My translator asked to remain anonymous and asked me to refer to her by her nom de guerre.

4. Although the Palestinian leadership usually refers to itself as the Palestinian National Authority, the term more common in the United States is Palestinian Authority. I will hence refer to the Palestinian Authority (PA).

5. Water is a crucial aspect to this conflict. Its scarcity, misallocation, use, and abuse is yet another point of conflict between the Israelis and Palestinians, especially those in the Gaza Strip. Israel declared all water resources state-owned following the 1967 war, severely restricting water access to the now 1.2 million Palestinians of the Gaza Strip, while simultaneously subsidizing and encouraging water consumption for six thousand Israeli settlers living in the Strip. The land appropriated for the settlements was also superior in terms of groundwater quality and quantity, a source of much anger amongst Gazan Palestinians to this very day. Stephanie Goeller, "Water and Conflict in the Gaza Strip," updated December 1997, last accessed 2 September 2002, www.american.edu/projects/mandala/TED/ice/GAZA.HTM.

6. Chris Hedges, "A Gaza Diary," *Harper's,* October 2001, 1–14.

7. Ibid., 11; see also the Palestinian Centre for Human Rights, "Fact Sheet: An Overview of the Gaza Strip," updated 18 March 2002, last accessed 2 September 2002, www.pchrgaza.org/facts/fact1.htm.

8. Ze'ev Schiff and Ehud Ya'ari, *Intifada* (New York: Simon and Schuster, 1990), 91.

9. Anonymous Palestinian student, interview with the author, 18 October 2002.

10. Hedges, "Gaza Diary," 12.

11. Philip Jacobson, "Home-Grown Martyrs of the West Bank Reap Deadly Harvest," *Sunday Telegraph,* 19 August 2001, 20.

12. UNRWA stands for United Nations Relief and Works Agency for Palestine Refugees in the Near East. UNRWA's Web site can be accessed at www.un.org/unrwa/.

13. "Annual Growth Rate of Registered Palestine Refugees and Female Percentage, 1953–2000," last accessed 29 August 2002, www.un.org/unrwa/pr/pdf/figures.pdf.

14. Abu Shanab's assertion that Israel's occupation of Gaza and the West Bank is illegal is questionable. Security Council Resolution 242 of 1967 does not require that Israel withdraw unilaterally from the occupied territories; it requires a negotiated settlement based on the principle of exchanging land for peace. But Abu Shanab is not the only person confused on this score. In March 2002, Secretary General Kofi Annan called Israel's occupation illegal, an assertion that was subsequently retracted by his spokesman. According to Sir Adam Roberts, professor of international relations at Oxford University, it is not the occupation per se that is illegal, but Israel's activities as an occupying power. Israel's attempts, since 1967, to colonize the occupied territories with its settlements is inconsistent with its obligation under Resolution 242 and with the rules of "belligerent occupation." See "Double Standards: Iraq, Israel, and the UN," *Economist,* 12 October 2002.

15. The full name of the militant wing is Izz-al-Din al-Qassam brigades.

16. Medical historians are uncertain about the nature of the disease; some suspect it may have been the first appearance of smallpox in the West. Rodney Stark, *The Rise of Christianity: A Sociologist Reconsiders History* (Princeton N.J.: Princeton University Press), 73; and Hans Zinsser, *Rats, Lice, and History* (1934; reprint, New York: Bantam, 1960).

17. Ibid.

18. Elaine Pagels, *Adam, Eve, and the Serpent* (New York: Vintage Books, 1988), 32.

19. *Passio Sanctarum Perpetuae et Felictas* 3, trans. H. Musurillo, in *The Acts of the Christian Martyrs* (Oxford, 1972), 106–13. Cited in Pagels, *Adam, Eve, and the Serpent,* 34.

20. Stark, *Rise of Christianity,* 181.

21. Herbert Musurillo, "The Martydom of Saints Perpetua and Felicta," last accessed 6 March 2003 www.pbs.org/wgbh/pages/frontline/shows/religion/maps/primary/perpetua.html.

22. Stark, *Rise of Christianity,* 181.

23. Pagels, *Adam, Eve, and the Serpent,* 48.

24. Stephanie Goeller, "Water and Conflict in the Gaza Strip," updated December 1997, last accessed 2 June 2002, www.american.edu/projects/mandala/TED/ice/GAZA.HTM.

25. Ladan Boroumand and Roya Boroumand, "Terror, Islam, and Democracy," *Journal of Democracy* 13.2 (April 2002): 7–8.

26. Ibid., 7.

27. Quoted in Gilles Kepel, *Muslim Extremism in Egypt: The Prophet and the Pharaoh* (London: Al Saqi Books, 1985), 41.

28. Rudolph Peters, *Jihad in Classical and Modern Islam* (Princeton: Markus Wiener Publishers, 1996), 44.

29. Boroumand and Boroumand, "Terror, Islam, and Democracy," 8–9.

30. See, for example, Howard Schneider, "Egypt's Brutal Response to Militants," *Washington Post,* 7 October 2001, A25.

31. Boroumand and Boroumand, "Terror, Islam, and Democracy," 5–6.

32. Ibid., 6.

33. Conversation with a young Palestinian man who was then in graduate school in the United States, September 2000.

34. Jerrold Post, "Thematic Analysis: The Terrorists in Their Own Words," unpublished manuscript, 79.

35. Ibid.

36. Ibid., 79–80.

37. See, for example, Leslie Susser, "Expulsion: Anatomy of a Decision," *Jerusalem Report,* 14 January 1993, 15, last accessed 3 September 2002, www.jrep.com/Info/10thAnniversary/1993/Article-1.html/.

38. Acting upon Executive Order 13224, the Bush administration froze the assets of several American-based Islamic charities in December 2001, including the Holy Land Foundation for Relief and Development (HLF), the Global Relief Foundation (GRF), and the Benevolence International Foundation (BIF). "Shutting Down the Terrorist Financial Network," updated 4 December 2001, last accessed 20 July 2002, www.ustreas.gov/press/releases/po841.htm. Amira, my translator, adds that Washington has attested that there is no direct evidence (at least none available to the public) that proves the link between HLF and GRF and the finance of terrorism. The U.S. government made the accusations and arrests based on what is known as secret evidence. The counterargument, Amira adds, is that providing funding to orphans is legitimate (which HLF claims to have done), and that it would be unfair to discriminate against orphans whose parents happen to be suicide bombers. An excerpt from a joint statement of major American Muslim organizations reads, "No relief group anywhere in the world should be asked to question hungry orphans about their parents' religious beliefs, political affiliations, or

legal status. Those questions are not asked of recipients of public assistance whose parents are imprisoned or executed in the United States, and they should not be a litmus test for relief in Palestine." See "Petition to the U.S. government to reverse its action against Palestinian Charity," last accessed 3 September 2002, www.petitionoline.com/hlf2001/petitionhtml/.

39. Johanna McGeary and David Van Biema, "Radicals on the Rise," *Time* 158, no. 26 (December 17, 2001): 50–57.

40. Ibid.

41. According to the Hamas Web site, Emad al-Alami is an elected member of the Hamas Political Bureau. He was arrested by Israel in 1988 on the charge of "performing information activities with the aim to externalize Hamas' achievements." He was released in 1990 and subsequently deported to Lebanon. See "Hamas Symbols and Leaders," Web site of Hamas, last accessed 3 September 2002, www.palestine-info.co.uk/hamas/leaders/index.htm#Emad.

42. McGeary and Van Biema, "Radicals on the Rise," 50–57.

43. In response to criticism of its funding activities, the embassy posted a statement on its Web site denying that it supports suicide bombing or terrorism, but did not remove these earlier press releases. See Web site of the Embassy of Saudi Arabia to the United States, last accessed 30 May 2002, www.saudiembassy.net/. Cited in David Tell, "The Saudi Terror Subsidy," *Weekly Standard* vol. 7, no. 35 (20 May 2002): 9.

44. Ariel Merari, lecture on suicide terrorism before the author's class on terrorism, Kennedy School of Government, Harvard University, Cambridge, Mass., 21 May 1999.

45. There are examples of individuals apparently acting on their own. In February 2001 a Palestinian bus driver deliberately ran over eight Israelis, apparently without organizational help. But these attacks take place in a societal context of suicide bombing as theater, with the *shaheed* the hero of the play. In some ways, as we shall see, society is now fulfilling some of the organization's role.

46. Merari, lecture on suicide terrorism; and Nasra Hassan, "An Arsenal of Believers; Talking to the 'Human Bombs,'" *New Yorker,* 19 November 2001, 36.

47. Yoram Schweitzer, "Suicide Terrorism: Development and Main Characteristics," in International Policy Institute of Counter-Terrorism (ICT), *Countering Suicide Terrorism: An International Conference* (Herzliyya, Israel: ICT, 2000), 82–83.

48. There are precedents, however, to the use of female suicide bombers in the Middle East. In March 1985, eighteen-year-old Sumayah Sa'ad drove a car loaded with dynamite into an Israeli military position in southern Lebanon, killing twelve Israeli soldiers and wounding fourteen others. Roughly two weeks later, on March 25, seventeen-year-old San'ah Muheidli drove a TNT-laden car into an IDF convoy, killing two soldiers and wounding two more. Amir Taheri, *Holy Terror: Inside the World of Islamic Terrorism* (Bethesda, Md.: Adler & Adler, 1987), 126–129, cited in Assaf Moghadam, "Palestinian

Suicide Bombings in the Israeli-Palestinian Conflict: A Conceptual Framework," in Reuven Paz, ed., Project for the Research On Islamist Movements, last accessed 28 April 2003, www.e-prism.org.

49. McGeary and Van Biema, "Radicals on the Rise," 50–57. An Israeli security official told Lelyveld, "These days they don't have any problem to recruit suicides. For every suicide they want, they have five, seven, ten volunteers." Joseph Lelyveld, "All Suicide Bombers Are Not Alike," *New York Times* magazine, 28 October, 2001, 17.

50. Abdul Hadi Palazzi, "Orthodox Islamic Perceptions of Jihad and Martyrdom," in International Policy Institute of Counter-Terrorism, *Countering Suicide Terrorism,* 69–70.

51. The nature of these gifts, which usually include seventy-two virgins, is now being questioned. Christoph Luxenberg, in *The Syro-Aramaic Reading of the Koran,* argues that the term *hur* (white raisin, a delicacy in the ancient Near East) was misconstrued as *houri* (virgins). Islamic tradition holds that the correct translation is "virgins," although the word in the earliest texts is *hur*. The ambiguities in the text are caused, in part, by the lack of vowels and diacritical dots. At least one respected dictionary of early Arabic reportedly defines *hur* as "white raisin," however. *Hur* is also translated as "white raisin" in ancient Aramaic, which is closely related to the language used in the earliest texts of the Koran. Luxenberg is a scholar in Germany of ancient Semitic languages. The name he publishes under, Luxenberg, is a pseudonym. Alexander Stille, "Scholars Are Quietly Offering New Theories of the Koran," *New York Times,* 2 March 2002, A1.

52. Emile Durkheim, *Suicide* (New York: Macmillan, 1966); and Gould et al., "Clustering of Attempted Suicide: New Zealand National Data," *International Journal of Epidemiology* 23.6 (1994): 1185–89. Both cited in David Cutler, Edward L. Glaeser, and Karen Norberg, "Explaining the Rise in Youth Suicide," *NBER Working Paper Series* (Cambridge: National Bureau of Economic Research, 2000).

53. Cutler, Glaeser, and Norberg, "Explaining the Rise in Youth Suicide."

54. See Raphael Israeli, "Islamikaze and Their Significance," *Terrorism and Political Violence* 9.3 (autumn 1997): 104. See also Moghadam, "Palestinian Suicide Bombings," 10–11.

55. Moghadam, "Palestinian Suicide Bombings," 36.

56. Ibid.

57. Cutler, Glaeser, and Norberg, "Explaining the Rise in Youth Suicide." On the culture of violence, see in particular Donatella della Porta and Mario Diani, *Social Movements: An Introduction* (Malden, Mass.: Blackwell, 1999); Donatella della Porta, *Social Movements, Political Violence, and the State: A Comparative Analysis of Italy and Germany* (Cambridge, England, and New York: Cambridge University Press, 1995); and Mark Juergensmeyer, *Terror in the Mind of God: The Global Rise of Religious Violence* (Berkeley: University of California Press, 2000).

58. Cutler, Glaeser, and Norberg, "Explaining the Rise in Youth Suicide."

59. Mark H. Moore et al., eds., *Deadly Lessons: Understanding Lethal School Violence* (Washington, D.C.: National Academy Press, 2002).

60. Alfonso Chardy, "Parents Defusing Children: Palestinians Seek Help to Keep Kids from Suicide Bombing," *Montreal Gazette,* 26 May 2002, A1.

61. James S. Robbins, "The Bombing Strategy," *National Review,* 15 May 2002; and Chardy, "Parents Defusing Children,"

62. Chardy, "Parents Defusing Children," A1.

63. Ibid.

64. Juergensmeyer, *Terror in the Mind of God,* 183.

65. The previous paragraphs based on a report published by the Middle East Media Research Institute (MEMRI): Yotam Feldner, " '72 Black-Eyed Virgins': A Muslim Debate on the Rewards of Martyrs," last accessed 3 September 2002, www.memri.org/.

66. Jack Kelley, "The Culture of Suicide Bombers," *USA Today,* 26 June 2001, 1A.

67. Moshe Sunder, "The Lost Garden of Eden," *Ma'ariv,* Weekend Section, 17 August 2001, 28, in Hebrew. Cited in Assaf Moghadam, "Palestinian Suicide Terrorism in the Second Intifada: Motivations and Organizational Aspects," *Studies in Conflict and Terrorism* 26.2 (March/April 2003), 73.

68. *Al-Quds Al-Arabi,* 11 May 2001. Cited in Feldner, " '72 Black-Eyed Virgins.' "

69. According to Human Rights Watch, 670 Palestinians were killed during the first two and a half years of the first Intifada—the most active years of the uprising. See Web site of Human Rights Watch, last accessed 2 September 2002, www.hrw.org/campaigns/israel/intifada-intro.htm/. According to the B'Tselem organization, the Israeli Information Center for Human Rights in the Occupied Territories, the number of Palestinian casualties reached 762 by the end of 1990, and 1,124 by 13 September 1993, the official ending date of the First Intifada. See "Palestinians Killed in the Occupied Territories (Including East Jerusalem) Since the Beginning of the Intifada (Dec. 9, 1987) Until the End of January 2002," Web site of B'Tselem, last accessed 3 September 2002, www.btselem.org/English/Statistics/Total_Casualties.asp.

70. Senior-ranking Israeli counterterrorism specialist, interview with the author, 8 August 1999.

Chapter 3: Demographics

1. Sadanand Dhume, "Islam's Holy Warriors: Laskar Jihad," *Masariku Network,* 26 April 2001.

2. The Bali bombing was carried out by Jamaah Islamiya, a rival group to Laskar Jihad, which works very closely with Al Qaeda. For a complete assessment see "Indonesia Backgrounder: How the Jamaah Islamiyah Terrorist Network Works," *ICG Asia Report,* no. 43, last accessed 13 January 2003, www.crisisweb.org/projects/showreport.cfm?reportid=845.

3. Timothy Mapes, "After Militancy, a New Threat? Indonesian Officials Fear Former Laskar Members May Stir Unrest," *Wall Street Journal,* 5 December 2002, A15.

4. "Indonesia's Terrorism Links," *Strategic Comments* 8, no. 4 (May 2002), International Institute for Strategic Studies (IISS).

5. Greg Barton, *Abdurrahman Wahid, Muslim Democrat, Indonesian President: a View from the Inside,* (Honolulu: University of Hawaii Press, 2002).

6. "Indonesia: The Search for Peace in Maluku," *ICG Asia Report,* no. 31, 8 February 2002, 16, last accessed 13 January 2003, www.crisisweb.org/projects/showreport.cfm?reportid=544.

7. "Indonesia: Lashkar Jihad Head Calls for Urgent Action to Curb Regional Conflict," *Media Indonesia,* quoted in *BBC Monitoring,* 9 August 2001.

8. Buddhism came to Indonesia sometime during the early centuries of the first millennium, soon after the arrival of Hinduism in the first century C.E. Many Buddhist monuments and temples were erected in Java during the Sailendra dynasty, which ruled a large section of Indonesia from the seventh to the ninth century. One of these is Borobudur in Yogyakarta, considered one of the most magnificent Buddhist stupas in the world. Both Hinduism and Buddhism were practiced as court religions from the ninth through the fourteenth centuries, when traders from India brought Islam to Indonesia. Today some 87 percent of Indonesia's population of 232 million people are Muslim, 10 percent are Christian, 2 percent are Hindu, and 1 percent is Buddhist. The Indonesian constitution established the concept of "Panscasila," in which all Indonesians must have a religion. The constitution does provide for religious freedom and belief in one supreme God. The government, however, recognizes only Islam, Catholicism, Protestantism, Buddhism, and Hinduism as legitimate religions. Atheism is forbidden, and by default, all Indonesians are required to profess one of these five officially sanctioned religions.

9. I have changed her name, but the name she gave me was equally as preposterously long.

10. Robert Hefner, e-mail communication, 12 May 2002.

11. Islamic scholar Robert Hefner explains that Salafism comes in a variety of forms, some of which are austerely puritanical but not particularly extremist as far as militancy and violence are concerned. Some Salafis, for example, are quite moderate and reject Wahhabism and, especially, a jihadi interpretation of Islam. Some observers prefer the term *neo-Salafis* to refer to extremists like Ja'far, since the term *Salafi* actually refers to an interest in basing one's life on that of the Prophet and his followers *(salaf),* and unaffiliated Muslims sometimes use the term.

12. He uses the term *Ahle Sunnah* for a particular reason. The term refers to the whole of mainstream Sunni Islam. It usually connotes moderate Islam as opposed to the harsh and violent version promoted by radicals like Ja'far. "Part of Ja'far's genius," Indonesia scholar Robert Hefner explains, "is that he appropriates the term to make it seem as if he is utterly mainstream, which he is not." Hefner, e-mail.

13. Abdul Hadi Palazzi, "Orthodox Islamic Perceptions of Jihad and Martyrdom," in International Policy Institute of Counter-Terrorism (ICT), *Countering Suicide Terrorism: An International Conference* (Herzliyya, Israel: ICT, 2000), 64–74.

14. According to a well-known hadith or tradition, the Prophet called the greater jihad "the jihad against one's soul, the jihad against our own limitations, our own defects." The lesser jihad is the jihad of the sword. Denial of this tradition is one of things that set followers of Wahhab apart. Cited in Palazzi, "Orthodox Islamic Perceptions," 64–74.

15. In the most recent elections of 1999, the majority of votes went to secular parties. Parties calling for the introduction of Shari'a across Indonesia received about 16 percent of the vote. "The Challenge for Moderate Islam: Islam in Indonesia," *Economist,* 22 June 2002, 64.

16. According to Robert Hefner, until the 1980s, the majority of Indonesian Muslims were relatively lax in their religious practice. This changed toward the end of the decade, however, when the country experienced an unprecedented Islamic resurgence. Despite the religious revival, the Islam favored by the majority is neither socially conservative nor politically radical. Most Indonesians continue to support religious pluralism and a Western-style democracy. Two-thirds of the people who voted for Megawati's Indonesian Democratic Party for Struggle consider themselves to be pious Muslims, although they have not voted for the imposition of Islamic law. Robert Hefner, "The Strategic Role and Political Future of Islam in Indonesia," (unpublished manuscript), March 2002.

17. Iannaccone limits the definition of strictness to a single attribute in his formal analysis: "The degree to which a group limits and thereby increases the cost of nongroup activities, such as socializing with members of other churches or pursuing 'secular' activities." Laurence R. Iannaccone, "Why Strict Churches Are Strong," *American Journal of Sociology* 99.5 (March 1994): 1182.

18. Strict religious communities simplify life by proclaiming an exclusive truth—a closed, comprehensive and eternal doctrine that provides answers to life's most troubling questions. Strict churches often limit secular pastimes and adherents' socializing with members of other churches or the secular world. Iannaccone, "Why Strict Churches Are Strong," 1180–1211.

19. Rodney Stark, *The Rise of Christianity* (San Francisco: Harper, 1997), 33.

20. Iannaccone's analysis shows a clear correlation between education, income level, and a tendency to donate money and to avoid socialization with those outside the church, and strictness. Iannaccone, "Why Strict Churches Are Strong," 1200. The research cited here only covers Jewish and Christian churches in the United States, however. See also Stark, *Rise of Christianity,* 33.

21. Gavin W. Jones and Chris Manning, "Labour Force and Employment during the 1980s," in Anne Booth, ed., *The Oil Boom and After: Indonesian Economic Policy and Performance in the Soeharto Era* (Kuala Lumpur: Oxford University Press, 1992), 363–410. Terence H. Hull and Gavin W. Jones, "Demographic Perspectives," in Hal Hill, ed., *Indonesia's New Order: The Dynamics of Socio-Economic Transformation* (Honolulu:

University of Hawaii Press, 1994), 123–78. Cited in Hefner, "Strategic Role and Political Future."

22. John Elster, *Ulysses Unbound* (Cambridge: University of Cambridge Press, 2000). Dixit and Nalebuf, for example, identify precommitment mechanisms to achieve self-regulation largely without enforcement by another. These include writing contracts, cutting off communication, and burning bridges behind you. Avinash Dixit and Barry Nalebuff, "Making Strategies Credible," in R. Zeckhauser, ed., *Strategy and Choice* (Cambridge, Mass.: MIT Press, 1991). See also Thomas Schelling, *The Strategy of Conflict* (Cambridge, Mass.: Harvard University Press, 1980).

23. Privileged young men living in largely traditional societies feel most acutely the tension between the simultaneous claims of modernity and tradition. This helps to explain why middle- and upper-middle class young men join or support militant groups in some parts of the Islamic world, for example, Egypt and Saudi Arabia.

24. The "traditionalists" are regarded as more liberal and more comfortable with the notion of a secular state, while the "modernists" tend to favor a greater role for Islam in government. "Indonesia: Next Steps in Military Reform," *ICG Asia Report* No. 24, 10 October 2001, 11. "Indonesia: Overcoming Murder and Chaos in Maluku," *ICG Asia Report* No. 10, 19 December 2000. Clifford Geertz, *The Religion of Java* (New York: Free Press, 1960).

25. Fabiola Desy Unidjadja, "Juwono Speaks on Maluku Tragedy," *Jakarta Post,* 15 July 2000. Quoted in *ICG Asia Report* No. 10, 19 December 2000, 13.

26. ICG reports that military cooperation with Laskar became more pronounced by June: "Credible witnesses saw military units providing covering fire for Muslim attacks on Christian neighborhoods in and around Ambon. Though the Police Mobile Brigade (Brimob), a paramilitary unit, had tended to side with Christians, they began to back down in the face of superior numbers and organization. On 21 June, Laskar Jihad fighters with support from soldiers overran Brimob headquarters in Ambon, effectively ending the police's role as a counterbalance." The civil emergency began with a frank admission from TNI spokesman Vice Air Marshal Graito Usodo that a significant proportion of the soldiers in Maluku were "emotionally involved" in the conflict. Roughly fourteen hundred potentially "contaminated" soldiers were rotated out of the province and steps were taken to reduce the likelihood of future involvement. See "Indonesia: Overcoming Murder and Chaos in Maluku," 10.

27. All the countries Ja'far mentions are part of the territory of Indonesia and Borneo (which is two-thirds Indonesian.) There are eight major islands or island groups. The largest land masses consist of Sumatra, Java, Kalimantan (also known as Borneo, whose capital city is Palangkaraya), Sulawesi (Celebes), and Irian Jaya (the western half of Papua New Guinea, where Sorong is located). The smaller islands fall into two main groups: the Moluccas (Maluku) to the northeast, and the lesser Sunda chain east of Bali.

28. *ICG Asia Report* No. 31, 8 February 2002, 2.

29. The province of Maluku is made up of thousands of small islands just north of East Timor. It is divided into roughly three regions, north Maluku, central Maluku, and south Maluku. However, the capital of Ambon and the northern municipality of Ternate, a recently independent Muslim province, are also considered their own regency. The general religious breakdown of Maluku is 57 percent Muslims and 43 percent Christian. Less than 1 percent of the population are "other." Each of the regencies has a distinct religious characteristic. North Maluku and Ternate have the highest percentages of Muslim population, 80 percent and 70 percent respectively. Central and Ambon have distinct religious populations, with 70 percent and 42 percent Muslims respectively. As you go to the southern regencies, the Christian majority grows, with close to 76 percent of the population.

30. "Indonesia: The Search for Peace in Maluku," *ICG Asia Report* No. 31, 8 February 2002, 2.

31. "Support for Ambon Martial Law Grows," *Jakarta Post,* 25 June 2000. Quoted in *ICG Asia Report* No. 31, 5.

32. Ibid.

33. "Indonesia: Overcoming Murder and Chaos in Maluku."

34. Ashutosh Varshney, "Ethnic Conflict and Civil Society: India and Beyond," *World Politics* 53.3 (April 2001).

35. Ibid.

36. *ICG Asia Report* No. 10.

37. "Settling the War of God and Gold," *Economist,* 2 February 2002.

38. Hefner, "Strategic Role and Political Future of Islam."

39. *ICG Asia Report* No. 31, 21.

40. Consultations with Jacqueline Babha, Chaim Kaufmann, Samantha Power, and Monica Toft, June 2002.

41. Laskar Jihad in "Berita Harian," 19 September 2000, quoted in "Indonesia: Overcoming Murder and Chaos in Maluku."

42. Lindsay Murdoch, "Brutal Religious War Leaves a Paradise Soaked in Blood," *Sydney Morning Herald,* 27 January 2001. Cited in *ICG Asia Report* No. 31. The article claims that the war began with a fight between a Muslim bus driver and a Christian passenger in January 1999. Within hours the fighting spread to all of the islands. Forced conversions must have happened within the following two years.

43. Sidney Jones, Testimony before the United States Commission on International Religious Freedom, 13 February 2001. Quoted in *ICG Asia Report* No. 31, 10.

44. Somini Sengupta, "Hindu Right Goes to School to Build a Nation," *New York Times,* 13 May 2002, A1.

45. Ibid.

46. Suresh Joshi, national coordinator for the educational wing of the voluntary service, says that his organization spends over a million dollars per year on its charitable projects, most of it focused on tribal peoples and *dalits.* Vidya Barati, an educational charity affiliated with the RSS, runs some twenty thousand low-cost private schools serving 2.4 million children throughout India. The organization has built over a thousand new schools every year in the last decade, all with the aim of returning the population to Hinduism. Sengupta, "Hindu Right Goes to School."

47. The Jamaat-i-Islami (JI), the largest of Pakistan's religious parties, was founded in 1941 by Maulana Abul Ala Maududi on the idea of the *ummah* (the Worldwide Community of Muslims) as an unadulterated and exclusive embodiment of the vision of Islam that would help purge the community of deviant behavior. The party would serve as a vehicle for propagating this vision and hence control Muslim politics of the time. Initially, Maududi opposed the Pakistan movement, arguing that Islam was a universal religion not subject to national boundaries nor subject to political principles. The JI changed its position once the decision was made to partition India on the basis of religion. In 1947 Maududi redefined the JI's purpose as the establishment of an Islamic state in Pakistan. To achieve this objective, the JI believed it was its duty to establish a political system in which decision making would be undertaken by a few pious people well versed in Islam.

48. *Tassawuf* means "Sufi." The Naqshbandi Order, an austere Sunni-Sufi order found in almost every Muslim political struggle around the world, was a precursor of Deobandism. Michael Rinder, e-mail communication with the author, 15 February 2002. But today, Deobandis of South Asia consider themselves very different from Sufis. In the Indonesian context, the Naqshbandis are a liberal force ardently opposed to jihadis. Hefner, e-mail.

49. "Laskar Jihad Worry Bogor Residents," *Jakarta Post,* 10 April 2000.

50. Jane Perlez, "Indonesia Arrests Leader of Militants in Islam Group," *New York Times,* 7 May 2002.

51. Derwin Pereira, "Arrest Me, Not Clerics, Says Indonesian VP," *Straits Times* (Singapore), 31 May 2002.

52. Robert Hefner argues that toward the end of his rule, beginning in 1994, Suharto began favoring the Islamists. He began providing funding and tactical support to radical militant organizations, in part because he saw them as allies against the democracy movement, which threatened his power. Testimony by Robert Hefner at a hearing of the East Asia and the Pacific Subcommittee of the House International Relations Committee, "South East Asia After 9/11: Regional Trends and U.S. Interests," Washington, D.C., 12 December 2001. Another view, expressed by a senior Singaporean official, is that Soeharto used the Islamists to fight communism.

53. Interview with Ja'far Umar Thalib in *Panji Masyarakat,* 26 April 2000. George Aditjondro, "Notes on the Jihad forces in Maluku," last accessed 8 August 2002, www.geocities.com/Choyse/PAPER/Aditjondro_l.htm.

54. Robert Cribb, "From Petrus to Ninja," in Bruce Campbell, ed., *Death Squads in Global Perspective: Murder with Deniability* (New York: St. Martin's Press, 2000). See also Dietrich Kebschull, "Indonesia: The Transmigration Program in Perspective" (Washington, D.C.: World Bank, 1986).

55. In Aceh, guerrillas in the Free Aceh Movement (GAM) are often accused of carrying out robberies and extortion—both of the population and the security services. The military provides bullets to a pro-government militia in central Aceh, although it is not clear whether as a policy or for personal enrichment.

56. *ICG Asia Report* No. 31, 26.

57. "TNI Urged to Probe Guns Used in Ambon Clashes," *Jakarta Post,* 10 December 1999. Cited in *ICG Asia Report* No. 31, 5.

58. Diego Gambetta, *The Sicilian Mafia* (Cambridge Mass.: Harvard University Press, 1995), 2.

59. "Around 2.5 million nonagricultural workers (3 percent of the labor force) were displaced during the crisis. Most of the displaced workers were wage employees and the job losses were most severe in the manufacturing and construction sectors. About three-quarters of the job losses were in the rural areas." IMF Country Report No. 02/154, *Indonesia: Selected Issues,* July 2002, 83. Figures for Indonesia's economic contraction vary from 12 to 17 percent of GDP. "The depth of the collapse in Indonesia, a 17 percent GDP contraction in 1998 according to the latest consensus forecast, if not unparalleled, is among the largest peacetime contractions since at least 1960." Joseph Stiglitz, "Must Financial Crisis Be This Frequent and This Painful?" World Bank representative at McKay Lecture, Pittsburgh, 23 September 1998.

60. John Gershman, "Is Southeast Asia the Second Front?" *Foreign Affairs,* Vol. 81, No. 4 July-August 2002. Also Hefner, "Strategic Role and Political Future of Islam."

61. Alan Krueger and Jitka Maleckova, "Education, Poverty, Political Violence and Terrorism: Is There a Cause Connection?" *NBER Working Papers,* 22 July 2002. The paper was pulled from a World Bank presentation in response to complaints about how the authors defined terrorism.

62. The study relies on the observation that the frequency of hate crimes went down during the American Depression and that wealthier and better-educated Palestinians are more likely to support terrorism in Israel. The problem with the first example is that hate crimes are hobbies, not full-time jobs, while the terrorism we worry about most today tends to be a full-time occupation. People are likely to have less time for their hobbies when money and jobs are scarce. And they are likely to be less discriminating about the nature of their employment, which might make a full-time paying job for a terrorist organization more attractive even as it made nonpaying hate crimes less appealing. Regarding the survey results in the West Bank and Gaza Strip, it is unclear whether wealthy and well-educated Palestinians would support attacks against Israeli civilians if the Palestinian National Authority had the means to attack the Israeli army. Thus, it is

not clear that wealth and education would be positively correlated with Palestinian support for terrorism if a military response were available. The study also makes use of data that purportedly show that Hezbollah fighters tended to be better educated and wealthier than their "civilian" counterparts. But a fifth of the militants were educated in Hezbollah schools. It is not the level of education that is likely to count, but the type. If children are educated in pro-jihadi schools, it is no surprise that the more educated they are, the more likely they are to join. Information about the militants' economic status was available for less than half the sample, and the authors of the study were forced to infer economic status from other indicators such as the militants' parents' professions. The economic findings are significantly less conclusive when militants' region of origin is taken into account.

63. Alan B. Krueger and Jitka Maleckova, "The Economics and the Education of Suicide Bombers: Does Poverty Cause Terrorism?" *New Republic,* 24 June 2002.

64. "Indonesia: Selected Issues," *IMF Staff Country Report* No. 00/132, International Monetary Fund, October 2000. Although this may be unpaid employment, families receive rewards if their sons become martyrs, and at least the boys have something to do while they are unemployed.

65. Joshua Angrist, "The Economic Returns to Schooling in the West Bank and Gaza Strip," *American Economic Review* 85.5 (December 1995): 1065–87, in Krueger and Maleckova, "Economics and the Education of Suicide Bombers."

66. Alexis de Tocqueville, *L'Ancien Régime et la Révolution* (Paris: Garnier Flammarion Philosophie, 1993); Harlan Cleveland, an American government official, coined the phrase "the revolution of rising expectations." When Cleveland was working in Asia in the 1950s, he observed that newfound political and social freedom, access to education and information, and the opening up of opportunities released enormous social energy and activity. Rising aspirations have spread across Europe, Africa, and Asia. Expectations, however, have outpaced actual human achievements. Rising aspirations, when not satisfied, can fuel resentment.

67. Rajiv Chandrasekaran, "Al Qaeda Feared to Be Lurking in Indonesia," *Washington Post,* 11 January 2002, A1. Ja'far claimed elsewhere that Al Qaeda had offered funding for his group, but he had refused it, saying he disapproved of bin Laden. Hefner reports that some Laskar Jihad members have had contacts with Al Qaeda for specific operations, but Ja'far's subsequent turning away from bin Laden is the more important development. Ja'far relies on funding from wealthy Indonesian industrialists, he says, and continued links with Al Qaeda could put that funding in jeopardy, at least for now. Testimony by Hefner, "South East Asia After 9/11." National Intelligence Agency chief Hendropriyono said that Al Qaeda was using a training camp in central Sulawesi, but he subsequently changed his mind. "Al-Qaeda 'Runs Camps on Island in Indonesia,'" *Financial Times,* 13 December 2001, 12.

68. Barry Desker and Kumar Ramakrishna, "Forging an Indirect Strategy in Southeast Asia," *Washington Quarterly,* spring 2002.

69. "Al Qaeda Trained in Indonesia," *Australian Financial Review,* 21 November 2001, 3.

70. *ICG Asia Report* No. 31, 19. See also Fabiola Desy Unidjadja, "International Training Camp in Poso 'Empty,'" *Jakarta Post,* 13 December 2001.

71. Hefner, "Strategic Role and Political Future of Islam," 50. Hefner considers Ja'far's post–September 11 view of Al Qaeda to be "revisionist."

72. *ICG Asia Report* No. 31, 20.

73. Gershman, "Is Southeast Asia the Second Front?"

Chapter 4: History

1. Isaiah Berlin, "The Origins of Israel" (1953), in Isaiah Berlin, *The Power of Ideas,* ed. Henry Hardy (London and Princeton: 2000), 143 Princeton University Press. The full quote of Berlin is: "I should like to begin with the strange fact that the State of Israel exists. It was once said by the celebrated Russian revolutionary, Alexander Herzen, writing in the mid–nineteenth century, that the Slavs had no history, only geography. The position of the Jews is the reverse of this. They have enjoyed rather too much history and too little geography. And the foundation of the State of Israel must be regarded as a piece of historical redress for this anomalous situation. The Jews have certainly had more than their share of history, or, as some might say, martyrology.'"

2. See Gershon Gorenberg, *The End of Days: Fundamentalism and the Struggle for the Temple Mount* (New York: Free Press, 2000), 170–71.

3. Midrash Tanhuma, written in the third century. Cited in Kanan Makiya, *The Rock* (New York: Pantheon, 2001), 283.

4. Chrysostom's Sermons, cited in J. Parkes, *The Conflict of the Church and the Synagogue,* 105–6. Cited in Lambert Dolphin, "The Destruction of the Second Temple," The Temple Mount in Jerusalem Web site, last accessed 11 January 2003, www.templemount.org/destruct2.html.

5. Cited in Makiya, *Rock,* 286.

6. Quoted in Ehud Sprinzak, *Brother Against Brother: Violence and Extremism in Israeli Politics from Altalena to the Rabin Assassination* (New York: Free Press, 1999), 261–62.

7. Other passages from the Old Testament that are frequently cited to show the Jewish connection to the land of Israel include Genesis 13:14–15, God said to Abram: "Lift up now thine eyes, and look from the place where thou art, northward and southward and eastward and westward; for all the land which thou seest, to thee will I give it, and to thy seed for ever." In Genesis 26:3, God confirms to Abraham's son Jacob, "Sojourn in this land, and I will be with thee, and will bless thee; for unto thee, and unto thy seed, I will give all these lands, and I will establish the oath which I swore unto Abraham thy father."

8. F. E. Peters, *Jerusalem: The Holy City in the Eyes of Chroniclers, Visitors, Pilgrims, and Prophets from the Days of Abraham to the Beginnings of Modern Times* (Princeton, N.J.: Princeton University Press, 1985), 76–84.

9. See, for example, Gorenberg, *End of Days,* 61–63. A group of rabbis, who included the

powerful head of the Merkaz ha-Rav Yeshiva, Rabbi Tzvi Yehuda Kook, issued a statement prohibiting Jews from stepping foot on the Temple Mount lest they would violate the purity of the compound. The rabbis argued that since the precise location of the Temple remained unknown, one might inadvertently step on the Holy of Holies, the chamber that held the Ark of the Covenant. Each Yom Kippur, the high priest—and only he—would enter the Holy of Holies, and call out the tetragrammaton, the name of God. All others were denied entrance to the Holy of Holies because they were considered impure.

10. The most relevant passage in the Koran is from verse 37:99 to verse 37:109. 102–5: "Then, when [the son] reached [the age of serious] work with him, he said: 'O my son! I have seen in a vision that I offer thee in sacrifice: now see what is thy view!' [The son] said: 'O my father! Do as thou art commanded: thou will find me, if Allah so wills, one of the steadfast!' So when they had both submitted [to Allah], and he had laid him prostrate on his forehead [for sacrifice], We called out to him: 'O Abraham! . . . Thou hast already fulfilled the vision!—thus indeed do we reward those who do right." All Koranic passages are taken from 'Abdullah Yusuf 'Ali, *The Meaning of The Holy Qur'an* (Brentwood, Md.: Amana Corporation, 1993).

11. See King Solomon's prayer at the dedication of the First Temple, I Kings 8:29–53.

12. Amidah is the core prayer in the Jewish prayer service and literally means "standing," since it is recited while standing. It is also known by the term Shmoneh Esreh, Hebrew for "eighteen," which stands for the number of benedictions included in the prayer on ordinary weekdays.

13. See, for example, David de Sola Pool, ed. and trans., under the direction of the Siddur Committee of the Rabbinical Council of America, *The Traditional Prayer Book for Sabbath and Festivals* (New York: Behrman House, Inc., 1960), 68. Most Jews interpret the instruction to rebuild the Temple as a mystical aspiration that they do not expect to carry out physically. But some interpret the instruction literally. For example, at the Ateret Cohanim yeshiva, located in Jerusalem's Old City, students are trained not to become Talmud scholars, but to practice Temple rites. Jewish law forbids Jews from ascending the Temple Mount, making active participation in building the Temple unlikely for most observant Jews, especially Orthodox ones. But some Jews believe that ascending the Temple Mount is permitted, now that the Temple is destroyed.

14. For messianic Jews, there is ample "proof" in the Old Testament that the End of the World will be accompanied by a battle over Jerusalem that will provoke divine intervention and, ultimately, a Day of Judgment. Zechariah 14:2–12 describes the scenario of the Last Days: "For I will gather all nations against Jerusalem to battle: and the city shall be taken, and the houses rifled, and the women ravished; and half of the city shall go forth into captivity. . . . And this shall be the plague wherewith the Lord will smite all the peoples that have warred against Jerusalem." In Daniel 12:1–2 of the Old Testament, Daniel predicts that the "time of the end" shall be a "time of trouble, such as there never was. . . . And many of them that sleep in the dust of the earth shall awake, some to everlasting life, and some to reproaches and everlasting abhorrence."

For Christians, the Apocalypse is most vividly described in the New Testament's Book of Revelation. Here too the End of Days is accompanied by a series of catastrophic events that culminates in the battle of Armageddon. "For they are the spirits of devils, working miracles, which go forth unto the kings of the earth and of the whole world, to gather them to the battle of that great day of God Almighty. . . . And he gathered them together into a place called in the Hebrew tongue Armageddon" (Revelation 16:14–16). All biblical passages from the New Testament are taken from *The New Testament: The King James Version* (Cambridge: The Bradley Press, 1954). At this junction, Jesus the warrior is believed to fight a coalition of all the kings. The devil is bound, while Jesus embarks on his millennial reign. "And I saw an angel come down from heaven, having the key of the bottomless pit, and a great chain in his hand. And he laid hold on the dragon, that old serpent, which is the Devil and Satan, and bound him a thousand years. . . . Blessed and holy is he that hath part in the first resurrection; on such the second death hath no power, but they shall be priests of God and of Christ, and shall reign with him a thousand years" (Revelation 20:1–6).

Islam too foresees the end of days, according to both the Koran and several hadith (sayings attributed to the Prophet). In the Koran, sura 14:48 (Ibrahim) states: "One day the Earth will be Changed to a different Earth, And so will be the Heavens, And [men] will be marshaled Forth, before Allah, the One, the Irresistible." In sura 18:8 (Al Kahf), it is written: "Verily what is on earth We shall make but as Dust and dry soil." As in Judaism and Christianity, the End of Days scenario in Islam is far from peaceful. According to several hadith, the Antichrist—a false Messiah by the name of *al-masih al-dajjal* will conquer the world in the last days. Again, the battle will take place in Jerusalem, where the false Messiah, who is Jewish, will be defeated by Jesus, who is regarded as a prophet in Islam. At the end, all human beings will be judged at the valley of Yehoshafat, near Jerusalem. See Gorenberg, *End of Days,* 44.

15. See "Biography of Chafetz Chaim," Torah.org Web site, last accessed 19 August 2002, www.torah.org/learning/halashon/ccbio.html.

16. For more on the assassin of Rabbi Meir Kahane, El-Sayyid Nosair, see Daniel Benjamin and Steven Simon, *The Age of Sacred Terror* (New York: Random House, 2002), 3–7.

17. See Jewish Defense League (JDL), "Introductory Message," JDL Web site, last accessed 11 January 2003, www.jdl.org/information/introductory_message.shtml.

18. Binyamin Kahane, the son of Rabbi Meir Kahane, was killed on 31 December 2000, along with his wife, in a drive-by shooting near the settlement of Ofrah in the West Bank.

19. "Kach and Kahane Chai," Web site of the International Policy Institute for Counter-Terrorism (ICT), Herzliyya, Israel, last accessed 11 January 2003, www.ict.org.il/.

20. Davan Maharaj, "Israel Says It Foiled Anti-Arab Plot," *Los Angeles Times,* 13 May 2002, 4.

21. "Kach and Kahane Chai."

22. Another version of the Pulsa di Nura appeared in the Israeli daily *Yediot Ahronot.* It reads: "Angels of destruction will hit him. He is damned wherever he goes. His soul will instantly leave his body . . . and he will not survive the month. Dark will be his path and God's angel will chase him. A disaster he has never experienced will beget him and all curses known in the Torah will apply to him." Dov Elboim, "The Murder Curse," *Yediot Ahronot,* 13 November 1995, quoted in Sprinzak, *Brother Against Brother,* 275.

23. Unfortunately, as Amos Elon points out, Michael Ben Yair published these words in an article that appeared in *Ha'aretz* in 2002, rather than in a legal brief when he was serving as attorney general. Michael Ben Yair, "The War's Seventh Day," *Ha'aretz,* 3 March 2002. Cited in Amos Elon, "Israelis and Palestinians: What Went Wrong?" *New York Review of Books,* 19 December 2002.

24. The Foundation for Middle East Peace, "Israeli Settlements in the Occupied Territories: A Guide," Special Report, March 2002, last accessed 11 January 2003, www.fmep.org/reports/2002/SR_March_2002.pdf.

25. For more information on Israeli settlement policy, see Foundation for Middle East Peace, "Israeli Settlements in the Occupied Territories."

26. Molly Moore, "On Remote Hilltops, Israelis Broaden Settlements," *Washington Post,* 13 December 2002, A1.

27. Ibid.

28. Ibid. For the view that Gaza and the West Bank are disputed rather than occupied territories, see "The Truth About 242," Web site of Christian Action for Israel, last accessed 11 January 2003, www.cdn-friends-icej.ca/un/242truth.html.

29. Moore, "On Remote Hilltops."

30. Ehud Sprinzak, "From Messianic Pioneering to Vigilante Terrorism," in *Inside Terrorist Organizations,* ed. David C. Rapoport (London: Frank Cass, 1988), 194–216.

31. Carmi Gillon, interview with the author, Tel Aviv, Israel, 8 August 1999.

32. Sprinzak, "From Messianic Pioneering to Vigilante Terrorism," 207.

33. Nadav Shragai, "Raising Funds for the Third Temple," *Ha'aretz,* 20 July 1999.

34. See Web site of Chai Vekayam, last accessed 19 August 2002, www.geocities.com/Heartland/Estates/2687/.

35. Yehuda Etzion, interview with the author, 4 August 1999.

Chapter 5: Territory

1. President Bush blamed Lashkar e Taiba (LET) for the attack, freezing its assets in response. David E. Sanger and Kurt Eichenwald, "Citing India Attack, U.S. Aims at Assets of Group in Pakistan," *New York Times,* 21 December 2001, A1. No group claimed responsibility for the attack, but India holds and Jaish-i-Muhammad responsible. LET categorically denies its relationship with Al Qaeda per unnamed LET member, e-mail communication, 11 January 2003. On the connection between LET and Al Qaeda, see, for example, Rohan Gunaratna, *Inside Al Qaeda: Global Network of Terror* (New York: Columbia University Press, 2002), 206. According to the Center for Defense Information (CDI), "The mercenary character of the group and the MDI's fraternal nature lead many Western agencies to believe that L[E]T members have participated in

other conflicts where Muslims were involved. Reports on L[E]T members fighting in Chechnya, Bosnia, parts of the Middle East, and the Philippines have emerged, possibly indicating close ties with Al Qaeda." "In the Spotlight: Lashkar-I-Taiba (Army of the Pure)," CDI Terrorism Project. Available on the CDI Web site, last accessed 14 January 2003, www.cdi.org/terrorism/lt.cfm. LET has spread its base of operations into the Middle East and is now recruiting operatives in Saudi Arabia, UAE, and Kuwait, according to a government agency that monitors the group closely. Jamaat-ud-Dawa is the formal name of LET's parent organization, formerly known as Markaz-Dawa-Wal-Irshad (MDI).

2. A former Pakistani official told me that these figures may be somewhat exaggerated, since the jihadi groups often give bloated statistics to show they are beating out the competition, but he felt the figures were in the right ballpark. Interview with former Pakistani official, 8 September 2002.

3. LET was designated a foreign terrorist organization (FTO) on 18 December 2001. It is the militant wing of a Wahhabi educational organization called Markaz-Dawa-Wal-Irshad (MDI), which changed its name formally to Jama'at ud Da'awa (Party of Preachers). Through a presidential decree, General Pervez Musharraf banned Jaish-i-Muhammad, LET, Sipah e Sahaba Pakistan (SSP), Tehrik e Jafria Pakistan (TJP), and Tehrik e Nifaz-e-Shariat-e-Mohammadi, and put under observation the Sunni Tehrik. *The News* (Lahore), 13 January 2002, last accessed 15 August 2002, www.oureffort2001.com/RESEARCH/TALIBAN/musharrafsspeech.htm. Two terrorist groups, Lashkar e Jhangvi and Sipah e Mohammad, were banned earlier, on 14 August 2001. Both are offshoots of SSP and TJP respectively. Available on the Web site of Rediff, last accessed 15 August 2002, www.rediff.com/news/2001/aug/14pak4.htm. Until then, the group had been sending me regular updates on its activities by e-mail, which then stopped.

4. Dexter Filkins, "As Pakistani's Popularity Slides, 'Busharraf' Is a Figure of Ridicule," *New York Times,* 5 July 2002, A1.

5. Rudyard Kipling, *Kim* (New York: Tom Doherty, 1999), 53.

6. This interview took place over two days on 24–25 February 1999.

7. "Donation for computer required." Formerly available on the group's Web site, last accessed 20 July 2002, www.markazdawa.org/englishweb/Donation%20for%20computer%20required.htm. The group used to provide its bank account numbers on its Web site, but since it was banned, that information is no longer provided. The general Web page for the group has also changed to www.jamatdawa.org. The Urdu-language page provides a link to a "Website fund," but it is "under construction."

8. Text of the 1972 Simla Agreement, last accessed 14 August 2002, www.jammu-kashmir.com/documents/simla.html

9. Eric S. Margolis, *War at the Top of the World: The Struggle for Afghanistan, Kashmir and Tibet* (New York: Routledge, 2001), 74.

10. India asserts that Pakistan incorporated the Northern areas under the federal government to make these areas distinct from the part of Kashmir now under Pakistani control.

The inference is that in case of any future bargain, Pakistan would claim the Northern areas to be undisputed and an integral part of Pakistan—a position India rejects. For details about the status of the Northern areas, see Victoria Schofield, *Kashmir in Conflict: India, Pakistan and the Unfinished War* (London and New York: I. B. Taurus, 2000), 179–81.

11. Margolis, *War at the Top of the World,* 77.

12. Ayesha Jalal and Sugata Bose, *Modern South Asia: History, Culture, Political Economy* (London and New York: Routledge, 1998), 226. See also Sumit Ganguly, *Conflict Unending: India-Pakistan Tensions Since 1947* (Washington, D.C.: Woodrow Wilson Center Press, 2001), 88.

13. Eqbal Ahmad, "Jihad International, Inc.," *Dawn,* 4 February 1998. Available on-line at the Dissident Voice Web site, last accessed 6 January 2003, www.dissidentvoice.org/Articles/EqbalJihadInc.htm.

14. Elizabeth Rubin, "Can Musharraf Reform Jihadi Culture?" *Christian Science Monitor,* 24 January 2002, 8.

15. According to a report by Bruce Riedel, a senior Clinton adviser, U.S. intelligence had reported to President Clinton that Pakistan had prepared its nuclear weapons for potential use. Isabel Hilton, "The General in His Labyrinth: Where Will Pervez Musharraf Lead His Country?" *New Yorker,* 12 August 2002, 42.

16. For a detailed analysis see Ashley J. Tellis, C. Christine Fair, Jamison Jo Medby, *Limited Conflict Under the Nuclear Umbrella: Indian and Pakistani Lessons from the Kargil Crisis* (Rand Corporation Publication, 2000), last accessed 19 January 2003, www.rand.org/publications/MR/MR1450.

17. Human Rights Watch, "Rights Abuses Behind Kashmir Fighting," 1999. Available on the Human Rights Watch Web site, last accessed 20 July 2002, www.hrw.org/press/1999/jul/kas0716.htm.

18. The phrase stands for extrajudicial killings—when military or law enforcement agencies kill suspects in stage-managed encounters. The official press releases announce that a person has been killed during encounters, though in practice unarmed civilians are killed in cold blood.

19. *Human Rights Watch World Report 2002,* 226. Available on the Human Rights Watch Web site, last accessed 21 August 2002, www.hrw.org/wr2k2/pdf/india.pdf.

20. Human Rights Watch, "Behind the Kashmir Conflict: Abuses by Indian Security Forces and Militant Groups Continue," 1999. Available on the Human Rights Watch Web site, last accessed 6 January 2003, www.hrw.org/reports/1999/kashmir/mil-abuses.htm.

21. Bureau of Democracy, Human Rights, and Labor, "Country Reports on Human Rights Practices—India," 4 March 2002. Available on the Department of State Web site, last accessed 6 January 2003, www.state.gov/g/drl/rls/hrrpt/2001/sa/8230pf.htm.

22. Schofield, *Kashmir in Conflict,* 198.

23. Human Rights Watch, "Kashmir: Attack on Civilians in Jammu Condemned," 16 July 2002, last accessed 20 July 2002, www.hrw.org/press/2002/07/kashmir0716.htm.

24. Nevertheless, Sayeed is widely believed to have taken cues from Pakistani intelligence agencies, especially Interservices Intelligence (ISI). For details see, Peter Chalk, "Pakistan's Role in Kashmir Insurgency," *Jane's Intelligence Review,* 1 September 2001.

25. Debt service obligations alone total nearly 50 percent of government expenditures. From 1999 to 2002, General Musharraf's lead economic wizard, Finance Minister Shaukat Aziz, has helped Pakistan's economy to revive to some extent, but Pakistan is still in no position to escape its dire debt situation in the foreseeable future. Since 11 September 2001, the United States has rewarded Pakistan's cooperation in the war against terrorism by helping the country reschedule its loans with international financial institutions. "US Assures Pakistan of Immediate Debt Relief," *Dawn,* 21 October 2001.

26. Pakistan's economy is currently suffering from a lack of foreign investment due to its unstable security environment. It also suffers from a heavy defense budget that consumes around 40 percent of its GDP. The country's economic outlook continues to be marred by its weak foreign exchange position, which relies on international creditors for hard-currency inflows. Foreign loans and grants provide approximately 25 percent of the government's revenue, but development projects and the education sector are only allotted roughly 5 percent of the GDP. For details, see S. Akbar Zaidi, *Issues in Pakistan's Economy* (Karachi: Oxford University Press, 2000).

27. In response to the Indian nuclear tests of May 11 and 13, 1998, Pakistan conducted two nuclear tests on May 28 and May 30, 1998. Experts argue that India conducted the tests in the belief that an overt nuclear capability would provide it with the status of a great power. The tests were also a show of strength vis-à-vis China and Pakistan and, in the words of George Perkovich, "reflected an overwhelming desire for global recognition and national pride." See Ganguly, *Conflict Unending,* 101; and George Perkovich, *India's Nuclear Bomb: The Impact on Global Proliferation* (California: University of California Press, 2001). Many Pakistani political analysts argued that Pakistan should have shown restraint by not conducting tit-for-tat tests and receiving international appreciation and financial support. Military and jihadi elements, however, exerted substantial pressure on the political leadership to conduct the tests. See Pervez Hoodbhoy, "Surviving South Asia's Nuclear Whirlpool," last accessed 6 January 2003, www.focusweb.org/focus/pd/sec/Altsec2/hoodbhoy.htm.

Two Pakistani nuclear scientists, Bashir-ud-din Mahmood and Abdul Majeed, were detained on 23 October 2001 in response to a U.S. request to question ten people with "specific knowledge" of Pakistan's nuclear weapons program. Both scientists were known to have close relations with the Taliban government in Afghanistan. Upon retirement two years ago, the two nuclear scientists started a charity that financed humanitarian and commercial activities in Afghanistan. Some directors of the charity, called Ummah Tameer-e-Nau, or Islamic Reconstruction, signed business deals with the Taliban, which included contracts to build hospitals, schools, and factories. Both scientists were released after a few weeks investigation, when no evidence of nuclear assistance to the Taliban regime was found. Presten Mendenhall, "Pakistan Releases Top Nuclear Scientist," MSNBC, 3 November 2002, last accessed 23 August 2002. www.msnbc.com/news/651022.asp. In an article published in the *Economist* in November 2001, however, one correspondent described how in the charity's house he found "several designs for a long thin balloon, something like a weather balloon, with lines and arrows indicating a suggested height of 10km (33,000 feet). There was also a sketch of a jet fighter flying towards the balloon alongside the words: 'Your days are limited! Bang.' " The correspondent suggested that the inhabitants of the house had worked on a plan to build a helium-powered balloon bomb carrying anthrax. See "Chilling Evidence in the Ruins of Kabul," *Economist*, 22 November 2001.

28. For a detailed perspective, see Luis Martinez and John Entelis, *The Algerian Civil War* (New York: Columbia University Press, 2000).

29. Harkat-ul-Mujahideen (HUM)—the "Holy Warriors Movement"—was the first Pakistani jihadi group to be listed by the U.S. Department of State as a foreign terrorist organization (FTO). The movement has been highly successful in guerrilla operations against Indian security forces in Kashmir, and it allegedly cooperated with the Pakistani army in the 1999 Kargil incursion. Some of HUM's activities, including the training of militants in Afghanistan, are widely believed to be partly funded by Osama bin Laden, the Saudi-born radical with whom the group maintains open ties. Fazlur Rahman Khalil, founder of the group, says that he met bin Laden early in the Afghan war. Khalil was a signatory to bin Laden's 1998 fatwa against the United States and a member of bin Laden's international network known as the International Islamic Front for Jihad against the Jews and Crusaders. HUM claims to be active in Bosnia, Chechnya, India, Burma, the Philippines, and Tajikistan. U.S. government officials allege that HUM has targeted Western military officials in Bosnia, and India accuses HUM of carrying out "dirty tricks," including murders in India on behalf of Pakistan's Interservice Intelligence Agency (ISI). (In turn, the ISI accuses India's intelligence agency of similar activities in Pakistan, usually in connection with sectarian or ethnic violence.) Harkat has changed its name several times. The various Harkat groups are suspected by the Department of State of carrying out a series of kidnappings and killings of Western tourists in Kashmir, as well as killing two American diplomats in Karachi in 1995 and four American oil company workers in 1997, also in Karachi. The hijackers of Indian Airlines flight IC 814 in December 1999 demanded the release of the group's chief ideologue, Maulana Masood Azhar, who was being held in an Indian prison, in exchange for freeing the hostage passengers and crew. After his release, Azhar formed a new Deobandi group, Jaish-i-Muhammad, which is more openly sectarian than HUM. A leader of a rival group told me in June that the ISI supports HUM, but Military Intelligence supports Jaish. The U.S. government added Jaish to its list of FTOs in October 2001.

30. This is a part of the rhetoric employed by most of the jihadi organizations on their Web sites. It has a historical reference as well—313 Muslims, in the battle of Badr in the times of Prophet Muhammad, defeated their opponents, who were reported to be around one thousand, according to Muslim historians.

31. Margolis, *War at the Top of the World.*

32. For details and history see, Schofield, *Kashmir in Conflict,* 87–91.

33. The World Bank Group, "Protecting Natural Resources in Northeastern Pakistan," June 2000, last accessed 24 August 2002, www.lnweb18.worldbank.org/sar/sa.nsf/Attachments/kashmir/$File/kashmir.pdf.

34. These figures were provided by the AJK government. The Indian official figures are notably different. The official Web site of India's Ministry of Home Affairs, last accessed 15 August 2002, maintains that a total of 29,488 persons have died during the insurgency, including 9,718 civilians, 14,356 militants, 2,358 foreign mercenaries, and 3,056 special forces personnel. For details, see www.mha.nic.in/annual%20report-4.htm#profile. Figures generally quoted by organizations like Human Rights Watch and Amnesty International also indicate that the total number of casualties are approximately forty thousand. However, none of these groups have provided a definitive figure. Addressing a conference in Boston in 2001, Yasin Malik, chairman of the pro-independence Jammu and Kashmir Liberation Front (JKLF), maintained that since the September 11 tragedy, there is a discernible increase in the intensity of the Indian forces' oppressive measures. According to him, an average twenty-five people are killed every day. See Hassan Abbas, "Kashmir as a Peace Bridge: An Idea Whose Time Has Come," *News* (Pakistan), 20 November 2001.

35. Ibid.

36. "Worldwide Refugee Information: India Country Report, 2002," U.S. Committee on Refugees, last accessed 6 January 2003, www.refugees.org/world/countryrpt/scasia/india.htm.

Part II: Holy War Organizations

1. Charles Heckscher and Joel Getzendammer, telephone interview with the author, 19 March 2002. The concept of strong and weak ties is developed by Mark Granovetter in "the Strength of Weak Ties," *American Journal of Sociology* 78: 1360–80.

2. As weapons technologies proliferate and continue to improve, it will be possible for leaderless resistors to carry out major attacks. The anthrax letter-attacks of fall 2001 are a foretaste of that future.

3. Joel Mowbray, "How They Did It: An 'evil one' confesses, and boasts," *National Review*, 23 December 2002; Vol. LIV, No. 24.

Chapter 6: Inspirational Leaders and Their Followers

1. James MacGregor Burns, *Leadership* (New York: Harper & Row, 1978), 4. I will say more about the requirement for moral action in Burns's conception of leadership.

2. Jennifer Gonnerman, "The Terrorist Campaign Against Abortion," *Village Voice*, 9 November 1998, 36; and Sharon Lerner, "The Nuremberg Menace," *Village Voice,* 10 April 2001, last accessed 21 August 2002, www.villagevoice.com/issues/0114/lerner.php.

3. Planned Parenthood, the Portland Feminist Women's Health Center, and several doctors brought suit against the American Coalition of Life Activists (ACLA), an antiabortion group, and a group of antiabortion activists, contending they had created a threatening environment that led to Slepian's murder. Although Horsley wasn't named as a defendant, the plaintiffs offered his Web site as evidence, since much of the initial information posted on the site was provided by ACLA. The plaintiffs won their lawsuit in February 1999 with a $107-million judgment, and Horsley's Internet provider immediately removed his Web site. But on March 28, 2001, the Ninth U.S. Circuit Court of Appeals unanimously reversed the verdict. See U.S. Court of Appeals for the Ninth Circuit, *Planned Parenthood of the Columbia/Willamette Inc.; Portland Feminist Women's Health Center; Robert Crist, M.D.; Warren M. Hern, M.D.; Elizabeth Newhall, M.D.; James Newhall, M.D., Plaintiffs-Appellees, and Karen Sweigert, M.D., Plaintiff, v. American Coalition of Life Activists; Advocates for Life Ministries; Michael Bray; Andrew Burnett; David A. Crane; Timothy Paul Dreste; Michael B. Dodds; Joseph L. Foreman; Charles Roy Mcmillan; Stephen P. Mears; Bruce Evan Murch; Catherine Ramey; Dawn Marie Stover; Charles Wysong, Defendants, and Monica Migliorino Miller, Donald Treshman, Defendants-Appellants,* Case No. 99-35320. See also Frederick Clarkson, "Journalists or Terrorists?" *Salon,* 31 May 2001, last accessed 21 August 2002, www.salon.com/news/feature/2001/05/31/nuremberg/index.html. Horsley's site, last accessed on 21 August 2002, can be found at www.christiangallery.com.

4. See the Web site of the Army of God, last accessed 5 August 2002, www.armyofgod.com

5. Attacks carried out in the name of the Army of God include bombings of an abortion clinic and a gay bar in Atlanta in 1997, and the bombing of an abortion clinic in Birmingham, Alabama, in 1998. These attacks are believed to have been perpetrated by Eric Robert Rudolph, who is also charged with the 1996 bombing at Atlanta's Centennial Olympic Park, in which one person was killed and over one hundred injured. In early November 2001, anthrax-hoax letters containing white powder and signed by the Army of God were delivered to private abortion clinics in Englewood and Fort Lee, New Jersey, and to more than 130 Planned Parenthood clinics in the United States.

6. Gonnerman, "Terrorist Campaign Against Abortion."

7. Jewish extremist groups in the United States that can be considered leaderless, are organized as networks, and recruit mainly through the Internet include Kach and Kahane Chai. Smaller radical Jewish organizations, such as the New Kach Movement or Noar Meir, can be regarded as offshoots of Kach and Kahane.

8. This system of organization, Beam claims, is almost identical to "the methods used by the committees of correspondence during the American Revolution." It is also similar in structure to communist revolutionaries' cells.

9. Louis Beam, "Leaderless Resistance," *Seditionist* 12 (February 1992). The essay received significantly more attention after Beam republished it and presented it to the Aryan National Congress in 1992. The essay is published on different Web sites at different times. See, for example, www.louisbeam.com/leaderless.htm, last accessed 30 August

2002, as well as www2.monet.com/~mlindste/ledrless.html, last accessed 30 August 2002. The idea, at least as practiced in America, was originally conceived by an American named Colonel Amoss in 1962, according to Louis Beam. Beam refined the concept in his essay "On Revolutionary Majorities," published in the *Inter-Klan* newsletter and *Survival Alert* 4 (1984) and again in *The Seditionist* in 1992.

Most published accounts have wrongly attributed the original idea to the 1992 essay. Mike Reynolds, interview with the author, 11 October 1999.

10. John Arquilla and David Ronfeldt, "The Advent of Netwar (Revisited)," in *Networks and Netwars* (Santa Monica: RAND, 2001). See also David Ronfeldt and John Arquilla, "Networks, Netwars and the Fight for the Future," available on the Web site of *First Monday: A Peer-Reviewed Journal*, last accessed 25 August 2002, www.firstmonday.dk/issues/issue6_10/ronfeldt/.

11. Al Qaeda is known to disseminate information, and possibly instructions for terror attacks, on Web sites. Several Web sites, including Al Qaeda's former Arabic Web site, alneda.com, have been hacked by U.S. authorities, although U.S. cyberterrorism experts acknowledge that the struggle against cyberterrorism is ongoing, as groups such as Al Qaeda can launch sites from different servers. Much of the information on these Web sites is written in Arabic and is often encrypted or scrambled in texts, and increasingly in digital photographs for better protection, a practice known as steganography. Steganography may well be the wave of the future, and a tactic that could someday permit "swarming" by groups such as Al Qaeda. See Jack Kelley, "Militants Wire Web with Links to Jihad," *USA Today*, 10 July 2002, 1A.

12. *Soldiers in the Army of God*, directed by Marc Levin and Daphne Pinkerson, Home Box Office, 2001.

13. Avram Goldstein, "Doctor Quits, Cites Antiabortion Threats," *Washington Post*, 4 November 1999, B1.

14. Ibid.

15. Ibid.

16. "Abortion Providers Decreased 14% between 1992 and 1996; Declines Mostly among Hospitals and Physicians' Offices Where Less Than 10% of Abortions Occur; All Measures of Abortion in the United States Now Lowest in 20 Years," news release, The Alan Guttmacher Institute, 1998, last accessed 11 January 2003, www.guttmacher.org/pubs/archives/newsrelease3006.html.

17. Neal Horsley's Nuremberg files were taken down for a while, but were available at the time of this writing. See Christian Gallery Web site, last accessed 24 August 2002, www.christiangallery.com/atrocity/aborts.html. Bob Lokey's site, last accessed 24 August 2002, can be found at www.alaweb.com/~savbabys/lokey1.html.

18. Lokey's Web site.

19. The pamphlet, last accessed 24 August 2002, is available at www.alaweb.com/~savbabys/holocaust.html.

20. "Incidents of Violence and Disruption Against Abortion Providers, 1977 to Present," Web site of the National Abortion Federation, last accessed 11 January 2003, www.prochoice.org/.

21. *Soldiers in the Army of God.*

22. Christian gallery Web site.

23. Ibid.

24. In a televised address to the American public on August 9, 2001, President George W. Bush said that he would allow federal taxpayer money to be used for research into stem cells from human embryos. The research conducted would be limited to those cells already extracted. Bush stressed that the government would not support the destruction of new embryos. See "Remarks by the President on Stem Cell Research," Bush Ranch, Crawford, Tex., 9 August 2001. The remarks are available on the Web site of the White House, last accessed 25 August 2002, www.whitehouse.gov/news/releases/2001/08/20010809-2.html.

25. Daniel Voll, "Neal Horsley and the Future of the Armed Abortion Conflict," *Esquire,* February 1999, 116.

26. Barbara Kellerman, "Leadership as a Political Act," in *Leadership: Multidisciplinary Perspectives,* ed. Barbara Kellerman (Upper Saddle River, N.J.: Prentice Hall, 1984), 81.

27. Sigmund Freud, *Moses and Monotheism* (New York: Vintage, 1967), 141, cited in Kellerman, "Leadership as a Political Act," 80.

28. Abraham H. Maslow, *Motivation and Personality* (New York: Harper and Row, 1987).

29. Benedict Anderson, *Imagined Communities: Reflections on the Origin and Spread of Nationalism* (London and New York: Verso, 1991).

30. Russel Hardin points out that most accounts of ethnic conflict assume that ethnic identity is given rather than "identified." Russel Hardin, *One for All: The Logic of Group Conflict* (Princeton, N.J.: Princeton University Press, 1995).

31. Burns's approach was revolutionary in that it defined leadership as an exchange that influenced both leaders and followers, rather than a trait of certain great men that, as some theorists suggested, might be brought out only in times of crisis. Burns's approach refines that of Edwin Hollander, who defined leadership as "a process, not a person," involving a relationship between leader and led. See discussion in Joseph C. Rost, *Leadership for the Twenty-First Century* (Westport, Conn.: Praeger, 1991), 61.

32. For a review of the literature on this point, see Rost, *Leadership for the Twenty-First Century.* Rost rejects Burns's view that leadership always entails the promotion of moral values, defining leadership as "an influence relationship among leaders and followers who intend real changes that reflect their mutual purposes." See Rost, *Leadership for the Twenty-First Century,* 102. For the view that Burns's insistence that leaders promote fundamental values is correct, see, for example, Ron Heifetz, *Leadership without Easy Answers* (Cambridge, Mass.: Harvard University Press, 1994). Heifetz defines leadership as "an

activity that fosters adaptive work and addresses the value conflicts that people hold." Heifetz argues that leaders use both formal and informal authority to find solutions that promote fundamental values (such as democracy, equality before the law, freedom).

33. See, for example, Rost, *Leadership for the Twenty-First Century,* 124–28, 165–67.

34. It strikes me that the only reasonable view is to admit that from a moral (as opposed to legal) perspective, we don't know when an embryo becomes a life that must be protected from harm. This becomes increasingly clear as the age of viability moves closer to conception as medical technology continues to improve. Thus, in making abortion legal and promoting it as a morally acceptable choice for those women who choose to undergo it, we are taking the consequentialist position that the evil of bringing an unwanted child into an overpopulated world is a lesser evil than abortion. Most difficult moral dilemmas require both deontological (rule-based) and ontological (consequence-based) considerations. The pro-choice position ought to admit these moral difficulties rather than glossing over them. The antiabortionists' position is cleaner and simpler: they admit no uncertainty about when life begins, and thus entertain no moral qualms about protecting innocent unborn by killing abortion providers. Where they go wrong, in my view, is in taking the law into their own hands, murdering the "murderers," rather than trying to change the law. Murdering doctors is not a last resort, a requirement for just war, and it puts institutions (such as the law) at risk that are an important part of our moral universe. For more, see the discussion in the introduction.

35. Alberto Melucci, *Challenging Codes: Collective Action in the Information Age* (Cambridge, England, and New York: Cambridge University Press, 1996), 293. This thesis is roundly rejected by a group of scholars who believe that the proclivity for "reciprocal altruism," to use Ann Florini's term, is even stronger than the drive to demonize the Other. But these scholars, whom I will call collective-action optimists, tend to focus on mobilizing groups to take actions that are nonviolent, like agreeing to set caps on pollution, rather than violent. Florini cites the work of Robert Wright and Richard Rorty as supporting this point of view. Ann Florini, *The Coming Democracy: New Rules for Running a New World* (Island Press, 2003).

36. Samuel Huntington, *The Clash of Civilizations and the Remaking of World Order* (New York: Simon and Schuster, 1996), 20.

37. Ronfeldt and Arquilla, "Networks, Netwars and the Fight for the Future."

38. Tony Perry, "Navy Takes a Scene out of Hollywood," *Los Angeles Times,* 27 November 2000, C1. Ronfeldt and Arquilla cite this article without quoting it.

39. Michael Bray, interview with the author, 25 April 1999.

40. He also shows me the New International Version, John 7:52–8:11, and others. There is indeed a footnote showing that the passage he refers to in *A Time to Kill* does in fact have notes explaining that later scholarship has questioned the authenticity of the passage, or in some cases the passage was omitted entirely.

41. *Soldiers in the Army of God.*

42. In March 2002, the *Sydney Morning Herald* reported that some forty-six families of Tulkarm received checks of $25,000 for each martyr, and $10,000 for each Palestinian shot by Israeli troops, from Saddam Hussein. See Paul McGeough, "Price of Martyrdom Becomes a Year's Rent," *Sydney Morning Herald*, 30 March 2002, 23.

43. Jerrold Post, "Terrorist Psycho-Logic," in *Origins of Terrorism: Psychologies, Ideologies, Theologies, States of Mind*, ed. Walter Reich (Washington, D.C.: Woodrow Wilson Center Press, 1998), 25.

44. Dave Grossman, *On Killing: The Psychological Cost of Learning to Kill in War and Society* (Boston, Mass.: Little, Brown & Co., 1995).

45. See Barbara Kellerman, "Introductory Remarks," in *Leadership: Multidisciplinary Perspectives,* ix–x.

46. Twentieth-century theorists have defined leadership variously as the ability to induce obedience, respect, loyalty, and cooperation (B. V. Moore, "The May Conference on Leadership," *Personnel Journal* 6 (1927): 127, cited in Rost, *Leadership for the Twenty-First Century,* 47); as a means of influencing others that depends on personality traits found only among a select few (see E. S. Bogardus, *Leaders and Leadership* [New York: Appleton-Century, 1934], 3–5, cited in Rost, *Leadership for the Twenty-First Century,* 47) or only under certain circumstances (see, for example, David M. Rosen, "Leadership Systems in World Cultures," in *Leadership: Multidisciplinary Perspectives,* 41); as a means for influencing a group to achieve its goals (see Rost, *Leadership for the Twenty-First Century,* 75); or as the process by which an organization achieves excellence (Rost, *Leadership for the Twenty-First Century,* 83. This is Rost's summary of the views expressed by Peters and Waterman, whose work inspired many subsequent studies, but who did not actually define leadership in their work. Leaders themselves disagree about what it is they do. Hitler defined leadership as coercion with a twist, observing that "whatever goal man has reached is due to his originality plus his brutality." Gandhi focused on the example the leader sets for others. "Clean examples have a curious method of multiplying themselves," he wrote. Truman defined a leader as "a man who has the ability to get other people to do what they don't want to do, and like it" (see Kellerman, "Leadership as a Political Act," 71). Several critical differences among these definitions are important for our purposes, however. Can leadership involve coercion, or must followers follow of their own volition? Are leaders necessarily "great men" with particular traits, such as charisma? Do the great men who lead us arise only in reaction to particular circumstances?

47. Inspirational leadership is different from Burns's transformational leadership in two senses. First, as noted in the text, it may promote immoral actions. Second, Burns emphasizes that it is the *realization* of mutual objectives that satisfies people's needs, but I would argue that in the save-the-babies movement, it is the *pursuit* of goals that satisfies participants' needs more than their achievement. Burns says that the "ultimate test of practical leadership is the realization of intended, real change that meets people's enduring needs." Burns, *Leadership,* 461. But here it is the pursuing of goals that meets the militants' needs, not the accomplishment of those goals. (In any case, it is often hard to assess the effectiveness of inspirational terrorist leadership. For example, it would be hard

for most observers to determine definitively whether a given terrorist act influenced the Messiah in some way.) Joseph Rost and others also reject the requirement that leadership involve the achievement of moral goals, but Rost goes a step further, arguing that charisma is a form of coercion "more consistent with the do-the-leader's-wishes conceptual framework than it is with the leadership-as-transformation framework." See Rost, *Leadership for the Twenty-First Century,* 85. The key difference between inspirational leadership as I am defining it here and Rost's "leadership for the twenty-first century" is that I assume that charisma can be used to inspire followers to take action they *want* to take, whether for spiritual, emotional, or material reasons, and that it may lead to more effective individuals and organizations; and that charisma can be part of inspirational leadership.

Unlike Rost, I do not see the function of charisma as only to persuade others to do the leader's bidding. Charismatic leaders often persuade us to do more of what we already want to do, but perhaps are too lazy or too afraid or too selfish to do without additional persuasion. Charisma, in my view, can make transformational or inspirational leaders more effective. It can, similarly, make transactional leaders or managers more effective.

48. Inspirational leaders are only successful when some of their followers also become leaders, while commanders do not necessarily aim to transform their subordinates into leaders. Inspirational leaders may be more effective if they have "great-man traits," especially charisma, but such traits are not required. From around 1910 and until World War II, the scientific study of leadership focused on the paradigm of "personality traits." The "traits theory" assumed that individuals who became leaders possessed a set of distinct characteristics, or traits, that separated them from followers. The aim of the scientific research on leadership during the period of the traits theory was to identify the characteristics of the individuals associated with leadership. See, for example, Martin M. Chemers, "The Social, Organizational, and Cultural Context of Effective Leadership," in *Leadership: Multidisciplinary Perspectives,* 93–94. See also James G. Hunt, "Organizational Leadership: The Contingency Paradigm and Its Challenges," in *Leadership: Multidisciplinary Perspectives,* 113–14.

Similarly, inspirational leaders may be more effective during political or social crises, but they may arise during crises that only they and their followers detect.

Inspirational terrorist leaders are bearers of symbolic goods. They use religion as a kind of technology. They use it to create community, thereby promoting altruism. They use it to promote the idea that joining a holy war and martyring oneself to a higher calling is a way to fulfill spiritual and emotional needs. Some try to create a kind of circle of grace in which all members see themselves as unusually good. The threat of being banished from that circle then deters defection. Inspirational terrorist leaders and their followers seek to alter the world as it currently exists (whether in a spiritual way such as bringing on Armageddon, or in a more material way, such as ending abortion or ousting U.S. troops from Saudi Arabia). Leaders and followers may aim to achieve multiple purposes, but they may not necessarily be pursuing an identical basket of goals. Some antiabortion activists are also racist and aim to provoke a race war, while others care only about abortion, for example.

49. Bob Lokey, telephone interview with the author, 1 May 1999.

50. "Paul Hill Speaks, Pro-Defending Life—A Reply to Credenda Agenda's 'Moving Beyond Profile,'" vol. 8, no. 5, issue no. 1, June 1997, last accessed 25 August 2002, www.trosch.org/bra/ph-v8_n5.htm.

51. Paul J. Hill, "Should We Defend Born and Unborn Children with Force," July 1993, last accessed 22 April 1999, www.trosch.org/bks/defnd-ph.htm.

52. Ibid.

53. Ibid.

54. Ibid.

55. Ibid.

56. Paul J. Hill, letter to the author, 26 April 1999.

57. National Abortion Federation, "2001 Year-End Analysis of Trends of Violence and Disruption Against Reproductive Health Care Clinics," Web site of the National Abortion Federation, last accessed 26 August 2002, www.prochoice.org/. The Web site also contains statistics of antiabortion violence since 1977.

58. Paul J. Hill, interview with the author, 27 April 1999, Florida State Prison. Journalist Steve Goldstein also participated in this interview.

59. *Soldiers in the Army of God.*

60. Ibid.

Chapter 7: Lone-Wolf Avengers

1. Alex Tizon and the *Seattle Times* Investigative Team, "John Muhammad's Meltdown," *Seattle Times,* 10 November 2002, A1.

2. Joseph S. Nye Jr., *The Paradox of American Power: Why the World's Only Superpower Can't Go It Alone* (Oxford and New York: Oxford University Press, 2002); see also, Joseph S. Nye Jr., "A Whole New Ball Game," *Financial Times,* 28 December 2002.

3. Robert O'Harrow Jr., "Kansi's Shadowy Stay in U.S. Leaves a Hazy Portrait," *Washington Post,* 3 March 1993, A1.

4. Ibid.

5. Ibid.

6. According to testimony by Special Agent Bradley J. Garrett of the FBI, who helped arrest Kansi in Pakistan, "He was especially furious over the Persian Gulf War and the treatment of Palestinians by Israel." Tim Weiner, "Pakistani Convicted of Killing 2 Outside CIA Headquarters," *New York Times,* 11 November 1997, A19.

7. John F. Burns, "Family of Pakistani in Killings at CIA Also Seeks a Motive," *New York Times,* 21 June 1997, A1.

8. Ibid. See also Tim Weiner, "FBI Men Sent to Karachi to Investigate the Slaying," *New York Times,* 13 November 1997, A10.

9. Weiner, "FBI Men Sent to Karachi," A10.

10. Mary Anne Weaver, "The Stranger," *New Yorker,* 13 November 1995, 59–72.

11. Jeff Stein, "Lone Gunmen," *Salon,* last accessed 4 September 2002, www.salon.com/news/1997/11/21news2.html.

12. See Weaver, "Stranger," 59–72.

13. On Kansi's denial, see Mir Aimal Kansi, interview with the author, Sussex One State Prison, Waverly, Va., 7 November 1999. The relatives of Kansi denied Abdullah Jan's involvement in court. See Burns, "Family of Pakistani in Killings," A1.

14. At the end of March 1993, the FBI reclassified Kansi as a suspected international terrorist. This move allowed the Department of State's Counter-Terrorism Rewards Program to raise the reward offered for information leading to the arrest of Kansi to $2 million. David B. Ottaway, "Frustrating the FBI," *Washington Post,* National Weekly Edition, 24–30 July 1995, 32, cited in "Central Intelligence Agency 1997—Capture and Trial of Kansi," Web site of The Literature of Intelligence: A Bibliography of Materials, with Essays, Reviews, and Comments, by J. Ransom Clark, Muskingum College, Ohio, last accessed 5 September 2002, www.intellit.muskingum.edu/cia1990s_folder/cia1997kansi.html. See also "Facing Justice," on-line *NewsHour* transcript, 18 June 1997, last accessed 4 September 2002, www.pbs.org/newshour/bb/law/june97/cia_6-18.html.

15. Zahid Hussain, "The Great Cover-Up," *Newsline,* July 1997, 18–24.

16. General Musharraf established the National Accountability Bureau in October 1999 to investigate major corruption cases involving politicians, government officials, money laundering, and loan write-offs to industrialists. Headed by a lieutenant general, the bureau employs many retired ISI officials, bankers, legal experts, and a few police officials.

17. The senior Pakistani official adds that the junior ISI official, who apparently received only $10,000 out of the total reward payment, came under investigation as his bank account statement showed an unusually large transaction, which NAB officials believed to be corruption money. As soon as the matter was reported to top NAB officials, the official adds, immediate instructions were given to close the case and destroy the interrogation reports. Senior Pakistani government official, interview with the author, 8 September 2002.

18. Burns, "Family of Pakistani in Killings," A1.

19. Senior Pakistani government official, interview.

20. According to the same senior Pakistani government official, who asked to remain unidentified, "The informed circles in Pakistan believe that Kansi was convinced that his father was ditched by the CIA after using him, and he sought revenge for this reason. It is believed by many that another of [Kansi's] close relatives was killed by the CIA in 1984 and may have also led Kansi to seek revenge." Former Pakistani spy chief Hamid Gul also suggested that Kansi might have had a personal motive in attacking CIA employees. See Stein, "Lone Gunmen."

21. Bill Baskervill, "Pakistani Who Killed CIA Agents in '93 Is Executed; Appeal Rejected; Reprisals Feared," *Boston Globe*, 15 November 2002, A2.

22. The material on James Dalton Bell summarizes Jessica Stern and Darcy Bender, "James Dalton Bell," forthcoming publication.

23. Jim Bell, "Assassination Politics," 3 April 1997, last accessed 30 August 2002, www.vader.com/ap.htm/.

24. Ibid.

25. For background information on diisopropyl fluorophosphate, see "EPA Chemical Profile: Isofluorphate," Web site of the Environmental Protection Agency, last accessed 30 August 2002, www.epa.gov/swercepp/ehs/profile/55914p.txt/. *Isofluorphate* is a synonym for *diisopropyl fluorophosphate*.

26. *United States of America v. James Dalton Bell*, 00-5172M, Complaint for Violation, 5.

27. Ibid., 97-5048M, Complaint for Violation, 4.

28. "Inco Nickel-Coated Carbon Fiber," Inco Material Safety Data Sheet, issued 31 March 1998, revised 4 January 1999.

29. James Bell, telephone interview with the author, 14 February 2000.

30. *United States of America v. James Dalton Bell*, 97-5048M, Complaint for Violation, 4.

31. United States District Court, Western District of Washington, 97-5047M, Inventory of Items Seized Under Authority of a Warrant, 1.

32. *United States of America v. James Dalton Bell*, 97-5048M, Complaint for Violation, 6.

33. Ibid., 5; and John Branton, "Judge Delays Bell's Sentencing," *The Columbian* (Vancouver), 21 November 1997, Section A.

34. John Painter Jr., "IRS Says Man from Tacoma Part of Plot," *Oregonian* (Vancouver), 21 November 1997; and Branton, "Judge Delays Bell's Sentencing."

35. *United States of America v. James Dalton Bell*, 97-5048M, Complaint for Violation, 5.

36. Ibid.

37. Ibid., 6.

38. Bell, telephone interview.

39. Ibid.

40. Ibid.

41. United States District Court, Western District of Washington, Application and Affidavit for Search, 97-5025M, 28 March 1997, 9. The issuance of such documents are "in violation of both the Criminal Code of Oregon, Section 162.355, Simulating Legal Process, and Title 26, U.S. Code, Section 7212(a), Corrupt Interference with the Administration of Internal Revenue Laws." Ibid., 7.

42. Ibid., 6.

43. *United States of America v. James Dalton Bell*, 00-5172M, Complaint for Violation, 4.

44. B. J. Berkowitz et. al, *Superviolence: The Civil Threat of Mass Destruction Weapons* (Santa Barbara: ADCON Corporation, 1972), 3–9, 4–4.

45. Robert S. Robins and Jerrold M. Post, *Political Paranoia* (New Haven: Yale University Press, 1997).

46. For more on how people can be simultaneously paranoid and correct about others' negative intentions, see Robins and Post, *Political Paranoia*.

47. As of 5 December 2001, eleven cases of inhalational and seven cases of cutaneous anthrax had been confirmed. There were four additional suspected cases of cutaneous anthrax. See "Update: Investigation of Bioterrorism-Related Anthrax, 7 December 2001," Web site of the Centers for Disease Control and Prevention (CDC), Morbidity and Mortality Weekly Report (MMWR), December 7, 2001 / 50(48), 1077–79, last accessed 30 August 2002, www.cdc.gov/mmwr/preview/mmwrhtml/mm5048a1.htm/.

The fatalities included a ninety-four-year-old woman from rural Connecticut, a hospital employee in New York City, both of whom are suspected to have contracted the disease from contaminated mail, two Washington, D.C., area postal workers, and a newspaper picture editor in Florida. Barbara Hatch Rosenberg, who has studied the anthrax case extensively, reports some uncertainty about the date the letter to the American Media Inc. office was mailed. See Barbara Hatch Rosenberg, "Analysis of the Anthrax Attacks," Web site of the Federation of American Scientists (FAS), last accessed 30 August 2002, www.fas.org/bwc/news/anthraxreport.htm/.

On 5 April 2002, CDC reported a case of suspected cutaneous anthrax in a worker who had been processing environmental samples for *Bacillus anthracis* in support of CDC investigations of the 2001 bioterrorist attacks in the United States. See "Public Health Dispatch: Update: Cutaneous Anthrax in a Laboratory Worker—Texas, 2002," Web site of the CDC, MMWR, June 7, 2002 / 51(22), 482, last accessed 30 August 2002, www.cdc.gov/mmwr/preview/mmwrhtml/mm5122a4.htm.

48. Jonathan Knight, "Crackdown on Hazardous Agents Raises Concern for Bona Fide Labs," *Nature* 414 (2001): 3–4.

49. Gerald Epstein, "Controlling Biological Warfare Threats: Resolving Potential Tensions among the Research Community, Industry, and the National Security Community," *Critical Reviews in Microbiology* 27, no. 4 (2001): 321–54.

50. The Antiterrorism and Effective Death Penalty Act of 1996, enacted on April 24, 1996, required HHS to regulate the transfer of "select agents." CDC came up with a list of twenty-four microbial pathogens and twelve toxins that, if transferred to another facility, would require registration with the CDC. Part 72 of Title 42 of Code of Federal Regulations. The act can be accessed through the Web site of the American Society of Microbiology (ASM), last accessed 30 August 2002, www.asmusa.org/pcsrc/pl104-132.pdf.

51. See "The Culture Collection in This World: WDCM Statistics, 27 November 2001," Web site of the World Data Centre for Microorganisms (WDCM), last accessed 30 August 2002, www.wdcm.nig.ac.jp/statistics2001.html/.

52. Michael Barletta, Amy Sands, and Jonathan B. Tucker, "Keeping Track of Anthrax: The Case for a Biosecurity Convention," *Bulletin of the Atomic Scientists* 58.3 (May/June 2002): 57–61. The purpose of these collections is to make cultures available for medical research. Hospitals use them to check the accuracy of diagnostic methods and instruments. Pharmaceutical companies use them to test the effectiveness of vaccines and other medical countermeasures. Universities use them in basic research.

53. Barry Kellman, "Biological Terrorism: Legal Measures for Preventing Catastrophe," *Harvard Journal of Law and Public Policy* 24.2 (spring 2001): 425–88.

54. *United States of America v. James Dalton Bell*, 00-5172M, Complaint for Violation, 10.

55. Ibid., 14.

56. Ibid., 1.

57. See Declan McCullagh, "Crypto-Convict Won't Recant," *Wired News*, 14 April 2000, last accessed 30 August 2002, www.wired.com/news/politics/0,1283,35620,00.html/. See also *United States of America v. Carl Johnson*, CR98-5393RJB, Superseding Indictment, last accessed 30 August 2002, www.cryptome.unicast.org/cryptome022401/cej040399.htm#Superseding, and *United States of America v. Carl Johnson*, CR98-5393RJB, Plea Agreement, last accessed 30 August 2002, www.cryptome.unicast.org/cryptome022401/cej040399.htm#PLEA/.

Chapter 8: Commanders and Their Cadres

1. "About CPL," Web site of the Center for Public Leadership, last accessed 7 January 2003, www.ksg.harvard.edu/leadership/aboutcpl.html.

2. In the American attack, which consisted of a barrage of seventy-nine Tomahawk missiles, some twenty-one people were killed and fifty wounded, according to Pakistani reports quoted by the *Washington Post*. Several Harkat operatives were killed during the raid. Eugene Robinson, "Reports of U.S. Strikes' Destruction Vary; Afghanistan Damage 'Moderate to Heavy'; Sudan Plant Leveled," *Washington Post*, 22 August 1998.

3. "Osama bin Laden—Links with Kashmiri Militants," undated Indian government document. Indian government officials, interviews with the author.

4. United States Department of State, *Patterns of Global Terrorism 2000*, last accessed 22 September 2002, www.usis.usemb.se/terror/rpt2000/.

5. Daniel McGrory, "Public Schoolboy Became bin Laden Pupil," *Times* (London), 4 October 2001.

6. Indian government official, interview with the author, 17 April 2002.

7. Daniel McGrory, "Public Schoolboy Became bin Laden Pupil."

8. Nick Fielding, "London Student Has Key Role in Terror Network," *Sunday Times*, 23 September 2001.

9. Indian government official, interview with the author, 17 April 2002.

10. The actual sum was 22 lakh. One lakh denotes one hundred thousand rupees. One hundred lakhs are one crore. In early January 2003, one dollar was worth approximately forty-eight Indian rupees.

11. In 1991, Azhar traveled with the vice chief of HUM, Maulana Farooq Kashmiri, to Saudi Arabia during Ramadan, raising some $6,500. After this, Azhar quit his teaching job to work full-time for HUM. In 1992, Azhar traveled again to Saudi Arabia on a fund-raising mission, this time with the assistance of Harkat's permanent representative in Saudi Arabia. He returned with $10,000. When he returned six months later in September 1993, a Saudi businessman secured an "Ekama" for Azhar, which allowed him to travel to Saudi Arabia anytime at will, without a visa. From there, he traveled to England, where he visited a number of towns where Pakistani expatriates tend to live. He raised some $30,000. While in England, Azhar arranged to acquire a fake Portuguese passport to travel freely throughout Europe on fund-raising missions. The next time he traveled, to Zambia, in February 1993, Azhar was able to raise another $30,000. From there he traveled to Saudi Arabia and Sharjah (UAE), where he collected an additional $12,000. In November 1993, Azhar traveled to Kenya at the behest of Somali Muslims, to lobby the Pakistani government to pull its troops out of the UN peacekeeping force there.

12. "Borderless Web of Killers," *Los Angeles Times,* 26 February 2002, Part 2, 12. See also Nick Fielding, "Pearl Murder Case Briton 'Was a Double Agent,'" *Sunday Times,* 21 April 2002.

13. Interrogation Report of Masood Azhar, Alias Adam Issa, undated document provided by Indian government.

14. Sipah e Sahaba Pakistan (SSP) was founded by Haq Nawaz Jhangvi, a cleric and leader of the Jamiat-ul-Ulema-e-Islami, in the mid-1980s to counter Shiite and Iranian influence in Pakistan. Pakistani Shiites believe that the SSP was the creation of Pakistan's Interservice Intelligence (ISI), at the behest of General Zia-ul-Haq. The SSP later gave rise to a number of breakaway groups, including Lashkar e Jhangvi (LEJ), formed in 1996. The SSP, which has been involved in sectarian killings for the past fifteen years, recently expanded its activities to Kashmir and became involved in the killing of Shiite political leaders in the region's Indian-administered part. It regularly issues a "hit list" of prominent Shiite professionals, scholars, and government servants. General Musharraf's government recently banned the SSP, along with its rival Shiite group Sipah-i-Mohammad. One of the SSP's leaders, Riaz Basra, who was killed by Pakistani law enforcement agencies in April 1999, was reportedly a part of the inner circle of Osama bin Laden. In certain areas of Punjab, the SSP is a relatively popular political party. Hassan Abbas, Shelley Cook, Brett Kenefick, Joseph Kopser, and Silbi Stainton, "Pakistani Terror," unpublished term paper written for the course "Non-State Threats to International Security," taught by Jessica Stern, Kennedy School of Government, Harvard University, 24 October 2001.

15. David Sanger and Kurt Eichenwald, "Reacting to Attack in India, U.S. Aims at Pakistan Group's Assets," *New York Times,* 21 December 2001, A1.

16. Brigadier Abdullah is in all likelihood an alias used to hide the identity of an important ISI official. Former high-ranking Pakistani government official, interview with the author, 8 September 2002.

17. Peter Finn and Dana Priest, "Weaker Al Qaeda Shifts to Smaller-Scale Attacks; Experts Say New Strategy Aims at Disruption," *Washington Post,* 15 October 2002, A1. According to Ajai Sahni of the Institute for Conflict Management in India, Mufti Nizamuddin Shamzai, the head of the Binori Mosque and Azhar's patron, had a long and close association with Mullah Omar, and is reported to have hosted bin Laden at the Mosque as far back as 1989. The JEM was one of the major facilitators of the Al Qaeda–Taliban relocation to Pakistan after September 11. Ajai Sahni, e-mail communication with the author, 2 February 2003.

18. Fielding, "Pearl Murder Case Briton."

19. Douglas Jehl, "Pakistan Is Willing to Give Up Suspect in Reporter's Death," *New York Times,* 27 February 2002, A1.

20. Ibid.

21. Jon Stock, "Inside the Mind of a Seductive Killer," *Times* (London), 21 August 2002, 4.

22. The humanitarian group said that Sheikh became ill and never left Croatia. According to Asad Khan, the charity's founder, Sheikh went alone to meet Bosnian refugees, who told him about rape and ethnic cleansing practiced by Serbs, claiming to be Orthodox Christians on a crusade against Islam. Aslan Cowell, "This Mild Schoolboy, Lost in the Islamic Inferno," *New York Times,* 11 October 2001, A4.

23. One prominent member of the SSG is Pakistan's president, Pervez Musharraf. Many leaders of the SSG have been trained in the United States, and the SSG is not known to be involved in training of jihadi groups.

24. Nasarullah Manzoor Langaryal was deputy commander of Harkat-ul-Jihad-e-Islami (HUJI). He was arrested in November 1993. His Indian interrogators describe him as a "highly motivated, committed, and fundamentalist militant." He fought Soviet forces in Afghanistan from 1983 until 1992 and got involved in militancy in Kashmir immediately after leaving Afghanistan. "List of disclosures made by mercenaries arrested in Jammu and Kashmir," undated Indian government document.

25. Ibid. "Profile of Ahmed Umar Saeed Sheikh," undated document provided by the Indian government.

26. Sayantan Chakravarty with Sheela Raval, Uday Mahurkar, and Hasan Zaidi, "Deadly Duo: The Dons of Terror," *BBC Monitoring,* 25 February 2002, 28. Undated documents provided by the Indian government.

27. Ibid.

28. Ibid. Suman K. Chakrabarti, "West Bengal: Prosecution Weakness," *India Today*, 5 August 2002, 30.

29. "Hijacker Received Money Through Pakistan," *The News*, 2 October 2001. Available on the Web site of *Karachi–The News International*, last accessed 22 September 2002, www.karachipage.com/news/Oct_01/100201.html. Undated Indian government documents supplied to author.

30. Chakravarty et al., "Deadly Duo," 28. Undated Indian government documents supplied to author.

31. Ibid. Paul Watson and Sidhartha Barua, "Worlds of Extremism and Crime Collide in Indian Jail," *Los Angeles Times*, 8 February 2002, 1.

32. Chakravarty et al., "Deadly Duo," 28.

33. Zahid Hussain and Daniel McGrory, "A World Apart, Two Young Men Twisted by One Idea," *Times* (London), 16 July 2002.

34. Tariq Ali, "Who Really Killed Daniel Pearl?" *Guardian*, 5 April 2002.

35. Senior Pakistani government official, interview with the author, 8 September 2002.

36. Isabel Hilton writes that a source told her that Saeed felt that he would escape the death penalty in return for his silence on operations that he claimed to have carried out in India for the ISI. Isabel Hilton, "The General in His Labyrinth," *New Yorker*, 12 August 2002. Robert Fisk writes that Saeed was not turned over to U.S. agents, and not tried publicly, out of fear that he might reveal links between the ISI and Al Qaeda. Robert Fisk, "The Murder of Daniel Pearl," *Independent*, 16 July 2002, 3.

37. Indian government official, interview with the author, 17 April 2002.

38. Robert Fisk, "U.S. Wary of Pakistan Intelligence Services' Links to Al-Qa'ida," *Independent on Sunday*, 21 August 2002.

39. Stimson's refusal in 1929 to expend State Department funds for cryptanalysis was made public in 1931 in a book by cryptanalyst Herbert O. Yardley, as a result of which the Japanese adopted machine ciphers more complex than the system employing simultaneous use of multiple codebooks that U.S. cryptanalysts had managed to crack before. David Kahn, "The Intelligence Failure of Pearl Harbor," *Foreign Affairs*, 70, No. 5 (winter 1991/1992): 138.

40. Erik Eckholm, "Qaeda Operative Is 'Hero' to Some in Pakistan," *New York Times*, March 2003, A13.

41. Syed Salahuddin, interview with the author, 2 August 2001.

42. Profile of Syed Salahuddin, supreme commander of Hizb-ul Mujahideen (HM), undated Indian government document.

43. Ibid.

44. See, for example, Jessica Tuchman Mathews, "Redefining Security," *Foreign Affairs* 68, No. 2 (spring 1989): 162–77. See also Ann M. Florini, ed., *The Third Force: The Rise*

of Transnational Civil Society (Washington, D.C. and Tokyo, Japan: Brookings Institution Press/Japan Center for International Exchange, 2000).

45. Taiba bulletin, 11 August 2000.

46. RAW is India's foreign intelligence agency, the counterpart to the ISI. The name stands for Research and Analysis Wing. Indian officials and scholars deny this claim.

47. Like the other interviewees for the Harvard project on leadership, this interviewee's identity cannot be revealed.

48. Interview with TUM-2, Pakistan, 2002. Interviewer: Muzamal Suherwardy.

49. Interview with HM-3, Pakistan, 2002. Interviewer: Muzamal Suherwardy.

50. Interview with HUM-1, Pakistan, 2002. Interviewer: Muzamal Suherwardy.

51. Interview with HM-3.

52. Interview with HUM-1.

53. Interview with HM-1, Pakistan, 2002. Interviewer: Muzamal Suherwardy.

54. Interview with JEM-2, Pakistan, 2002. Interviewer: Muzamal Suherwardy.

55. Of all *madrassahs,* only roughly 10–15 percent espouse an extremist ideology. Jessica Stern, "Pakistan's Jihad Culture," *Foreign Affairs,* November/December 2000, 115. See also P. W. Singer, "Pakistan's Madrassahs: Ensuring a System of Education, Not Jihad," Brookings Institution Analysis Paper #14, November 2001. Available on the Web site of the Brookings Institution, last accessed 7 January 2003, www.brook.edu/dybdocroot/views/papers/singer/20020103.htm.

56. This conversation took place in Moinhuddin Haider's office in Islamabad on 8 June 2000, but Haider has publicly promoted this plan many times since.

57. "Elections in Pakistan," Electionworld.com, last accessed 7 January 2003, www.electionworld.org/election/pakistan.htm.

58. David Rhode, "Turning Away from U.S., Pakistan's Elite Gravitate toward Islamic Religious Parties," *New York Times,* 13 October 2002, A8; see also David Rhode, "Pakistani Fundamentalists and Other Opponents of Musharraf Do Well in Elections," *New York Times,* 11 October 2002, A13.

59. Interview with TUM-2. Others responded more simply "education."

60. Interview with JEM-2.

61. Ibid. JEM-2 had worked for several militant groups: Harkat e Jihadi Islami, Harkat-ul-Mujahideen, and Harkat-ul-Ansar. He is now closest to Jaish-i-Muhammad.

62. Interview with HUM-1.

63. Firdous Syed, e-mail communication, 22 February 2003.

64. E-mail communication with Ajai Sahni, 2 February 2003. Interviews with Indian government officials in Delhi, January 2003.

Chapter 9: The Ultimate Organization: Networks, Franchises, and Freelancers

1. Hidaya Rubea Juma, mother of Khalfan Khamis Mohamed. Quoted in "Special Assignment," SABC Africa News, date unavailable.

2. R. F. Burton, *Zanzibar: City, Island, and Coast* (London: Tinsley Brothers, 1872), 117. The monsoon between December and February blows north, northeast from the Arabian peninsula and the west coast of India, then reverses direction in April. This remarkable pattern of winds made oceangoing trade possible long before overland commerce was possible. Michael F. Lofchie, *Zanzibar: Background to Revolution* (Princeton, N.J.: Princeton University Press, 1965), 21.

3. Lofchie, *Zanzibar*, 24. Lofchie explains the Persians mingled completely with the Africans and were no longer detectable as a separate group. Arabs, who arrived later, became the upper classes in Zanzibar, while immigrants from the Indian subcontinent were traders, and the Africans became the lowest class.

4. Nathalie Arnold, telephone conversation with the author, 14 October 2002. Bruce McKim, telephone conversation with the author, 16 October 2002; Human Rights Watch, "Tanzania: 'The Bullets Were Raining'—The January 2001 Attack on Peaceful Demonstrators in Zanzibar," *Human Rights Watch Report*, 14, no. 3 (A) (April 2002), last accessed 16 October 2002, www.hrw.org/reports/2002/tanzania/zanz0402.pdf.

5. Arnold, telephone conversation. McKim, telephone conversation.

6. Quoted in "Pemba Island," All About Zanzibar Web site, last accessed 16 October 2002, www.allaboutzanzibar.com/indepth/guidebook/pb00-01-11.htm.

7. Alice Werner, *Myths and Legends of the Bantu* (London: Cass, 1968). See chapter 16, "Doctors, Prophets, and Witches," available in the book's on-line version at the Najaco Web site, last accessed 16 October 2002, www.najaco.com/books/myths/bantu/16.htm.

8. Werner, *Myths and Legends*. See also John. E. E. Craster, *Pemba: The Spice Island of Zanzibar* (London: T. F. Unwin, 1913). Werner believes that some of these stories reflect the prejudices of white Christians.

9. *United States of America v. Usama bin Laden, et al.,* S(7) 98 Cr. 1023 (27 June 2001), 8321.

10. Ibid., 8324–25.

11. This material summarizes Federal Bureau of Investigation, FD-302a, of Khalfan Khamis Mohamed, 10/5–7/99 at Cape Town, South Africa. Marked "particularly sensitive." This document was entered into evidence at Mohamed's trial.

12. Ibid.

13. Ibid.

14. Ibid.

15. *United States of America v. Usama bin Laden* (27 June 2001), 8327–28.

16. Ibid., 8329.

17. Ibid., 8328.

18. This material summarizes Federal Bureau of Investigation, FD-302a.

19. *United States of America v. Usama bin Laden,* (2 May 2001), 5437.

20. This material summarizes Federal Bureau of Investigation, FD-302a.

21. Ibid.

22. *United States of America v. Usama bin Laden,* (28 June 2001), 8431.

23. Ibid., 8431–32.

24. Ibid., (3 July 2001), 8740.

25. Quoted in Benjamin Weiser and Tim Golden, "Al Qaeda: Sprawling, Hard-to-Spot Web of Terrorists-in-Waiting," *New York Times,* 30 September 2001, 1B4.

26. Testimony of Jerrold Post, *United States of America v. Usama bin Laden* (27 June 2001), 8311–62.

27. Excerpts of the Al Qaeda training manual are available at the Web site of the U.S. Department of Justice, last accessed 14 January 2003, www.usdoj.gov/ag/training manual.htm.

28. See the Web site of the U.S. Department of Justice, last accessed 11 October 2002, www.usdoj.gov/ag/manualpart1_1.pdf, "Declaration of Jihad (Holy War) against the Country's Tyrants—Military Series," First Lesson, p. 13 (translated version).

29. Ibid., Second Lesson, pp. 15–20 (translated version).

30. Copies of the more extensive *Encyclopedia of the Afghan Jihad* were found from Al Qaeda members arrested in Asia, the Middle East, and Europe. The *Encyclopedia* covers tactics, security, intelligence, handguns, first aid, explosives, topography, land surveys, and weapons, and has been compiled since Soviet troops withdrew from Afghanistan in 1989. Originally designed as a record of the Afghan fighters' knowledge and experience in guerrilla warfare, it gradually came to include terrorist tactics, as Al Qaeda developed into a terrorist organization. A work of several thousand pages written and translated over five years, the *Encyclopedia* also appeared in CD-ROM in 1996. For more information on the *Encyclopedia of the Afghan Jihad,* see Rohan Gunaratna, *Inside Al Qaeda: Global Network of Terror* (New York: Columbia University Press, 2002), 70.

31. Well-known members of the Shura Council include Muhammad Atef, an Egyptian who served as military commander and was reportedly killed in Afghanistan in late 2001; and Ayman al-Zawahiri, a surgeon who runs the Egyptian Islamic Jihad, responsible for the 1981 assassination of President Anwar el-Sadat of Egypt. For other members of the council, see testimony of Jamal Ahmad al-Fadl, *United States of America v. Usama bin Laden* (6 February 2001), 204–07.

32. Ibid., 204–214.

33. Bruce B. Auster et al., "The Recruiter for Hate," *U.S. News & World Report,* 31 August 1998, 48.

34. The explosion blew a hole in the fuselage, and only an extraordinary flight performance by the pilot enabled an emergency landing at Naha airport in Okinawa. Gunaratna, *Inside Al Qaeda,* 175.

35. For more on the Bojinka Plot—also known as Oplan Bojinka—see Simon Reeve, *The New Jackals: Ramzi Yousef, Osama bin Laden and the Future of Terrorism* (Boston, Mass.: Northeastern University Press, 1999), 71–93; and Gunaratna, *Inside Al Qaeda,* 175–77.

36. Judy Aita, "U.S. Completes Presentation of Evidence in Embassy Bombing Trial: Defense Expected to Begin Its Case April 16," The Washington File, Office of International Information Programs, U.S. Department of State, April 2002, last accessed 2 October 2002, www.usinfo.state.gov/regional/af/security/a1040558.htm.

37. See, for example, *United States of America v. Mokhtar Haouari,* S(4) 00 Cr. 15 (3 July 2001), 630–35 (www.news.findlaw.com/cnn/docs/haouari/ushaouari70301rassamtt.pdf).

38. Peter L. Bergen, *Holy War, Inc.: Inside the Secret World of Osama bin Laden* (New York: Free Press, 2001), 185.

39. Statement for the Record of J. T. Caruso, Acting Assistant Director, Counter-Terrorism Division, Federal Bureau of Investigation (FBI), on Al-Qaeda International Before the Subcommittee on International Operations and Terrorism Committee on Foreign Relations, United States Senate, Washington, D.C., 18 December 2001. Available on the Web site of the FBI, last accessed 2 October 2002, www.fbi.gov/congress/congress01/caruso121801.htm.

40. For a summary of President Musharraf's 12 January 2002 speech, see "Musharraf Declares War on Extremism," BBC News Online, 12 January 2002, last accessed 18 October 2002, www.news.bbc.co.uk/1/hi/world/south_asia/1756965.stm. See also "Pakistan's Leader Comes Down Hard on Extremists," CNN.com, 12 January 2002, last accessed 18 October 2002, www.cnn.com/2002/WORLD/asiapcf/south/01/12/pakistan.india/.

41. See, for example, "Confessions of an Al-Qaeda Terrorist," *Time,* 23 September 2002, 34.

42. Kit R. Roanet, David E. Kaplan, Chitra Ragavan, "Putting Terror Inc. on Trial in New York," *U.S. News & World Report*, 8 January 2001, 25.

43. See, for example, Bergen, *Holy War, Inc.,* 80.

44. Gunaratna, *Inside Al Qaeda,* 58–59.

45. Peter Baker, "Defector Says bin Laden Had Cash, Taliban in His Pocket," *Washington Post,* 30 November 2001, A1. See also Molly Moore and Peter Baker, "Inside Al Qaeda's Secret World; bin Laden Bought Precious Autonomy," *Washington Post,* 23 December 2001, A1.

46. Baker, "Defector Says bin Laden Had Cash," A1.

47. Osama bin Laden established the International Islamic Front in a statement calling for a jihad against the Jews and Crusaders on 23 February 1998. Signatories other than Osama bin Laden were Ayman al-Zawahri, leader of Egypt's Jihad group, Rifai Taha, head of Egypt's Gama'a al-Islamiya, Mir Hamza, secretary general of Pakistan's Ulema Society, and Fazlul Rahman, head of the Jihad Movement in Bangladesh. Other organizations whose membership in the IIF has been publicized include the Partisans Movement in Kashmir (Harkat ul-Ansar), Jihad Movement in Bangladesh, and the Afghan military wing of the "Advice and Reform" commission led by Osama bin Laden. last accessed 21 March 2003, http://www.satp.org/satporgtp/usa/IIF.htm.

48. Gunaratna, *Inside Al Qaeda,* 57–58.

49. *United States of America v. Usama bin Laden* (4 June 2001), 7007.

50. Gunaratna, *Inside Al Qaeda,* 60.

51. "Exclusive Interview: Conversation with Terror," *Time,* 11 January 1999, available on-line at *Time Asia* last accessed 8 October 2002, www.time.com/time/asia/news/interview/0,9754,174550-1,00.html.

52. Pamela Constable, "Bin Laden Tells Interviewer He Has Nuclear Weapons," *Washington Post,* 11 November 2001, A32.

53. Ayman al-Zawahiri, *Knights under the Prophet's Banner,* chap. 11. Excerpts of the book were translated by FBIS. See "*Al-Sharq al-Awsat* Publishes Extracts from Al-Jihad Leader Al-Zawahiri's New Book," *Al-Sharq al-Awsat* (London), 2 December 2001, in FBIS-NES-2002-0108, Document ID GMP20020108000197.

54. James Risen, "Question of Evidence: A Special Report; To Bomb Sudan Plant, or Not: A Year Later, Debates Rankle," *New York Times,* 26 October 1999, A1.

55. Testimony by Ahmed Ressam, *United States of America v. Mokhtar Haouari* (5 July 2001), 620–22.

56. See, for example, Peter Finn, "Five Linked to Al Qaeda Face Trial in Germany; Prosecutors Focus on Alleged Bombing Plans," *Washington Post,* 15 April 2002, A13.

57. This manual was found in the house of a Libyan Al Qaeda member who lived in Manchester, England. Benjamin Weiser, "A Nation Challenged: The Jihad; Captured Terrorist Manual Suggests Hijackers Did a Lot by the Book," *New York Times,* 28 October 2001, A8.

58. The manual was part of the so-called *Encyclopedia of the Afghan Jihad,* a seven-thousand-pages-long collection, which used to consist of ten volumes, of guidelines for terrorist attacks against targets worldwide. See Gunaratna, *Inside Al Qaeda,* 70. See also Mark Boettcher, "Evidence Suggests Al Qaeda Pursuit of Biological, Chemical Weapons," CNN.com, 14 November 2001, last accessed 8 October 2002, www.cnn.com/2001/WORLD/asiapcf/central/11/14/chemical.bio/.

59. On 13 March 2001, Italian authorities bugged a conversation in which a Milan-based Al Qaeda cell led by a Tunisian, Essid Sami Ben Khemais, spoke of "an extremely efficient liquid that suffocates people" and that was to be "tried out" in France. The liquids, one cell member was overheard saying, could secretly be placed in tomato cans and would be dispersed when the cans were opened. See Peter Finn and Sarah Delaney, "Al Qaeda's Tracks Deepen in Europe; Surveillance Reveals More Plots, Links," *Washington Post,* 22 October 2001, A1. See also "Disturbing Scenes of Death Show Capability with Chemical Gas," CNN.com, 19 August 2002, last accessed 8 October 2002, www.cnn.com/2002/US/08/19/terror.tape.chemical/index.html.

60. Barton Gellman, "Al Qaeda Near Biological, Chemical Arms Production," *Washington Post,* 23 March 2003, A1.

61. *United States of America v. Usama bin Laden,* (7 February 2001), 357–365.

62. Gunaratna, *Inside Al Qaeda,* 36.

63. *United States of America v. Usama bin Laden* (19 June 2001), 7464.

64. "Report Links bin Laden, Nuclear Weapons," *Al-Watan al-Arabi,* 13 November 1998; available from FBIS, Document ID FTS19981113001081. Quoted in Kimberly McCloud and Matthew Osborne, "CNS Reports: WMD Terrorism and Usama bin Laden," Web site of the Center for Nonproliferation Studies, Monterey Institute of International Studies, last accessed 8 October 2002, www.cns.miis.edu/pubs/reports/binladen.htm. The November report in *Al-Watan* followed that in another Arabic newspaper, the London-based *Al-Hayat,* which declared that bin Laden had already acquired nuclear weapons. "An Aide to the Taliban Leader Renews His Refusal to Give Information on Nuclear Weapons to bin Laden from Central Asia," *Al-Hayat,* 6 October 1998, quoted in McCloud and Osborne, "CNS Reports." See also Joseph, "Chemical Labs Show Al Qaeda Still Active."

65. Joseph, "Chemical Labs Show Al Qaeda Still Active."

66. Steven Erlanger, "Lax Nuclear Security in Russia Is Cited as Way for bin Laden to Get Arms," *New York Times,* November 12, 2001.

67. Kamran Khan and Molly Moore, "2 Nuclear Experts Briefed bin Laden, Pakistanis Say," *Washington Post,* 12 December 2001, A1. See also Peter Baker and Kamran Khan, "Pakistan to Forgo Charges Against 2 Nuclear Scientists; Ties to Bin Laden Suspected," *Washington Post,* 30 January 2002, A1.

68. Ibid.

69. David Albright, Kathryn Buehler, and Holly Higgins, "Bin Laden and the Bomb," *Bulletin of the Atomic Scientists,* January/February 2002, 23.

70. ICT, "Al-Qa'ida (The Base)," International Policy Institute for Counter-Terrorism (ICT), Herzliyya, Israel, last accessed 9 October 2002, www.ict.org.il/inter_ter/orgdet.cfm?orgid=74.

71. "Bin Laden's Martyrs for the Cause: Thousands of Terrorists. Dozens of Cells. One Mission," *Financial Times,* 28 November 2001, 17.

72. The information about the European recruiters is taken from *The Recruiters,* produced by Alex Shprintsen, edited by Annie Chartrand, June 2002, CBC News, Canada. A summary of the documentary is available on the Web site of the Canadian Broadcasting Corporation, last accessed 10 October 2002, www.cbc.ca/national/news/recruiters/network.html/.

73. "Bin Laden's Martyrs for the Cause," 17.

74. Michael Powell, "Bin Laden Recruits with Graphic Video," *Washington Post,* 27 September 2001, A19.

75. "Alliance Says It Has Found a School run by a Titan of Terrorism," *New York Times,* 1 December 2001; Jane Perlez, "School in Indonesia Urges 'Personal Jihad' in Steps of Bin Laden," *New York Times,* 3 February 2002.

76. The German Bundeskriminalamt, the Federal Criminal Agency, estimates the number of militant Islamic trainees at Al Qaeda training camps at 70,000. See "Bin Laden's Martyrs for the Cause," 17. The CIA estimates the number at 110,000. Quoted in Gunaratna, *Inside Al Qaeda,* 8. Of the 6–7 million Al Qaeda supporters, some 120,000 are willing to take up arms. Estimates of the Central Intelligence Agency, quoted in Gunaratna, *Inside Al Qaeda,* 95.

77. Gunaratna, *Inside Al Qaeda,* 8.

78. Walter Pincus and Vernon Loeb, "Former Recruits Provide Best Knowledge of Camps; Intelligence on Targeted bin Laden Training Sites Sketchy," *Washington Post,* 8 October 2001, A16.

79. Department of Defense, Interrogation Report of John Walker Lindh, JPWL-000389-000407, 402–03. Declassified 5 March 2002.

80. Bryan Preston, "Inside Al Qaeda's Training Camps," *National Review,* 1 October 2002. Available at the Web site of National Review Online, last accessed 19 October 2002, www.nationalreview.com/comment/comment-preston100102.asp.

81. *"Al-Sharq al-Awsat* Publishes Extracts." Parts one through eleven of serialized excerpts from Egyptian Al-Jihad Organization leader Ayman al-Zawahiri's book, "Knights under the Prophet's Banner" (FBIS translated text, henceforth: Ayman al-Zawahiri, "Knights under the Prophet's Banner Part: 1).

82. Ibid. Part I

83. Ibid. Part XI

84. Ibid. Part XI

85. Department of Defense, Interrogation Report of John Walker Lindh.

86. Interview with HM-1, Pakistan, 2002. Interviewer: Muzamal Suherwardy.

87. Lashkar e Taiba public affairs officer, interview with the author, 3 August 2001. This interview was attended by a Pakistani journalist who writes under a pseudonym for tehelka.com, an electronic newspaper published in India. He wrote an article that highlighted this surprising admission.

88. Frederick R. Karl, Introduction to Joseph Conrad, *The Secret Agent,* (New York and London: Penguin, 1983 edition).

89. Franz Fanon, *The Wretched of the Earth* (New York: Grove Press, 1963), 94.

90. Ibid., 195–97.

91. See John Esposito, *Unholy War: Terror in the Name of Islam* (New York: Oxford University Press, 2002), 43.

92. Ibid.

93. Based on interviews in the United States, Lebanon, Gaza, Israel, Pakistan, and Indonesia, 1998–2001.

94. Andrew Higgins and Allan Cullison, "Saga of Dr. Zawahiri Sheds Light on the Roots of Al Qaeda Terror: How a Secret, Failed Trip to Chechnya Turned Key Plotter's Focus to America and bin Laden," *Wall Street Journal,* 2 July 2002.

95. Neil MacFarquhar, "Islamic Jihad, Forged in Egypt, Is Seen as bin Laden's Backbone," *New York Times,* 4 October 2001, B4.

96. Lawrence Wright, "The Man behind bin-Laden," *New Yorker,* 16 September 2002, 77.

97. See, for example, Ahmed Rashid, "The Taliban: Exporting Extremism," *Foreign Affairs,* November/December 1999; Ahmed Rashid, *Jihad: The Rise of Militant Islam in Central Asia* (New Haven, Conn.: Yale University Press, 2002); and Gunaratna, *Inside Al Qaeda,* 168–72.

98. C. J. Chivers, "Uzbek Militants Decline Provides Clues to U.S.," *New York Times,* 8 October 2002, A15.

99. Ibid.

100. Ibid.

101. Ibid.

102. Bernard Lewis, "License to Kill," *Foreign Affairs,* November/December 1998, 15.

103. John F. Burns, "Bin Laden Taunts U.S. and Praises Hijackers," *New York Times,* 8 October 2001, A1. A transcript of President Bush's 20 September 2002 speech is available at the Web site of the White House. "Address to a Joint Session of Congress and the American People," United States Capitol, Washington, D.C., 20 September 2002, last accessed 19 October 2002, www.whitehouse.gov/news/releases/2001/09/20010920-8.html.

104. Judith Miller, "Bin Laden's Media Savvy: Expert Timing of Threats," *New York Times,* 9 October 2001, B6.

105. "UK-Based Paper Notes Al-Qa'ida Military Training on Internet Site, Encyclopedia," in *Al-Sharq al-Awsat* (London), 16 February 2002. Available in FBIS-NES-2002-0216, Article ID GMP20020216000057.

106. *Thaqafat al-Jihad*, placed by "OBL2003." http://members.lycos.co.uk/himmame/vb/printthread.php?threadid=1881. Printed (in translation) in Reuven Paz, editor, The Project for the Research of Islamist Movements (PRISM), Occasional Papers, Volume 1 (2003), Number 3 (March 2003).

107. *Bayan li-Taliban yu'akid an bin Laden taliq walam yu'taqal* (A statement by Taliban confirms that Bin Laden is free and has not been arrested). See on-line in: http://www.o-alshahada.net/vb/printhead.phb?s=&threadid=82. Printed (in translation) in Reuven Paz, editor, The Project for the Research of Islamist Movements (PRISM), Occasional Papers, Volume 1 (2003), Number 3 (March 2003).

108. Dee Hock, *Birth of the Chaordic Age* (San Francisco: Berrett Koehler Publishers, 1999).

109. Virtual networks of leaderless resisters make sense for groups that will be satisfied with the kind of attacks that can be carried out by small groups or individuals acting on their own. The mission is openly communicated, but detailed plans are not discussed with the leadership of the movement or among groups. The need for secrecy and the need for inspirational leaders to be able to plausibly deny their knowledge of past or present plots distort the communication flow.

110. Manuel Castells, *The Rise of the Network Society* (Cambridge, Mass.: Blackwell Publishers, 1996). Quoted in Joel Garreau, "Disconnect the Dots," *Washington Post*, 17 September 2001, C1.

111. Richard Wolfe, Carola Hoyos, and Harvey Morris, "Bin Laden's Wealth Put in Doubt by Saudi Dissidents," *Financial Times*, 24 September 2001, 5.

112. Paul McKay, "The Cost of Fanatical Loyalty," *Ottawa Citizen*, 23 September 2001, A8.

113. Barry Meier, "'Super' Heroin Was Planned by bin Laden, Reports Say," *New York Times*, 4 October 2001, B3.

114. For more information on the remittance system of *hawala*, see the Web site of Interpol, last accessed 7 January 2003, www.interpol.int/Public/FinancialCrime/MoneyLaundering/hawala/default.asp#2.

115. Douglas Frantz, "Ancient Secret System Moves Money Globally," *New York Times*, 3 October 2001, B5. See also Judith Miller and Jeff Gerth, "Business Fronts: Honey Trade Said to Provide Funds and Cover to bin Laden," *New York Times*, 11 October 2001, A1.

116. *United States of America v. Usama bin Laden* (26 February 2001), 1415.

117. Interviews with Laskar Jihad, Jakarta, 9 August 2001, and Yogyakarta, 11 August 2001; questionnaires administered in Pakistan.

118. Mark Hosenball, "Terror's Cash Flow," *Newsweek*, 25 March 2002, 28.

119. See for example, "Hate Literature Blitz Planned by Neo-Nazi Groups to Coincide with Jewish Holidays and 9/11," Anti-Defamation League (ADL) press release, 27 August 2002. Available at the ADL Web site, last accessed 13 January 2003, www.adl.org/PresRele/ASUS_12/4148_12.asp. The flyer can be viewed at the National Alliance Chicago Web site, last accessed 13 January 2003, www.natallchicago.com/Human-Shields2.pdf.

120. See the Web site of the World Church of the Creator, last accessed 13 January 2003, www.creator.org/.

121. The article can be found in German at www.deutsches-reich.de/, last accessed 17 March 2003.

122. Louis Beam, "Battle in Seattle: Americans Face Off the Police State," last accessed 14 January 2003, www.louisbeam.com/seattle.htm.

123. See Web site of the American Revolutionary Vanguard, last accessed 14 January 2003, www.attackthesystem.com/islam.html.

124. Thanassis Cambanis and Charles M. Sennott, quoting Magnus Ranstorp et al., "Fighting Terror: Going After the Network Cells; Qaeda Seen Still Dangerous," *Boston Globe,* 6 October 2002, A17.

125. In December 2001, Singapore authorities arrested fifteen Islamist militants who had plotted to bomb U.S. targets, including naval vessels in Singapore. The commander of the group was an Indonesian based in Malaysia named Ruduan Isamuddin (known as Hambali), whom Senior Minister Lee Kuan Yew referred to as Bashir's "right-hand man." The same group was accused of planning to bomb U.S. embassies in Southeast Asia on the anniversary of September 11. Details of the plot, and the relationship of Jamaah Islamiyah to Al Qaeda, were revealed to U.S. investigators by Al Qaeda's regional manager in Southeast Asia, Omar al Faruq.

Jamaah Islamiyah has been involved in a series of failed attempts to attack Western targets in Singapore, and information about its planned attacks led the United States to shut embassies in Southeast Asia on several occasions. The group has also attempted several times to assassinate Megawati. Singaporean investigators have learned about how JI functions from the operatives they took into custody in December 2001 and August 2002. Several JI members had been trained in Al Qaeda camps in Afghanistan. Others were trained in Mindanao by the Moro Islamic Liberation Front. JI leaders were instructed by Al Qaeda to stay away from mainstream Muslim life in Singapore to avoid drawing attention to themselves. They were not active in *madrassahs.*

126. A good example of a broad mission statement is the one that was used to mobilize participants in the Battle of Seattle. Groups opposed the World Trade Organization (WTO) for multiple reasons. American unions, supporters of Ralph Nader, and environmentalists were on the same side for completely different reasons. They demanded that WTO members adopt mandatory standards regarding pollution and protecting workers—in the case of the unions, because it would help them compete with their third-world rivals, and in the case of the environmentalists and "Naderites" because it would reduce worldwide emissions and promote workers' health. Developing countries opposed the WTO because they feared it would impose precisely those standards, which would help rich companies in the West at the expense of the poor in the third world.

127. The Battle of Seattle is perhaps the best example of an operation that succeeded despite the inherent difficulties of surmounting this problem. Individuals came to Seattle for their own reasons.

128. Ronfeldt and Arquilla argue, in contrast, that swarming is the ideal approach for networked terrorist organizations. But I argue that the requirement for secrecy will make large-scale swarming difficult for terrorist organizations, absent impenetrable communication systems. For their argument, see David Ronfeldt and John Arquilla, "Networks, Netwars, and the Fight for the Future," last accessed 15 August 2002, www.firstmonday.dk/issues.issue6_10/ronfeldt/.

Chapter 10: Conclusion/Policy Recommendations

1. While several of the militants interviewed for this book felt they communicated with God, none felt satisfied that God was taking care of their grievances, at least not in real time.

2. See Karen Armstrong, *The Battle for God* (New York: Alfred Knopf, 2000), 199.

3. The relationship of many religious terrorists in the Arab and Islamic world to modernism is contradictory: on the one hand, they feel threatened by it, while on the other hand there is a feeling of being left behind, almost an inferiority complex. A recent Gallup Poll conducted in Arab and Islamic countries revealed that Muslims overwhelmingly cited technology, computers, and knowledge when asked what they liked most about the West. In Iran, Kuwait, Indonesia, Jordan, and Morocco, more respondents chose the category "Technology/Computers/Expertise/Knowledge" than any other category when asked what they liked most about the West. In Iran, Indonesia, and Kuwait, the percentage of people who said that technology was what they liked best about the West was above 52 percent. The Gallup Organization, "The 2002 Gallup Poll of the Islamic World." Surveys can be purchased at the Web site of the Gallup Organization, last accessed 10 December 2002, www.gallup.com/poll/summits/islam.asp.

4. It is important to point out that macro-level studies of the root causes of terrorism, however, can only get us so far. When it comes to terrorism, the actions of a single individual, together with chance, can make a big difference. Consider the chain of events set in play by the assassination of Archduke Franz Ferdinand by a Serb terrorist. Moreover, we would need a lot more data at all these levels before we could identify root causes of terrorism. My goal here is to identify possible risk factors that have emerged from my interviews, which, when more data are collected, could be measured systematically (even if root causes probably cannot be identified). It is also important to realize that different terrorist movements attract different sorts of persons; and that, as Martha Crenshaw points out, internal bargaining and interactions both inside and outside the group must be taken into account, in addition to individual terrorist psychologies. Martha Crenshaw,

"Thoughts on Relating Terrorism to Historical Context," in Martha Crenshaw, ed., *Terrorism in Context* (University Park: Pennsylvania State University Press, 1995). The chapters in this volume provide an excellent introduction to the variety of individual, social, and political factors that have given rise to terrorism.

5. The term *bad neighborhood* was used by Michael Ignatieff. See Michael Ignatieff, "Intervention and State Failure," *Dissent,* winter 2002, 115–123.

6. Valerie Hudson and Andrea Den Boer, "A Surplus of Men, a Deficit of Peace: Sex Ratios in Asia's Largest States," *International Security* 26, no. 4 (spring 2002): 5–38. See also the Central Intelligence Agency's World Factbook 2002 for recent data, last accessed 13 January 2003, www.odci.gov/cia/publications/factbook/. Some studies suggest that a high ratio of males to females occurred at the time of the Vikings. See, for example, Carol Clover, "The Politics of Scarcity: Notes on the Sex Ratio in Early Scandinavia," *Scandinavian Studies* 60, (1988): 147–88. However, as Judith Jesch, professor of Viking studies at the University of Nottingham, explains, not enough evidence exists to support such a claim as far as the Vikings and other ancient societies are concerned. She adds that there are a variety of reasons for privileging males in the sources, such as a higher likelihood for burials of high-status males to be observed in the archaeological record, and the clear preference for males in memorial inscriptions. Jesch adds that there is also little evidence that Viking society was more violent than its counterparts in other contemporary societies. Judith Jesch, e-mail correspondence with the author, 14 January 2003.

7. Robert Bates, interview with the author, 8 October 2002. James D. Fearon and David D. Laitin, "Ethnicity, Insurgency, and Civil War," *American Political Science Review* 97, No. 1 (February 2003): 75–90.

8. Robert J. Barro, "Inequality, Growth, and Investment," *NBER Working Papers* 7038, National Bureau of Economic Research, Inc., 1999.

9. Paul Collier and Anke Hoeffler, "On the Economic Causes of Civil War," *Oxford Economic Papers* 50 (1998).

10. Without surveying a large number of Indonesians, it is not possible to disaggregate the partial effects of the economic downturn and/or Soeharto's fall (which in any case was also precipitated by the crisis), or other factors not yet identified.

11. For more on the idea of cultural humiliation, see Bernard Lewis, *What Went Wrong? Western Impact and Middle Eastern Response* (Oxford and New York: Oxford University Press, 2002).

12. Michael Kimmel and Abby L. Ferber, "White Men Are This Nation: Right-Wing Militias and the Restoration of Rural American Masculinity," *Rural Sociology* 65, no. 4 (2000): 582–604.

13. Military psychologist Dave Grossman estimates that two percent of men take pleasure from killing. Dave Grossman, *On Killing* (New York: Little Brown, 1995). Psychiatrist Jerrold Post argues that terrorists are driven to commit acts of violence as a consequence of psychological forces, and "that their special psycho-logic is constructed to rationalize acts they are psychologically compelled to commit." In Jerrold M. Post,

"Terrorist Psycho-Logic: Terrorist Behavior as a Product of Psychological Forces," in Walter Reich, ed., *Origins of Terrorism: Psychologies, Ideologies, Theologies, States of Mind* (Washington, D.C.: Woodrow Wilson Center Press, 1998). For a discussion of male genetic predisposition to violence, see Dale Peterson and Richard Wrangham, *Demonic Males: Apes and the Origins of Human Violence* (Boston, Mass.: Houghton Mifflin, 1996). See also Grossman, *On Killing*.

14. Anti-U.S. feelings intensified after the attack and subsequent occupation of Iraq. Four percent of Saudis have a favorable opinion of the United States, according to a poll conducted by Zogby International in March 2003. Ninety-seven percent of those polled in Saudi Arabia believed that the threat of terrorism against America would increase after the war in Iraq. The findings were only marginally better elsewhere in the Arab world. Only 6 percent in Jordan and Morocco, 8.8 percent in the United Arab Emirates, 13 percent in Egypt and one out of three in Lebanon had a favorable opinion of the United States. These findings represent an enormous drop in sympathy for the United States. Only a year earlier, a Zogby poll revealed that youth, in particular, were far more sympathetic to the United States, when the main complaint was U.S. policy in regard to Israel. In March 2003, there was widespread perception in the Islamic world that U.S. government policies were anti-Muslim across the board. John Zogby explains, "We have lost a lot of good will for a long time." He is especially concerned about angry youth. "The problem is that someone will reach them and organize them and that does not bode well for the United States." E-mail communication with John Zogby, 23–24 April 2003. Discussion with Ashraf Jehangir Qazi, Pakistan's ambassador to the United States, 23 April 2003. Discussion with Youssef Ibrahim, 16 April 2003. See also Pew Global Attitude Project, "What the World Thinks in 2002: How Global Publics View Their Lives, Their Countries, the World, America," 53–72. The full survey report is available at the Web site of the Pew Research Center for the Public and the Press, last accessed 10 December 2002, www.people-press.org/reports/files/report165.pdf.

15. It is unlikely that Al Qaeda could have managed the sophisticated embassy bombings and the September 11 attacks without the Egyptians' help. Al Qaeda established operational cooperation with the Egyptian Islamic Jihad (EIJ) by 1989 and later absorbed the group, co-opting its leaders into high ranks within Al Qaeda. Egypt has long been at the center of the Muslim religious revival movement, and hence the Egyptian alumni brought with them a substantial amount of experience in guerrilla and terrorist operations. EIJ also provided a ready-made officer corps for Al Qaeda.

16. See, for example, the 2002 Arab Human Development Report, and in particular chapter 2, "The State of Human Development in the Arab Region," and chapter 6, "Using Human Capabilities: Recapturing Economic Growth and Reducing Human Poverty." Available at the Web site of the United Nations Development Programme (UNDP), last accessed 10 December 2002, www.undp.org/rbas/ahdr/bychapter.html. For data on individual countries in the Middle East, see the MENA region Web page, available at the Web site of the World Bank, last accessed 10 December 2002, www.lnweb18.worldbank.org/mna/mena.nsf.

17. See, for example, Michael L. Ross, "Does Resource Wealth Cause Authoritarian Rule?" paper presented at Yale University, 10 April 2000, available at the Web site of Yale University, last accessed 11 December 2002, www.yale.edu/leitner/pdf/ross.pdf. See also Jeffrey Sachs and Andrew M. Warner, "Natural Resource Abundance and Economic Growth," *Development Discussion Paper No. 517a,* 1995, Harvard Institute for International Development, Cambridge, Mass.; Carlos Leite and Jens Weidmann, "Does Mother Nature Corrupt? Natural Resources, Corruption, and Economic Growth," *IMF Working Paper WP/ 99/85,* 1999; and Paul Collier and Anke Hoeffler, "On Economic Causes of Civil War," *Oxford Economic Papers 50,* 1998. Barro argues that democracies take hold only after a certain degree of wealth is achieved. Robert J. Barro, "Democracy and Growth," *NBER Working Paper 4909,* National Bureau of Economic Research, Inc., 1994.

18. Thomas Carothers, "The End of the Transition Paradigm," *Journal of Democracy* 13.1 (2002): 5–21. See also Fareed Zakaria, "The Rise of Illiberal Democracy," *Foreign Affairs,* November/ December 1997.

19. Abdelaziz Testas, "The Roots of Algeria's Religious and Ethnic Violence," *Studies in Conflict and Terrorism* 25, no. 3 (May–June 2002): 161–84.

20. According to Ahmed Rashid, some observers believed that the victory of the Islamists was sponsored by the Pakistani military in an attempt to gain concessions from the United States, to "ensure that the West does not question continued military rule and keep the Kashmir issue on the boil, ensuring a predominant role for the army." See Ahmed Rashid, "EU Condemns 'Flawed' Pakistan Elections," *Daily Telegraph,* 14 October 2002, 15.

21. Paul Schulte, "Reflections on watching *The Battle of Algiers* for the first time after September 11th," unpublished manuscript. Also see, J-L Marret, "Terrorism and Its Raison d'Être: What Makes Terrorists Tick?" unpublished manuscript, Foundation for Strategic Research, Paris.

22. Ron Scherer and Alexandra Marks, "Gangs, Prison: Al Qaeda Breeding Grounds?" *Christian Science Monitor,* 14 June 2002, last accessed 13 January 2003, www.csmonitor.com/2002/0614/p02s01-usju.html. See also David E. Kaplan et al., "Made in the U.S.A.," *U.S. News & World Report,* 10 June 2002, 17. For an assessment of French Islamists, see Marret, "Terrorism and Its Raison d'Être."

23. Sometimes purchasers are not required to show ID when purchasing guns from private vendors at gun shows. A known member of Hezbollah was found to have purchased weapons with the intent to ship them to Lebanon. A suspected member of Al Qaeda was found to have frequently bought and sold weapons at shows and was convicted on various weapons charges. Agents of the Provisional IRA were found to have purchased firearms to ship them back to Ireland hidden in toys. Susan Page, "Terrorists Use Gun Shows, McCain Says," *USA Today,* 28 November 2001, 14A. Jonathan Cowan, president of Americans for Gun Safety, letter to the author, 1 March 2002.

24. Martha Crenshaw, "The Effectiveness of Terrorism in the Algerian War," in Crenshaw, *Terrorism in Context,* 475 n. 4.

25. Michael Ignatieff, "The Lesser Evil: Political Ethics in an Age of Terror," The Gifford Lectures, 2002/2003.

26. Milton J. Bearden, "War of Secrets: When Playing the Field, the Game Gets Rough," *New York Times,* 8 September 2002, 4.

27. In April 2002, for example, Chris Patten, the European Union's external affairs commissioner, was quoted saying, "I see the U.S. unilateralist temptation as one of the central problems, perils, challenges, and opportunities confronting the English-speaking peoples of today." Roy Watson, "EU Calls on U.S. to Resist Unilateralist 'Temptation,'" *Times* (London), 1 May 2002. With regard to the death penalty, German and French authorities, for example, have been reluctant to turn over evidence in their possession that could help the United States in its trial against Zacarias Moussaoui, who is suspected of having conspired with the nineteen September 11 hijackers to commit acts of terrorism. Moussaoui, if convicted, could face the death penalty in the United States. See, for example, James Harding, "Handover of Evidence Condemned in Trial of Alleged Terrorist," *Financial Times,* 29 November 2002, 8. See also Peter Finn, "Germany Reluctant to Aid Prosecution of Moussaoui," *Washington Post,* 11 June 2002, A1. For an overview of the European Union's law against the death penalty, see "EU Law and Policy Overview: EU Policy on the Death Penalty," Web site of "The European Union in the U.S.," last accessed 6 March 2003. www.eurunion.org/LEGISLAT/DeathPenalty/deathpenhome.htm.

28. While at the Metropolitan Correctional Center in Manhattan, Mamdouh Mahmud Salim, along with Khalfan Khamis Mohamed, discussed in chapter 9, stabbed two prison guards, sprayed hot sauce, and stabbed one prison guard in the eye, while attempting to escape captivity. Benjamin Weiser, "Traces of Terror: The Inmate; Suspect Admits to a Plot," *New York Times,* 7 September 2002, 7.

29. Rohan Gunaratna, Debriefing of John Walker Lindh, Virginia, 25–26 July 2002.

30. Peter Maass, "Dirty War," *New Republic,* 11 November 2002, 18.

31. Ibid.

32. "Liberation Tigers of Tamil Eelam," in *Terrorism: Questions and Answers,* Council on Foreign Relations, last accessed 13 January 2003, www.terrorismanswers.com/groups/tamiltigers.html.

33. "LTTE Attack Paralyzes Sri Lanka's International Airport," International Policy Institute for Counter-Terrorism (ICT), July 24, 2001. Available at the ICT Web site, last accessed 13 January 2003, www.ict.org.il/spotlight/det.cfm?id=644.

34. James Q. Wilson, *Political Organizations* (Princeton, N.J.: Princeton University Press, 1995). Members stop caring about the purported mission as much as they care about their jobs. Over time, grievance shifts to greed—whether for political power, money, attention, or the satisfactions of leadership. Groups whose grievances have in large measure been addressed, but whose organizations persist, include ETA, the Basque separatist organization, and both sides in the dispute in Northern Ireland, which have turned what was a kind of holy war into a business. Other groups that we have discussed switch mis-

sions to ensure the group's survival. It is unlikely that Mubarak's replacement with a religious leader would satisfy his enemies today, for example.

35. Most of the world's nuclear weapons stockpiles and the world's stockpiles of weapons-usable materials (both military and civilian) are concentrated in the five nuclear weapons states acknowledged by the nuclear Non-Proliferation Treaty (NPT): the United States, Russia, China, Great Britain, and France. Additional nuclear weapons or components exist in Israel, India, and Pakistan, and enough civilian plutonium for many nuclear weapons also exists in Belgium, Germany, Japan, and Switzerland. Some "twenty tons of civilian highly enriched uranium (HEU) exist at 345 operational and shut-down civilian research facilities in fifty-eight countries, sometimes in quantities large enough to make a bomb." See Matthew Bunn, John P. Holdren, and Anthony Wier, "Securing Nuclear Weapons and Materials: Seven Steps for Immediate Action," Project on Managing the Atom, Belfer Center for Science and International Affairs, John F. Kennedy School of Government, Harvard University, May 2002. Available on-line at the Web site of the Nuclear Threat Initiative, last accessed 11 December 2002, www.nti.org/e_research/securing_nuclear_weapons_and_materials_May2002.pdf. See also Ashton Carter et al., "Soviet Nuclear Fission: Control of the Nuclear Arsenal in a Disintegrating Soviet Union," Center for Science and International Affairs, Harvard University, 1991; and Ashton B. Carter and William J. Perry, *Preventive Defense: A New Security Strategy for America* (Washington, D.C.: Brookings Institution Press, 1999).

36. "The National Security Strategy of the United States," September 2002, available at the White House Web site, last accessed 13 January 2003, www.whitehouse.gov/nsc/nss.pdf.

37. For an article that examines the complexity of the double-standard debate, see "Double Standards—Iraq, Israel and the UN," *Economist,* 12 October 2002.

38. While the view of economists like Dani Rodrik and Joe Stiglitz, who believe that free trade harms some countries, are debated by many of their mainstream colleagues, few would argue with the proposition, put forward by Amartya Sen, that the industrialized West is benefiting more from globalization than the developing world.

39. Susan Nieman, *Evil in Modern Thought: An Alternative History of Philosophy* (Princeton, N.J.: Princeton University Press, 2002), 287.

INDEX